2010 International Conference on Optical MEMS and Nanophotonics

(OPT MEMS 2010)

Sapporo, Japan
9 – 12 August 2010

IEEE Catalog Number: CFP10MOE-PRT
ISBN: 978-1-4244-8926-8

**Copyright © 2010 by the Institute of Electrical and Electronic Engineers, Inc
All Rights Reserved**

Copyright and Reprint Permissions: Abstracting is permitted with credit to the source. Libraries are permitted to photocopy beyond the limit of U.S. copyright law for private use of patrons those articles in this volume that carry a code at the bottom of the first page, provided the per-copy fee indicated in the code is paid through Copyright Clearance Center, 222 Rosewood Drive, Danvers, MA 01923.

For other copying, reprint or republication permission, write to IEEE Copyrights Manager, IEEE Service Center, 445 Hoes Lane, Piscataway, NJ 08854. All rights reserved.

***This publication is a representation of what appears in the IEEE Digital Libraries. Some format issues inherent in the e-media version may also appear in this print version.

IEEE Catalog Number: CFP10MOE-PRT
ISBN 13:   978-1-4244-8926-8

Additional Copies of This Publication Are Available From:

Curran Associates, Inc
57 Morehouse Lane
Red Hook, NY 12571 USA
Phone: (845) 758-0400
Fax: (845) 758-2633
E-mail: curran@proceedings.com
Web: www.proceedings.com

TABLE OF CONTENTS

Ultra-High Efficiency Solar Cel Development Activity in "SOLAR QUEST", the International Research Center for Global Energy and Environmental Technologies 1
Yoshiaki Nakano

Mems Scanner Enabled Real-time Depth Sensitive Hyperspectral Imaging 3
Youmin Wang, Sheldon Bish, Ashwini Gopal, James W Tunnell, Xiaojing Zhang

Vertical Comb-Drive MEMS Mirror with Sensing Function for Phase-Shift Device 5
Kentaro Oda, Kyohei Terao, Takaaki Suzuki, Hidekuni Takao, Ichiro Ishimaru, Fumikazu Oohira

Array of Cat's Eye Retro-Reflectors with Modulability for an Optical Identification System 7
Keng-hsing Chao, Chun-da Liao, Jui-che Tsai

Magnetic Actuated Moems Resonant Biosensor Array 9
Erman Timurdogan, Sezin Nargul, Halil Kavakli, Erdem Alaca, Hakan Urey

Development of an Integrated Microsystem for the Multiplexed Detection of Protein 11
Ciara K. O' Sullivan

Tunable Optofluidic Micro-iris 13
Philipp Müller, Nils Spengler, Wolfgang Mönch, Hans Zappe

Single Chamber Adaptive Membrane Lens with Integrated Actuation 15
Jan Draheim, Florian Schneider, Tobias Burger, Robert Kamberger, Ulrike Wallrabe

Eye-shaped Coplanar Variable Liquid Lens Using Spherical Polymer Encapsulation 17
Jae Yong An, Ji Hwan Hur, Sung Kil Lee, Jong-Hyun Lee

Dynamic Response of Dielectric Liquid Microlens 19
Chih-Cheng Yang, Yih-Ching Wang, J. Andrew Yeh

Optical Scanning with Mems In-plane Vibratory Gratings and Its Applications 21
Guangya Zhou, Yu Du, Kelvin K.L. Cheo, Hongbin Yu, Fook Siong Chau

A Large Rotational Angle Micromirror Based on Hypocycloidal Electrothermal Actuators for Endoscopic Imaging 23
Xiaojing Mu, Yingshun Xu, Janak Singh, Nanguang Chen , Hanhua Feng, Guangya Zhou, Aibin Yu, Chee Wei Tan, Kelvin Wei Sheng Chen, Fook Siong Chau

Translatory Mems Actuator with Extraordinary Large Stroke for Optical Path Length Modulation 25
Thilo Sandner, Thomas Grasshoff, Harald Schenk

Polymer-mems Torsion Mirror with Large 27
Dzung Viet Dao, Satoshi Amaya, Susumu Sugiyama

Low Cost and Large Deflection Angle Polymer Mems Mirror Using Glass Substrate 29
Osamu Sasaki, Takaaki Suzuki, Kyouhei Terao, Hidekuni Takao, Fumikazu Oohira

Phosphor Sensors Using Mechanoluminescence 31
Chao-Nan Xu

A New Fabrication Technique for Integrating Silica 33
Karen E. Grutter, Anthony M. Yeh, Susant K. Patra, Ming C. Wu

Four-mask Process Based on Spacer Technology for Scaled-down Lateral NEM Electrostatic Actuators 35
Daesung Lee, W. Scott Lee, Subhasish Mitra, Roger T. Howe, H. S. Philip Wong

Nanostructured Origamitm Folding of Patternable Resist for 3d Lithography 37
Se Young Yang, Hyung-ryul Johnny Choi, Martin Deterre, George Barbastathis

Absorbent Liquid Immersion Angled Exposure for 3D Photolithography 39
Hironori Kubo, Shinya Kumagai, Minoru Sasaki

Nanobiodevice Based Single Cell Imaging for Cancer Diagnosis and in Vivo Imaging for Stem Cell Therapy 41
Yoshinobu Baba

Pulsed Laser Triggered High Speed Fluorescence Activated Microfluidic Switch 43
Ting-Hsiang Wu, Yue Chen, Sung-Young Park Soojung Claire Hur, Dino Di Carlo, Eric Pei-Yu Chiou

A Laser Driven Optofluidic Device for High-Speed and Precise Volume-Controlled Droplet Generation on Demand 45
Sung-Yong Park, Ting-Hsiang Wu, Yue Chen, Pei Yu Chiou

Reshaping Gold Micro and Nano Structures with Polarization Dependent Photothermal Annealing 47
Fan Xiao, Pei Yu Chiou

Near-field Plasmonic Enhancement Via Nanogratings on Hollow Pyramidal Aperture Probe Tip 49
Yuyan Wang, Yu-Yen Huang, Kazunori Hoshino, Ashwini Gopal, Xiaojing Zhang

Surface Optomechanics: Mechanical Whisperinggallery Modes in Microspheres 51
John Zehnpfennig, Matthew Tomes, Tal Carmon

Silicon Integrated Electronic-Photonic ICs 53
Patrick Lo, Dim-Lee Kwong

High Power THZ Photoconductive Antenna Using Localized Surface Plasmon Resonance 55
Sang-Gil Park, Yongje Choi, Minwoo Yi, Jun-Hyuk Choi, Kyung-Hwan Jin, Jong-Chul Ye, Jaewook Ahn1, Ki-Hun Jeong1

Control of Solid-state Lasers Using an Intra-cavity Mems Micro-mirror 57
W. Lubeigt, J. Gomes, G. Brown, A. Kelly, V. Savitski, D. Uttamchandani, D. Burns

Linear MEMS Micromirror Array for UV-NIR Femtosecond Pulse Shaping 59
Stefan M. Weber, Jerome, Extermann, Wilfried Noell, Fabiio Jutzi, Sebastien Lani, Denis Kiselev, Luigi Bonacina, Nico F. De Rooij, Jean-Pierre Wolf

Tunable Optical Diffusers for High-power Laser Applications Based on Magnetically Actuated Membranes 61
Jonathan Masson, Andreas Bich, Wilfried Noell, Reinhard Voelkel, Kenneth J. Weible, Nico F. de Rooij

Plasmonics for Ultrasensitive Biomolecular Nanospectroscopy 63
Hatice Altug, Ahmet A. Yanik, Ronen Adato, Serap Aksu, Alp Artar, Min Huang

A Photoelectrophyscial Capacitor with Direct Solar Energy Harvesting and Storage Capability 65
Chi-Wei Lo, Chensha Li, Hongrui Jiang

Remote Switching of Cellular Activity Using Light Through Quantum Dots 67
Katherine Lugo, Xiaoyu Miao, Fred Rieke, Lih Y. Lin

Mechanically Tunable Coupled Photonic Crystal Nanocavities 69
Xiongyeu Chew, Guangya Zhou, Fook Siong Chau, Jie Deng, Xiaosong Tang, Yee Chong Loke

Ring-resonator Reflector with a Waveguide Crossing 71
Wei Shi, Raha Vafaei, Miguel Angel Guillen Torres, Nicolas A. F. Jaeger, Lukas Chrostowski

A MEMS Digital Microshutter (DMSTM) for Low-power High Brightness Displays 73
J. Lodewyk Steyn, Timothy Brosnihan, John Fijol, Jignesh Gandhi, Nesbitt, Hagood IV, Mark Halfman, Steve Lewis, Richard Payne, Joyce Wu

Flexible Display System Based on Mems Fabry-perot Interferometer 75
G.Tortissier, C.-Y. Lo, H. Fujita, H. Toshiyoshi

Low Voltage Electrostatic 90 Turning Flap for Reflective Mems Display 77
Fabio Jutzi, Francois Gueissaz, Wilfried Noell, Nico F. de Rooij

Gan Pitch-variable Grating Fabricated on Si Substrate 79
H. Sameshima, T. Tanae, F. Hu, K. Hane

Synchronized Laser Scanning of Multiple Beams by Mems Gratings Integrated with Resonant Frequency Fine Tuning Mechanisms 81
Yu Du, Guangya Zhou, Kelvin Koon Lin Cheo, Qingxin Zhang, Hanhua Feng, Fook Siong Chau

Fluorometric Bio-Sniffer (Biochemical Gas Sensor) with UV-LED Light for Fomaldehyde Vapor as VOC (Volatile Organic Chemical) 83
Tomoko Gessei, Gen Itabashi, Yuki Suzuki, Daishi Takahashi, Takahiro Arakawa, Hiroyuki Kudo, Kohji Mitsubayashi

High-precision Optical & Fluidic Micro-bench for Endoscopic Imaging 85
Niklas Weber, Hans Zappe, Andreas Seifert

A Fully Integrated Thermo-pneumatic Tunable Microlens 87
Wei Zhang, Khaled Aljasem, Hans Zappe, Andreas Seifert

A Tunable Liquid Lens with Extended Depth of Focus 89
Jingran Kang, Guangya Zhou, Hongbin Yu, Fook Siong Chau, Haiqing Chen

Micromachined Two Dimensional Lens Scanner with Large Aperture Beam 91
Hyeon-Cheol Park, Cheol Song, Ki-Hun Jeong

Large-size Infrared Reflow Microlens Based on Stacked layers 93
Takuro Aonuma, Shinya Kumagai, Minoru Sasaki

Optical Waveguide Devices for Bioanalysis 95
James S Wilkinson

Submicron silicon waveguide Mach-Zehnder interferometer using micro electro-mechanical phase-shifter 97
T. Ikeda, Y. Kanamori, K. Hane

Mems-actuated Waveguide Phase Modulators 99
Chun-Che Chang, Wei-Chao Chiu, Jiun-Ming Wu, Ming-Chang M. Lee Jia-Min Shieh

Inertial Force Sensor Using Optical Mach-zehnder Interferometer and Multi Mode Interferometer 101
Masato Suzuki, Gou Kawai, Kouji NishiokTomokazu Takahashi, Seiji Aoyagi, Yoshiteru Amemiya, Masataka Fukuyama, Shin Yokoyama

Pure Piston Motion of Optically Flat Micromirrors in a Fully Programmable Micro Diffraction Grating 103
R. Lockhart, R.P. Stanley, M. Tormen

Lamellar Grating Based Mems Fourier Transform Spectrometer 105
Huseyin R. Seren, N. Pelin Ayerden, Jaibir Sharma, Sven TS Holmstrom1, Thilo Sandner, Thomas Grasshoff, Harald Schenk, Hakan Urey

A Study on Color-tunable MEMS Device based on Plasmon Photonics 107
Taelim Lee, Akio Higo, Hiroyuki Fujita, Yoshiaki Nakano, Hiroshi Toshiyoshi

A Mixed-signal Analysis for Tilted MEMS Torsion Mirror Devices 109
Satoshi Maruyama, Akio Higo, M. Nakada, K. Takahashi, T. Takahashi, M. Mita, Hiroyuki Fujita, Yoshiaki Nakano, and Hiroshi Toshiyoshi

A 2-axis MEMS Scanner for the Landing Laser Radar of the Space Explorer 111
M. Mita, T. Mizuno, M. Ataka, H. Toshiyoshi

Vacuum Operation Characteristics of Two-dimensional Micro-mirror 113
Hoang Manh Chu, Kazuhiro Hane

A Two-axis Hybrid Mems Scanner Incorporating Electrothermal and Electrostatic Actuators 115
Gordon Brown, Li Li, Ralf Bauer, Jinsong Liu, Deepak Uttamchandani

Mems Scanning Mirror Used As an Laser External Modulator for Photoacoustic Spectroscopy 117
Li Li, Graham Thursby, George Stewart, Deepak Uttamchandani

Torsional Mirror Driven by a Cantilever Beam Integrated with 1x10 Individually Biased Pzt Array Actuator for VOA Application 119
Kah How Koh, Takeshi Kobayashi, Chengkuo Lee

Design, Fabrication, and Package of MEMS-based Image Stabilizer for Photographic Cell Phone Applications 121
Jin Chern Chiou, Chen-Chun Hung, Chun-Ying Lin

Development of A 2x2 Optical Switch Using Bi-stable Solenoid-based Actuators 123
Bonnie Tingting Chia, Cheng-Wen Ma, Bo-Ting Liao, Sun-Chih Shih, Yao-Joe Yang

Fabrication and Evaluation of Piezoelectric Drive Type 2-axis Tilt Control Device Using Epitaxial PZT Thin Film 125
Katsuya Ozaki, Daisuke Akai, Kazuaki Sawada, Makoto Ishida

Compliant Scanning Micromirror Actuated with a Displacement Amplification Mechanism 127
Tzung-Ming Chen, Florian Schneider, Ulrike Wallrabe

Multilayer Piezoelectric Ceramic Actuator for Laser Scanner 129
Jae-Sung Song, In-Sung Kim, Soon-Jong Jeong, Min-Soo Kim

Droplet-based Lateral Tunable Optofluidic Microlens Array with Pneumatic Control 131
Ye Liu, Hongrui Jiang

Excellent fault tolerance of a MEMS optically differential reconfigurable gate array 133
Hironobu Morita, Minoru Watanabe

Batch Fabrication of Flowable Colorimetric Pressure Sensing Particles Via Surface Micromachining 135
S. Chalasani, Y. Xie, Y. Zeng, C. H. Mastrangelo

Enhanced Contrast of Wavelength Selective Mid-IR Detector Stable against Temperature Change 137
Katsuya Masuno, Shinya Kumagai, Kohji Tashiro, Masaru Hori, Minoru Sasaki

Fabrication and Verification for the Micro Holographic Optical Pickup 139
Jin Chern Chiou, Kuan Chou Hou

A Novel Fabrication Method of the Micro Cube Beam-Splitter with Optical Surface Roughness 141
Kuo-Yung, HungYing-Chuan, Chen Shih-Hao, Huang Yun-Ju Chuang

Dynamic trapping and release of multiple particles in a polarized optical vortex 143
Baile Zhang, George Barbastathis

Design and Fabrication of Large Fiber-mode-matched Three-dimensional Adiabatic Tapered Couplers for Integrated Optics 145
Chun-Wei Liao, Yao-Tsu Yang, Sheng-Wen Huang, and Ming-Chang M. Lee, Pi-Yao Lin, Chao-Min Chou, Jia-Ming Shieh

Fabrication of LED Based Ultra Slim Optical Pointing Device 147
Jae Young Joo, Do-Kyun Woo, Sun Sub Park, Sun-Kyu Lee

Fast Atom Beam-based Fabrication of High-efficienct Blazed Grating Using Slanting Angle Control of a Substrate 149
ChaBum Lee, Kazuhiro Hane, Sun-Kyu Lee

X-ray Imaging Test for a Single-stage Mems X-ray Optical System 151
Ikuyuki Mitsuishi, Yuichiro Ezoe, Kensuke Ishizu, Teppei Moriyama, Yoshitomo Maeda, Takayuki Hayashi, Takuro Sato, Makoto Mita, N Y. Yamasaki, K. Mitsuda, Mitsuhiro Horade, Susumu Sugiyama, Raul E. Riveros, Taylor Boggs, Hitomi Yamaguch, Yoshiaki Kanamori, Kohei Morishita, Kazuo Nakajima, Ryutaro Maeda

Improvement of GaN crystalline quality on nanoscale patterned sapphire substrates 153
Yu-Sheng Lin, J. Andrew Yeh

The Morphological Control of MEH-PPV Films on an Ito Electrode for Hybrid Solar Cell Fabrication 155
Quynh Nhu Nguyen Truong, N.T.N. Truong, C. Park, Jae Hak Jung

Photonic MEMS Vibrating at X-band Rates (11 GHz) 157
Matthew Tomes, Tal Carmon

Gyroscopic Optomechanics 159
Xingyu Zhang, Matthew Tomes, Tal Carmon

Measurements of Light Fields Emerging from Fine Amplitude Gratings 161
Myun-Sik Kim, Toralf Scharf, Hans Peter Herzig

Nanoscale Epitaxial Growth of GaN on Freestanding Circular GaN Grating 163
Yongjin Wang, Fangren Hu, Kazuhiro Hane

Preparation of Anodic Aluminum Oxide Nano-Template using Al/Si Substrate for Large Area LED Applications 165
Lan Shen, Doogwook Kim, Bonggi Min, Zhengbin Gu, Chinho Park

Polarization Control of InAs Quantum Dot Semiconductor Laser using External Light Injection Technique 167
P.C. Peng, R.L. Lan, S.T. Hsu, H.H. Lu, G. Lin, H.C. Kuo, G.R. Lin, J.Y. Chi

Experimental Demonstration of the Vernier Effect using Series-Coupled Racetrack Resonators 169
Robi Boeck, Nicolas A.F. Jaeger, Lukas Chrostowski

Polarization Independent Grating Coupler for Silicon-on-insulator Waveguides 171
Chun-Chia Chiu, Ding-Wei Huang

An Approach for Modeling Photonic Crystal Circuits 173
Yih-Bin Lin, Rei-Shin Chen, Ju-Feng Liu, Han-Bin Lin

Design of High Transmission Broadband 90-degree Bends for Two Dimensional Cubic Photonic Crystals 175
Yih-Bin Lin, Cheng-Ru Li, Rei-Shin Chen, Jung-Young Su

Design and Fabrication of Dielectric Nanostructured Luneburg Lens in Optical Frequencies 177
Satoshi Takahashi, Chih-hao Chang, Se Young Yang, George Barbastathis

Nonlinear Kerr Effect Aperiodic Luneburg Lens 179
Hanhong Gao, Satoshi Takahashi, Lei Tian, George Barbastathis

Configuration Analysis of Sensing Element for Microcantilever Sensor Using Dual Nano-ring Resonator 181
Bo Li, Fu-Li Hsiao, Chengkuo Lee

Investigation of Strain Sensitivity of Photonic Crystal Nanocavity for Mechanical Sensing 183
Bui Thanh Tung, Dzung Viet Dao, Susumu Sugiyama

Magnetic Response of Continuous Split Ring Structures at Visible Frequencies 185
Yi-Hao Chen, Alex F. Kaplan, L. Jay Guo

An adaptive objective for optical motion correction in MRI 187
F. Schneider, J. Draheim, T. Burger, J. Maclaren, M. Herbst, M. Zaitsev, R. Bammer, U. Wallrabe

Mems-based X-ray Optics for Future Astronomical Missions 189
Yuichiro Ezoe, Ikuyuki Mitsuishi, Kensuke Ishizu, Teppei Moriyama, Kazuhisa Mitsuda, Noriko Y. Yamasaki,
Takaya Ohashi, Mitsuhiro Horade, Susumu Sugiyama, Raul E. Riveros, Taylor Boggs, Hitomi Yamaguchi,
Yoshiaki Kanamori, Nicholas T. Gabriel, Joseph J. Talghader, Kohei Morishita, Kazuo Nakajima, Ryutaro Maeda

Large Electrostatically and Electromagnetically Actuated Mirror System for Space Applications 191
Dara Bayat, Caglar Ataman, Benedikt Guldimann, Sebastian Lani, Wilfried Noell, Nico F. de Rooij

Micromirror Arrays Designed and Tested for Space Instrumentation 193
Frederic Zamkotsian, Michael Canonica, Kyrre Tangen, Patrick Lanzoni, Emmanuel Grassi, Rudy Barette,
Christophe Fabron, Severin Waldis, Wilfried Noell, Nico de Rooij, Laurent Marchand, Ludovic Duvet

Commercialization of Self-Assembled Quantum-Dot Lasers: From Optical Communication to Consumer Electronics 195
M Sugawara

Author Index

(Plenary)

Ultra-High Efficiency Solar Cell Development Activity in "SOLAR QUEST", the International Research Center for Global Energy and Environmental Technologies

Yoshiaki NAKANO

Research Center for Advanced Science and Technology, The University of Tokyo

4-6-1 Komaba, Meguro-ku, Tokyo, 153-8904, Japan

Phone: +81-3-5452-5150, Fax: +81-3-5452-5151, e-mail: nakano@rcast.u-tokyo.ac.jp

The research and development of innovative solar cells with high conversion efficiency and low production cost is important for achieving the long-term goal of "Cool earth 50", in which we aim at 50% reduction in the emission of greenhouse gas in the entire world by 2050 in order to prevent further global warming. The primary target of this project is the dramatic improvement in a photovoltaic conversion efficiency, which will innovate energy cycle in the world in a manner that reduces the use of fossil fuels, and thus eliminating the emission of greenhouse gas, in cooperation with the innovations in energy storage and delivery.

The Research Center for Advanced Science and Technology (RCAST) of the University of Tokyo has organized a research consortiam named "SOLAR QUEST" in 2008 and started next generation solar cell development activity under the support of NEDO (see Fig. 1). This project aims at the solar cells achieving a conversion efficiency of more than 40%. Our approach to such a high efficiency starts from the improvement of III-V multi-junction solar cells by introducing a novel material for each cell that realizes an ideal combination of bandgaps and lattice-matching (see Fig. 2). Further improvement of the multi-junction cells incorporates quantum structures such as stacked quantum wells and quantum dots, which allows higher degree of freedom in the design of the bandgap and the lattice strain. Highly controlled arrangement of either quantum dots or quantum wells allows the coupling of the wavefunctions, and thus forms intermediate bands in the bandgap of a host

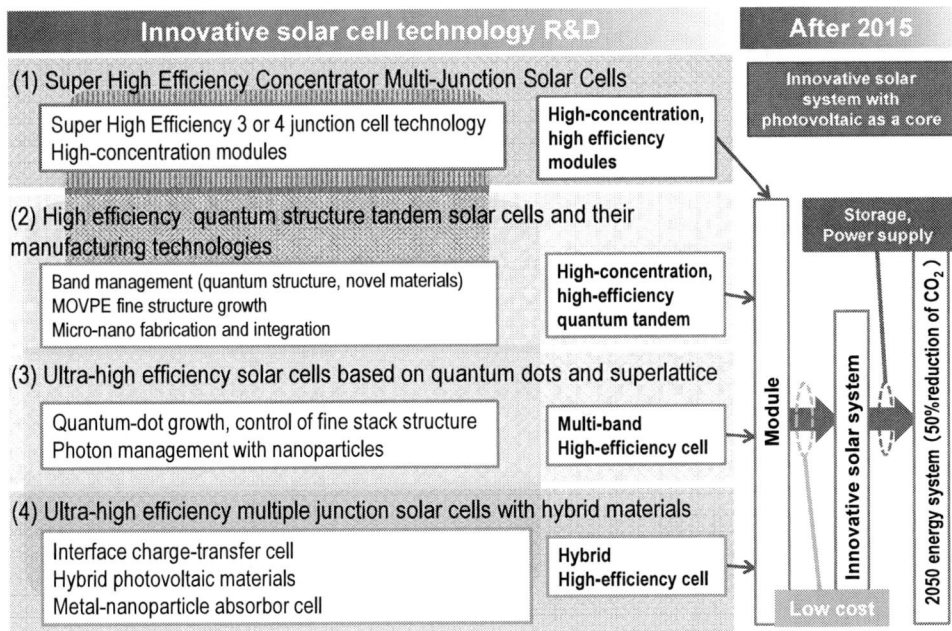

Fig. 1. *Organization of the NEDO "Post-silicon Solar Cells for Ultra-High Efficiencies" Project*

material, which allows multiple photon absorption that will theoretically leads to a conversion efficiency exceeding 50% (see Fig. 3). In addition to such improvements in the efficiency of III-V cells, novel approach that enhances photon absorption efficiency by the use of plasmon or energy transfer using novel materials, microfabrication technology for the integrated high-efficiency cells, and the development of novel material systems that realizes high efficiency and low cost at the same time.

In addition to such intensive research and developments, we also promote international research cooperation and provide an opportunity for the exchange of information among domestic and international organizations.

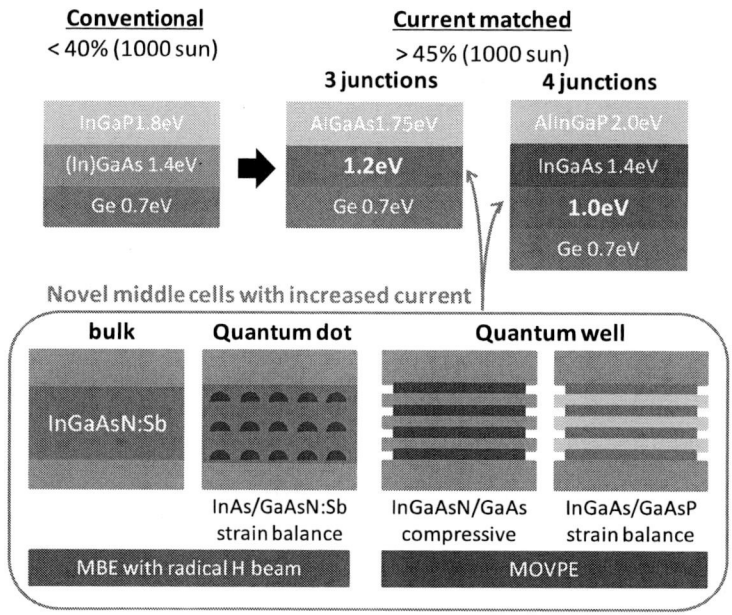

Fig. 2. Strategy toward 45% and higher efficiency in III-V compound semiconductor multiple junction solar cells investigated in this project.

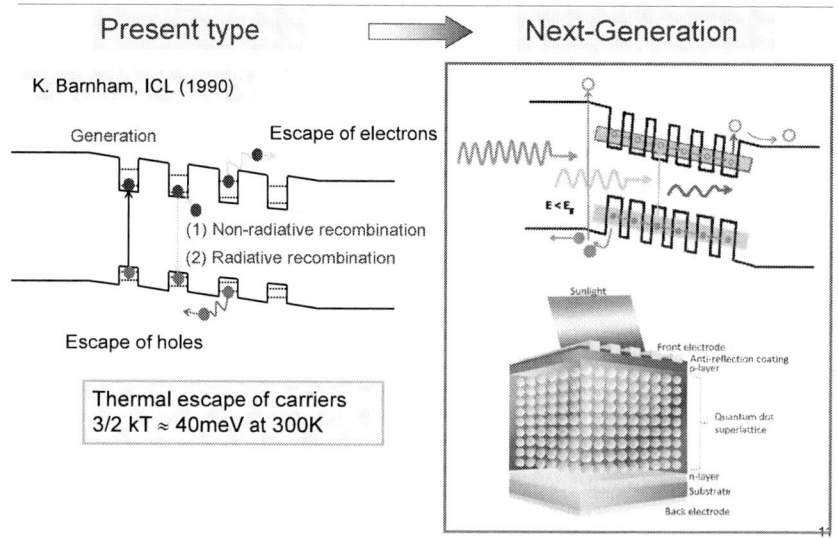

Fig. 3. III-V compound semiconductor intermediate band solar cell based on quantum dots and superlattice studied in this project.

MEMS SCANNER ENABLED REAL-TIME DEPTH SENSITIVE HYPERSPECTRAL IMAGING

Youmin Wang, Sheldon Bish, Ashwini Gopal, James W Tunnell and Xiaojing Zhang

Department of Biomedical Engineering

The University of Texas at Austin, Austin, Texas, USA

Tel: (512) 475-6872; Fax: (512) 232-4275; E-mail: jtunnell@mail.utexas.edu, John.Zhang@engr.utexas.edu

Abstract:

We demonstrate a hyperspectral and depth sensitive optical scattering diffusion imaging microsystem, where fast scanning is provided by a CMOS compatible 2-axis MEMS mirror. By using lissajous scanning patterns, large field-of-view (FOV) of 1.2cm x 1.2cm images with lateral resolution of 100µm can be taken at 1.3 frame-per-second (fps). Hyperspectral and depth-sensitive images were acquired on phantom samples consisting of quantum dots (QDs) patterned at various depths in PDMS, showing 6 nm spectral resolution and 0.43 wavelength per second acquisition speed. Images were also acquired on biological sample of porcine epithelium with QDs placed underneath the surface.

1. INTRODUCTION

The clinical need for noninvasive skin cancer detection devices has led to intensive commercial and research efforts on developing novel instruments on imaging, spectroscopy or the combination of both [1]. In particular, portable hyperspectral screening device is highly desirable for sensitive biopsy-free characterization of diseases in situ and precision guided microsurgery. This will not only save the costly millions of biopsies performed every year, but also to make the diagnosis and treatment for diseases like melanoma more convenient and timely. A single-point spectral diagnosis (SDx) device has been developed for the noninvasive diagnosis of skin cancer using optical fiber based probe placed in gentle contact with the skin surface [2]. The high accuracy of the preliminary results demonstrate the potential of SDx for the non-invasive, real-time diagnosis of skin cancer; however, these results suffer from inaccuracies due to "sampling error" inherent in point measurements. While SDx provides detailed physiological data that can be useful for diagnosis, it does not allow for mapping or imaging of skin sites.

In this paper, we report the design and characterization of a MEMS scanner based hyperspectral imaging system combining the advantages of both functional imaging and spectroscopy. Integrating the CMOS compatible dual-axis micromirror, we have already demonstrated imaging modalities, including a handheld forward-imaging confocal microscope capable of sub-micrometer lateral resolution [3] and OCT [4]. The new system is capable of hyperspectral and depth-sensitive imaging with adjustable lateral and spatial resolution.

2. MEMS HYPERSPECTRAL SYSTEM

The implementation of the hyperspectral imaging system is shown in Figure 1.

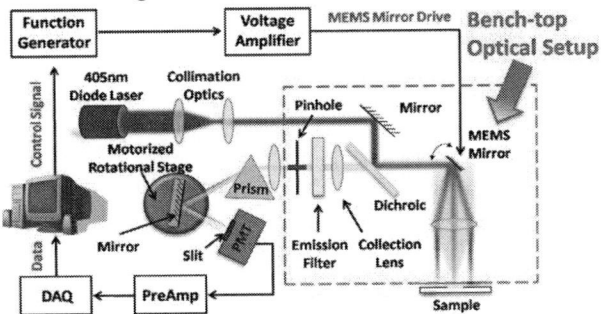

Figure 1: Schematic of the MEMS HSI system.

The excitation laser beam is deflected by the MEMS micromirror and focused onto the sample. The emitted fluorescence signal from the sample is collected by the micromirror simultaneously and focused onto the plane of a pinhole for depth sensitive imaging, and then dispersed by a prism for hyperspectral imaging, before being acquired by the photomultiplier tube (PMT) (Hamamatsu H9858).

The MEMS micromirror, shown in Figure 2, was fabricated using CMOS compatible process [5]. The maximum optical deflection angles are 22° and 12° respectively, at resonant frequencies of 2.67 kHz and 1.19 kHz for the inner and outer rotation axes.

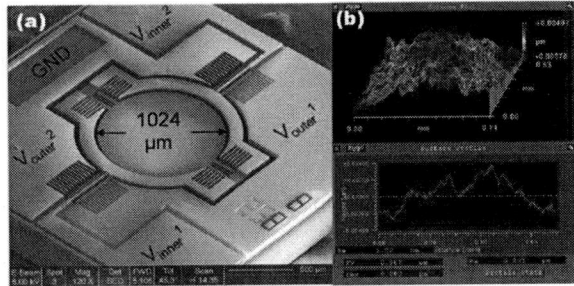

Figure2: SEMs of CMOS compatible microscanner. (a) Scanning electron micrograph of micromirror.(b) Roughness measurement showing a 8nm roughness in average

3. EXPERIMENTAL RESULTS

We characterized the MEMS hyperspectral imaging systems using both PDMS-QD phantom and biological samples such as porcine epitheliem. A multilayer QDs PDMS phantom sample is fabricated for evaluating the quality of depth sensitive and hyperspectral imaging simultaneously (Figure 3). Micro-contact printing (µCP) is used to pattern QDs on the surfaces of a multi-stack of PDMS thin layer. Each spin-coated PDMS thin-layer is 200µm thick, infusd with titanium dioxide to simulate scattering in tissue. The sample contained a 3x3 QDs patterns array, each depth with 3 different colors.

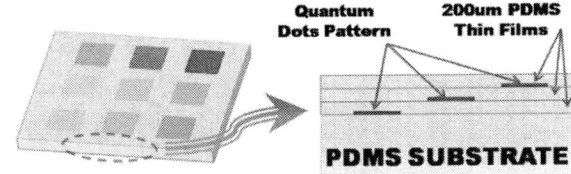

Figure 3: Schematic of the µCP fabricated QDs multilayer PDMS sample. Left: Isometric View. Right: Side View.

By actuating the MEMS micromirror scanning in a lissajous pattern, the lateral resolution and the field of view (FOV) of the real-time imaging system is experimentally determined as ~100µm and 1.2cm by 1.2cm, while rendering at a frame rate of 0.8 frames per second (fps). Using our real-

978-1-4244-8926-8/10 $26.00 © 2010 IEEE

time HSI system on the PDMS-QD phantom, we took 30 images of different wavelengths through the spectrum within 70s (Figure 4). Therefore 0.43 wavelength per second acquisition speed was obtained. The 100μm slit placed in front of the PMT guarantees a spectral resolution of 8nm.

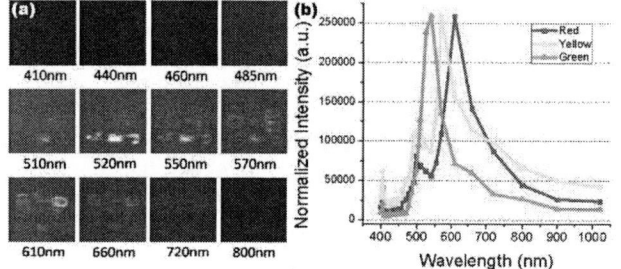

Figure 4: MEMS HSI for PDMS-QD phantom imaging. (a) PDMS sample under 12 selected wavelengths from the 30 total acquired wavelengths. (b) Normalized spectrum of 3 featured positions on the PDMS sample.

Using the images of 3 peak wavelengths for each color, the pseudocolor image was merged and rendered (Figure 5a), which shows good preserving of features when comparing to the mosaic image from commercial microscope (Figure 5b).

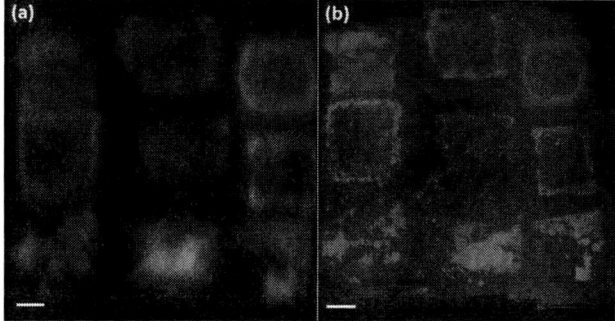

Figure 5: Comparison of images acquired using MEMS HIS and Olympus microscope. (a) Pseudocolor image merged from 3 peak wavelength images. (b) Microscope image using Olympus BX51 microscope. Scale bars are 1mm.

For acquiring depth sectioned images; a 100μm pinhole is translated in the focusing plane as a collection aperture for light returning from the sample (Figure 1).

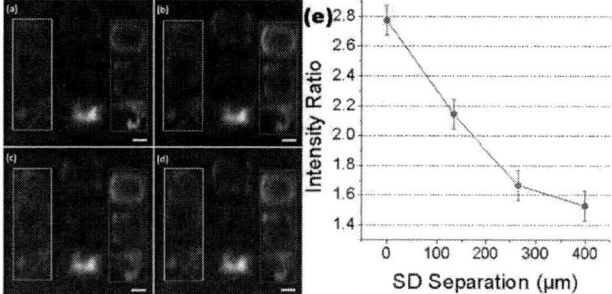

Figure 6: (a)-(d): Depth sampled images for four different SD separations. Fluorescence intensities within selected regions of the image were calculated to obtain ratios of shallow to deep intensities. Red and yellow outlined regions contain shallow (200μm) and deep (600μm) QDs respectively. (e) Mean intensity ratios of shallow vs. deep quantum dot stamps over 0 to 399μm SD separations. Scale bar is 1mm.

Source-detector (SD) separation is the radial distance between the illumination spot and the image pinhole, and will be proportional to the mean sampling depth of the collected photons. In Figure 6 our different SD separations were performed, showing the depth sectioning capability in terms of the ratios of fluorescence intensities between the shallowest and deepest layers.

Figure 6e implies the diminishing trend of the intensity ratios with increasing SD separation. This implies that fluorescence emanating from more deeply implanted quantum dots account for a higher percentage of the overall collected light as SD separation increases, as indicated by previous studies [6]. Here we have demonstrated that by applying SD separation up to 400μm, the selectivity of the deepest layer increased by 1.83 times.

Besides, the biological imaging potential was demonstrated using 2 wavelength quantum dots placed under the surface of ex vivo porcine epithelium (Figure 7). The spectrums of two QDs injected cites were measured. Pseudocolor image is also acquired from hyperspectral images using the same way as in Figure 4. This shows the potential of using bio-conjugating agents for hyperspectral fluorescence imaging. Hyperspectral resolution and imaging speed of 6nm was demonstrated here for the biological imaging.

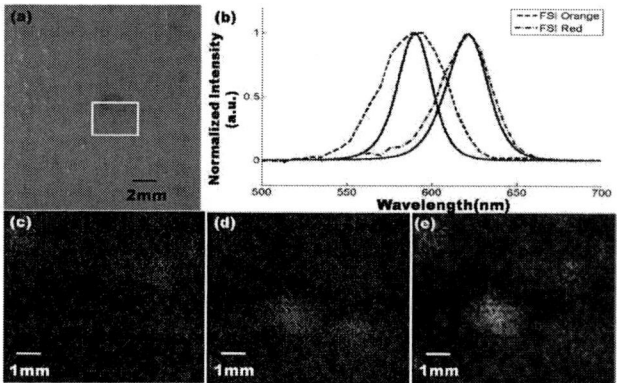

Figure 7: Biological Sample of porcine epithelium with QDs placed underneath the surface. (a) Camera image of sample, MEMS HSI scan area is delineated by the white box. (b) Emission spectrum of quantum dots used. Bold curve derived using USB4000 spectrometer, broken line curves derived from the hyperspectral imaging system (25 points from 550-700nm). (c)-(d) De-mixed image acquired at the peak wavelength of orange and red QDs. (e) Pseudocolor image merged from (c), (d) after thresholding despeckling denoise.

4. CONCLUSIONS

We demonstrated a real-time functional spectral imaging system capable of depth-sensitive and hyperspectral imaging using a dual-axis MEMS micromirror. Spatial FOV of 1.2cm x 1.2cm and resolution of 100um are obtained at a frame rate of 1.3fps. Hyperspectral resolution of 6nm and imaging speed of 0.43 wavelength per second were obtained for both PDMS phantom and biological sample.

The range of depth sectioning can be tuned by adjusting applying source-detector separation.

5. REFERENCE

[1] YG Patel, et al., J Biomedical Optics, 2007, 12, 3.
[2] N Rajaram, et al., Appl Opt 49(2):142-152, 2010
[3] K.Kumar, et al., Biomed. Microdev., 2009, 12, 2.
[4] K.Kumar, et al., J of Optics, 10, 044013 (7pp), 2008
[5] K.Kumar, et al., MEMS 2009, Sorrento, Italy
[6] SA Burgess et. al. OSA Tech. Digest March 16-19, 2008

Vertical Comb-Drive MEMS Mirror with Sensing Function for Phase-Shift Device

Kentaro Oda*, Kyohei Terao*, Takaaki Suzuki*, Hidekuni Takao*, Ichiro Ishimaru* and Fumikazu Oohira*

* Department of Intelligent Mechanical Systems Engineering, Faculty of Engineering, Kagawa University
2217-20, Hayashi-cho, Takamatsu, Kagawa, 761-0396, Japan
Tel +81-87-864-2341, Fax +81-87-864-2341, E-mail: s09g506@stmail.eng.kagawa-u.ac.jp

Abstract

We aim to achieve the phase-shift device which is a key component of two-dimensional Fourier spectrometer using the MEMS mirror fabricated by the micro fabrication technology. This mirror maintains parallel movement to the reference plane toward the vertical direction with a high precision. Therefore, in this study, the vertical electrostatic comb-drive actuator and displacement capacitive sensors were fabricated monolithically on one chip, and they were arrayed in 4 directions of the movable mirror. We fabricated the MEMS mirror that can move toward the vertical direction with sensing the tilting angle. As the result, we confirmed that the vertical comb-drive MEMS mirror with the sensing function for the phase-shift device could be realized.

Keywords : MEMS mirror, SOI, vertical comb-drive actuator, capacitive sensor, phase-shift, Deep-RIE

1. Introduction

The MEMS mirrors fabricated by the micro fabrication technology which are small size and low electric power have been widely applied for many fields such as optical switch, laser scanner and attenuator etc. [1] We aim to apply the MEMS mirror for not only tilting use such as the optical switch etc. but also the optical sensing use such as the spectrometer with a phase-shift method [2] by the vertically driven movable mirror. In the former study, [3] we showed that the fabricated MEMS mirror can be applied to the spectrometer with phase-shift method, but the fabricated MEMS mirror is difficult to maintain the parallel movement to the reference plane toward the vertical direction with a high precision, so the necessity of an On-Chip sensor and the feedback control to make the high precision and the parallel drive vertical to the reference plane (fixed mirror) has become clear. In this study, the vertical electrostatic comb-drive actuators were fabricated on one chip, and they were arrayed in 4 directions of the movable mirror. We mainly evaluated the sensor characteristics of the fabricated MEMS mirror.

2. Principle measurement

Fig.1 shows the principle configuration of the spectrometer with a phase-shift method. A light from an object is changed to a parallel light, and the half of the parallel light is reflected at the area of the movable mirror and the other half of the light is reflected at the area of the fixed mirror. In this case, when the movable mirror is moved, the phase difference is generated at the half area of reflected light. This generates the interference between the different phase lights when the two lights converge at the imaging surface. The interferogram is a summation of the interference intensity of each object light spectrum. Therefore, the relative intensity of each wavelength can be calculated by Fourier-transform analysis of the obtained interferogram. This principle is two-dimensional Fourier spectroscopy with phase-shift interferometer between the object lights. Hence, by using all object lights converged by an objective lens, the two-dimensional spectroscopy for a very week light can be obtained.

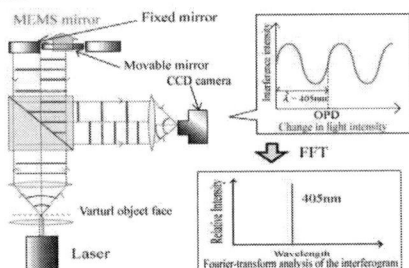

Fig.1 Principle configuration of the spectrometer with phase-sift method

3. Configuration of MEMS mirror

In this study, the vertical electrostatic comb-drive actuator is used in the actuator part and the differential capacitance detection method [4] is used in sensor part as shown in fig.2. The sensor consists of two kinds of comb teeth, A TYPE (The height of the mobile comb tooth is the half of the fixed comb tooth) and B TYPE (The height of the mobile comb tooth is the double of the fixed comb tooth) as shown in fig.3. The capacitance change in the horizontal direction is canceled, and the capacitance change in only the vertical direction can be the output value by the differential detection method. It is possible to detect the vertical displacement and the inclination angle as shown in table.1 by designing the sensors in 4 directions of a movable mirror as shown in fig.2.

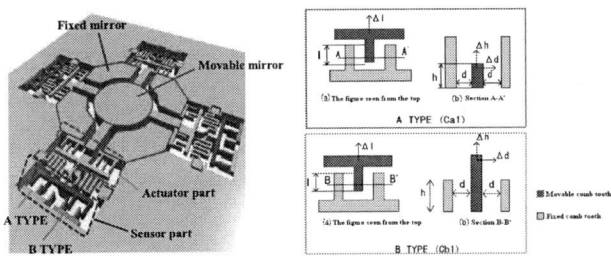

Fig.2 The configuration of the MEMS mirror   Fig.3 The configuration of a set of comb tooth part

Table.1 Output of capacitive sensor according to the drive direction

	I	II	III	IV	V	VI
①	−	+	−	+	0	0
②	−	+	0	0	+	−
③	−	+	+	−	0	0
④	−	+	0	0	−	+

I :Down II :Up III :Front inclination IV :Back inclination V :Left inclination VI :Right inclination
+ : Capacitance increase − : Capacitance decrease 0 : No change

4. Design and Fabrication of MEMS mirror

The capacitance change between the comb teeth of the sensor was analyzed using the finite element analysis software "ANSYS". The sensitivity 0.062 [pF/μm] and measuring range 10 [μm] which is appropriate to the spectroscopic measurement can be obtained using the comb tooth interval 4 [μm] as shown in fig.4. The designed configuration of the device is shown in fig.5. [5] The actuator part, the sensor part, the movable mirror and the fixed

978-1-4244-8926-8/10 $26.00 © 2010 IEEE

mirror were fabricated monolithically as shown in fig.6.

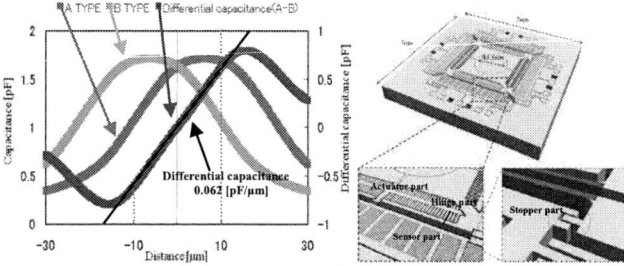

Fig.4 The analysis result of the sensor part

Fig.5 Designed configuration of a device

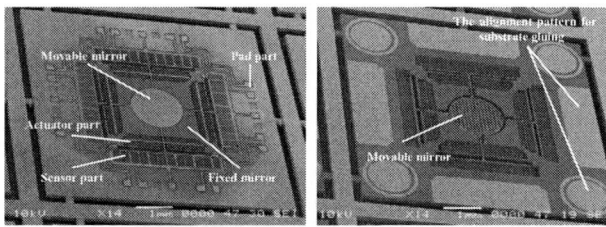

Fig.6-a The surface side   Fig.6-b The back side

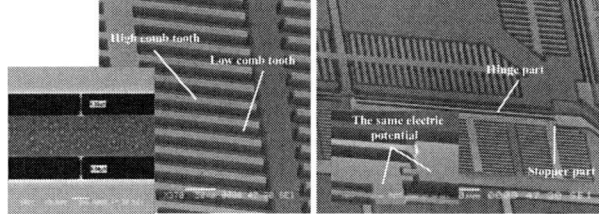

Fig.6-c The sensor part   Fig.6-d The stopper part

5. Characteristics of MEMS mirror

The static drive characteristic of the fabricated MEMS mirror with a sensing function was examined. By applying the same voltage to 4 actuators and the differential capacitance change in each sensor was measured. The experimental result is shown in fig.7. First, it was confirmed that 4 inner sensors showed the good differential capacitance characteristic which was almost same sensitivity as the designed value in fig.7.

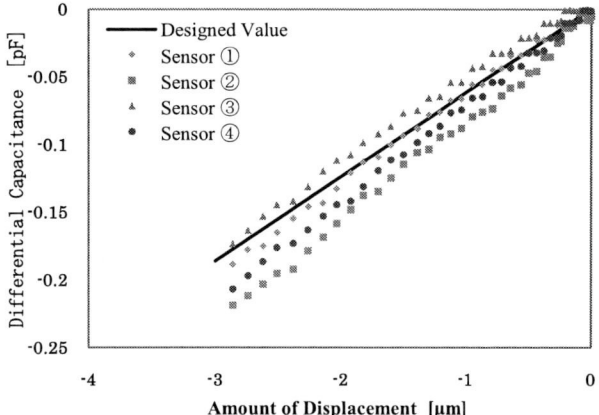

Fig.7 The measurement result of 4 inner sensors

Next, by applying the same voltage to 4 actuators, the vertical amount of the displacement was measured by the laser displacement sensor (measured value) which was used as the outside sensor. By using 4 inner sensors, the vertical amount of the displacement was measured from the differential capacitance change. The experimental result is shown in fig.8. The amount of the displacement of 3 [μm] was obtained in about 40 [V] which is the same as the designed value. It was confirmed that 4 inner sensors showed almost the same amount of displacement the

characteristic as the outside laser sensor.

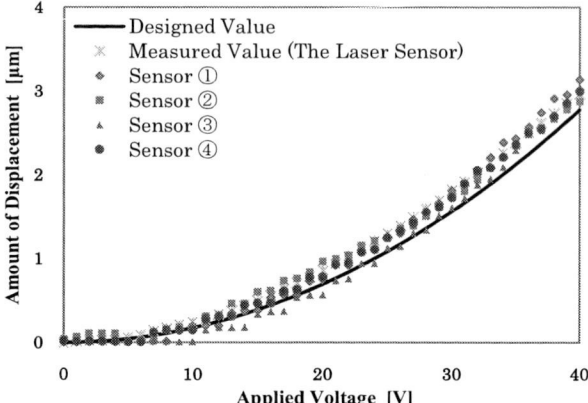

Fig.8 The comparison of outside laser sensor and 4 inner sensors

Finally, by applying the voltage to the actuator (fig.2-②), the movable mirror is tilted intentionally (table.1-VI), and the tilting angle was measured by 4 inner sensors. The experimental result is shown in fig.9. It was confirmed that 4 inner sensors characteristic can distinguish tilting angle and the tilting direction from each output of 4 inner sensors as shown in fig.9.

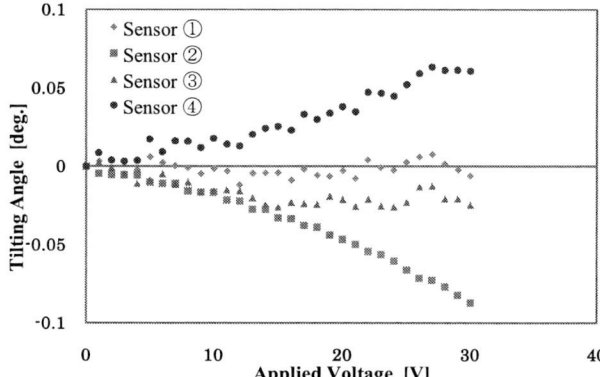

Fig.9 The detection of tilting angle and the direction of the MEMS mirror

6. Conclusions

The vertical comb-drive MEMS mirror with a sensing function was designed and fabricated for the purpose of applying to the spectroscopic sensing with a phase-shift method. The evaluation of the moving and sensing characteristic was performed. The amount of displacement of 3 [μm] was obtained in about 40 [V], which is the same as the designed value. It was confirmed that 4 inner sensors showed almost the same amount of displacement the characteristic as the outside laser sensor. Then it was confirmed that 4 inner sensors characteristic can distinguish tilting angle and the tilting direction from each output of 4 inner sensors. From these results, it was shown that this vertical comb-drive MEMS mirror with the sensing function has feasibility for the phase-shift devices which can control the vertical amount of the displacement and the tilting angle of the mirror.

References

[1] H. Obi, H. Fujita, H. Toshiyoshi, "Design for high stable electrostatic torsion mirror with vertical-comb actuators", Proceedings of IEICE General Conference 2004, pp.242, 2004
[2] Y. Inoue, I. Ishimaru, T. Yasokawa, K. Ishizaki, M. Yoshida, M. Kondo, "Variable phase-contrast fluoresce spectrometry for fluorescent stained cell", Appl Phys Lett, Vol.89, No.12, pp.121103-121103-3, 2006
[3] D. Inoue, F. Oohira, K. Yamamoto, M. Kondo, T. Harada, I. Ishimaru, G. Hashiguchi, M. Hosogi, "Application of vertical comb-drive MEMS mirror for a phase-shift device", Proceedings of the 24th sensor symposium, pp.367-370, 2007
[4] H. Hamaguchi, K. Sugano, T. Tsuchiya and O. Tabata, "A Differential Capacitive Three-Axis SOI Accelerometer Using Vertical Comb Electrodes", Transducer'07, Vol.2, pp.1483-1486, 2007
[5] K. Oda, F. Oohira, I. Ishimaru, T. Suzuki, "Vertical Comb-Drive MEMS Mirror with a Sensing Function for Phase-Shift Devices", Proceedings of the 13th Intelligence Mechatronics Workshop, pp.299-304, 2008

978-1-4244-8926-8/10 $26.00 © 2010 IEEE

ARRAY OF CAT'S EYE RETRO-REFLECTORS WITH MODULABILITY FOR AN OPTICAL IDENTIFICATION SYSTEM

Keng-hsing Chao, Chun-da Liao, and Jui-che Tsai

Graduate Institute of Photonics and Optoelectronics and Department of Electrical Engineering

National Taiwan University

No. 1, Sec. 4, Roosevelt Rd., Taipei 10617, Taiwan

Tel: +886-2-3366-3700 Ext. 247, Fax: +886-2-3366-3686, E-mail: jctsai@cc.ee.ntu.edu.tw

Abstract

In this paper, we present an array of cat's eye retro-reflectors with modulability for an optical identification (ID) system. The modulability is obtained through the use of a smart film, a polymer dispersed liquid crystal device (PDLC). Switching of the image pattern of retro-reflected light is demonstrated. Such a dynamic image can function as the "key" so that a high-security, high-privacy identification system can be constructed.

Keywords: Retro-reflector, cat's eye, optical ID

1. INTRODUCTION

Radio frequency identification (RFID), which has been demonstrated to have numerous applications, is a technology often seen in our daily life. Examples of RFID applications include product identification, in a supply chain, making transactions with VISA payWave, etc. However, it still exhibits several weaknesses. For example, an RFID system can easily experience interference caused by the surrounding electromagnetic waves in the environment. And the coding can only be done in the time domain, limiting the security of RFID. Moreover, the RF signals are basically radiating electromagnetic waves and may be easily intercepted and eavesdropped, which also compromises the security of this technology.

Using light waves to carry the information can resolve these issues. An optical beam, particularly a laser beam, can be directional, making the signal difficult to be intercepted. This greatly enhances the system security. Moreover, optical signals are not affected by surrounding electromagnetic waves. Therefore, a more accurate identification process can be achieved. The barcodes seen in nowaday supermarkets are an example of the optical ID. However, such a barcode scatters, instead of reflecting, the scanning laser beam sent from the reader, and, therefore, the advantage of the directionality of the laser beam is not taken. Also, due to the radiating property of scattered light, the distance between the barcode tag and reader has to be kept short for enough optical power to arrive at the reader.

With an optical retro-reflector, the laser beam can be reflected directly back along its incoming path, regardless of the incident angle. This way, the returned signal to the reader can be directional, decreasing the chances of being eavesdropped; and adding the ability to modulate light intensity to the retro-reflector will further accomplish the goal of remote interrogation/identification. The signal strength is, of course, greater than that of the scattered light received by the reader in a barcode-based optical ID system. Moreover, by employing an array of optical retro-reflectors with modulability, the signal coding can be performed in both the time and spatial domains. This means a dynamic image can be generated and used as the

"key" for the identification (Fig. 1). An optical ID system with enhanced security and privacy can then be constructed.

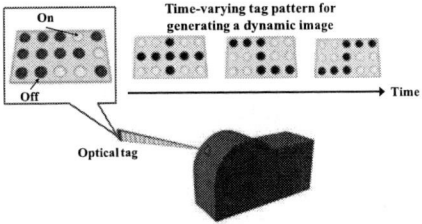

Fig. 1. Concept: an optical identification system using a dynamic image as the "key".

One of the most commonly used retro-reflectors is the corner cube retro-reflector (CCR), which consists of three orthogonal mirrors [1]. However, the fabrication process of MEMS CCRs with modulability, particularly arrays of them, is complicated, normally requiring several steps of delicate assembly. Cat's eye retro-reflectors [Fig. 2(a)] are also often seen [2]. I. Glaser *et al* demonstrated a smart card with an array of "cat's eyes", all of which were to be switched on or off together by a reflective LC modulator [3]. However, their so-called "cat's eye" intrinsically is not a cat's eye retro-reflector as the back reflecting surface is flat instead of being concave. This would significantly reduce the acceptance angle. Also, their purpose of having an array is to increase the light accepting area of the smart card, not to generate any optical image or pattern. As a whole, the smart card acts as a single switchable "quasi" retro-reflector, capable of only time-domain coding.

In this paper, we present an array of real cat's eye retro-reflectors with modulability for an optical identification system. The modulability is obtained through the use of a smart film, a polymer dispersed liquid crystal device (PDLC). Switching of the image pattern of retro-reflected light is demonstrated.

2. DEVICE DESIGN AND FABRICATION

A simple cat's eye retro-reflector consists of a front focusing lens and a back concave reflecting mirror, as shown in Fig. 2(a). Assuming a homogeneous cat's eye material, R_1+R_2 has to be equal to the focal length of the micro lens to achieve optimal retro-reflection [2]. For our device, a smart film, which can modulate the light transmittance, is

978-1-4244-8926-8/10 $26.00 © 2010 IEEE

sandwiched between the arrays of micro lenses and concave micro mirrors [Fig. 2(b)].

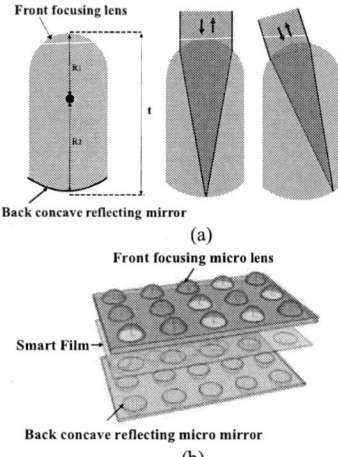

Fig. 2. (a) A simple cat's eye retro-reflector and (b) exploded 3-D view of our array of cat's eye retro-reflectors with modulability.

The micro lenses are made of the photoresist AZ4620 and fabricated on a 150-μm-thick glass substrate (D263 T Borosilicate Thin Glass). Thermal reflow process is used to form the required spherical lens shape. The back concave mirrors are fabricated with the same fabrication process, except that an additional silver coating is needed on the photoresist after the dome shape is formed with the thermal reflow step. The part with the concave mirrors is then flipped, and the smart film and module of micro lenses are attached in order. Fig. 3 shows the device photos taken by an optical microscope. For a nominal device, the diameter of the cat's eye and the pitch of the array are 300 μm and 1 mm, respectively, and the radii of curvature of the micro lens and back concave micro mirror are 447 μm and 730 μm.

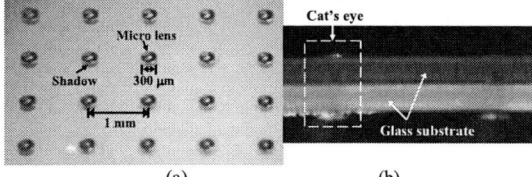

Fig. 3. Device photos: (a) a micro lens array only, and (b) a cat's eye array device with no smart film and air gap and without silver coating on the concave mirror part.

3. EXPERIMENTS

Fig. 4 is the experimental setup. A lens with a 4-cm focal length is placed in front of the 632.8-nm He-Ne laser. The laser beam is first focused, then diverges, and finally hits the sample with an incident of 17° and a beam diameter of 1.2 cm. The retro-reflected light is reflected by the beam splitter, and then passes through the imaging lens so that an image is formed on the observation plane/screen.

We first test a cat's eye array without the smart film. The spacing between the upper (micro lens) and lower (concave mirror) modules is 0.14 mm. As shown in Fig. 5(a), an array of red bright spots, resulting from the retro-reflected light from the cat's eye device, can be observed on the observation screen. Devices with spacing far off the ideal value or some misalignment between the upper and lower modules are also tested for comparison. No bright spot can be observed as the conditions for retro-reflection are not met.

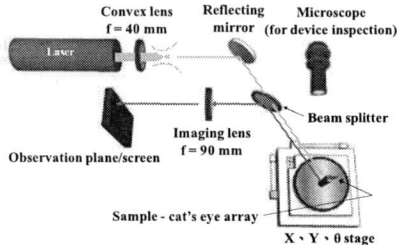

Fig. 4. Experimental setup

We next experiment on the device with the smart film. The smart film is purchased from the Chiefway Optronics Co., LTD [4]. The light transmittance is >75% under the clear mode (65 V applied voltage) and is <6% in the opaque mode (no applied voltage). The response time is around 10 msec for switching from the clear to opaque mode, and is around 2 msec for the reverse switching. The switching of the image pattern is demonstrated by Fig. 5(b)-(c). Fig. 5(b) is the image when the voltage is applied, and an array of red light spots is observed. Once the voltage is removed, the smart film becomes opaque and the light spots disappear [Fig. 5(c)].

Fig. 5. (a) Observed image with a cat's eye device of no smart film. (b)-(c) Demonstration of the switching of the image pattern, showing the images taken when the applied voltage to the smart film is (b) on and (c) off.

4. CONCLUSIONS

We have reported a device with an array of cat's eyes with modulability. The switching of the image pattern of the retro-reflected light has been demonstrated. This device can be used as the optical ID tag of an identification system. Future devices will incorporate the function of independent control over individual cat's eyes. Thanks to the retro-reflection property, such an optical identification system can possesses high security and the capability of long-distance interrogation/identification.

ACKNOWLEDGEMENTS

This work was supported by National Science Council of Taiwan under Grants NSC 98-2221-E-002-090 and NSC 96-2221-E-002-165-MY3, and Excellent Research Projects of National Taiwan University, 98R0062-07.

REFERENCES

[1] C. S. Chang et. al., "A novel addressable switching micro corner cube array for free-space optical applications," *Proc. of MEMS 2003*, pp. 279- 282.
[2] A. Lundvall et. al., "High performing micromachined retroreflector," *Optics Express*, vol. 11, pp. 2459-2473, Oct. 6 2003.
[3] I. Glaser et. al., "Optical smart card using semipassive communication," *Optics Letters*, vol. 31, pp. 712-714, Mar. 15 2006.
[4] http://www.smart-film.com.tw/index.asp

MAGNETIC ACTUATED MOEMS RESONANT BIOSENSOR ARRAY

Erman Timurdogan, Sezin Nargul, Halil Kavakli, Erdem Alaca, Hakan Urey

Koç University, Istanbul-Turkey

ABSTRACT

A biosensor platform is utilized in liquid for detection of His protein. T-shaped thin film nickel cantilevers are actuated magnetically using self-oscillation principle and the sensing s based on embedded diffraction gratings on the cantilevers with interferometric sensitivity. Experiments are performed n a custom 1ml liquid chamber and 5.7ng/ml detection sensitivity is demonstrated. Method is label-free, robust and allows multiplexing. It also doesn't require any electrical connections to the MEMS sensor chip. Therefore, it is suitable for real-time multi analyte screening using a disposable chip in a portable device.

MAGNETIC THIN FILM ACTUATED CANTILEVER BIOSENSOR AND ADVANTAGES

The resonant microcantilevers are widely investigated as AFM tips, gas and mass detectors. Mass sensing ability of he microcantilever arrays are utilized as detection of analyte mass and biosensors.

We report recent progress in our biosensor technology that utilize magnetic actuation with thin Nickel films and diffraction grating based readout using embedded diffraction gratings.[1] Fig. 1 illustrate the MEMS cantilevers and Fig. 2 illustrate the actuation and optical readout principle. Thin Ni cantilevers are actuated by an external electrocoil at resonance. This actuation technique eliminates problems such as stiction and pull-in and allows liquid operation.

Some of the key advantages of the label-free sensor echnology include:

- There are no electrical connections on the MEMS sensor, allowing for a completely disposable sensor;
- Optical detection method is suitable for multiplexed operation;
- Operation both in air and liquid is possible;
- The sensor is more robust compared to Laser Doppler Vibrometer and other optical readout techniques;
- Read-out is not affected by the refractive index variations due to dynamic mode operation that monitors the frequency instead of deflection.

l

SENSOR CHARACTERİZATİON

Microcantilever arrays are fabricated with single mask ithography. Fabrication flow of double layer (Au-Ni) microcantilever arrays is explained in detail in a previous work. [1] The biological mass is sensed through the shift in he resonance frequency of microcantilevers which is monitored optically. Precision of the detection and SNR are mproved by magnetic optimization, self-excitation of

cantilevers, and by detecting the 1st order diffracted light that reject bias and other noise.

The sensitivity of the biosensor is improved by increasing the vibration frequency, which can be achieved either by decreasing the dimensions of the micro cantilever or using higher order vibration modes. [2] Fig. 3 shows the resonant peaks and Q-factor measured both in air and in liquid environments.

Since proposed biosensing platform is envisioned as a multi-analyte, low cost, frequency multiplexed system. Direct excitation with a signal generator is not desired. Self-excitation is used to eliminate signal generator and increase SNR at the same time.[3,4] Brownian motion of the cantilevers is monitored using a photodiode (PD) placed at the 1st order diffraction angle. The PD intensity is amplified and phase-shifted to achieve resonant deflection and used as a drive signal.

Fig. 4 shows the custom fluidic chamber with 1mL sample volume that is used in the biosensing experiments, which also illustrate the details of the magnetic drive coil and magnet configuration.

Fig. 5 shows label-free detection of the interaction between hexahistidine-tagged human K-opioid receptor membrane protein and anti His antibody demonstrated in liquid. After surface is saturated with HKor Antigen which is verified with SEM images, Anti His Antibody is immobilized on the surface with different concentrations. A reference measurement is performed without the DSP linker. A minimum detectable antibody concentration of 5.7ng/ml was calculated using the slope of the frequency shift vs. concentration curve and the variation in the reference measurements.

The measurement principle is shown to impart immunity to environmental noise, facilitate operation in liquid media and bring about the prospect for further miniaturization of actuator and readout leading to a portable biochemical sensor.

REFERENCES

[1] A. Ozturk, et al., "A magnetically actuated resonant mass sensor with integrated optical readout", Photonics Technology Letters, 2008

[2] F. Lochon, et al. "An alternative solution to improve sensitivity of resonant microcantilever chemical sensors: comparison between using high-order modes and reducing dimensions", Sensors and Actuators B Chemical, 2004

[3] D. Ramos, et al., "Phototermal self-excitation of nanomechanical resonators in liquids", Applied Physics Letter, 2008

[4] J. Tamayo, et al., "Underlying mechanisms of the self-sustained oscillation of a nanomechanical stochastic resonator in a liquid", Physical review B, 2007

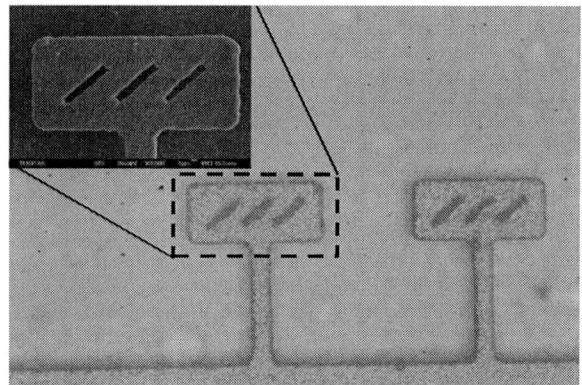

Figure 1. Fabricated cantilever array under light microscope and inset shows the SEM image of the particular cantilever (5x50 tail piece and 35x18 head piece in um, and thickness 1.1um)

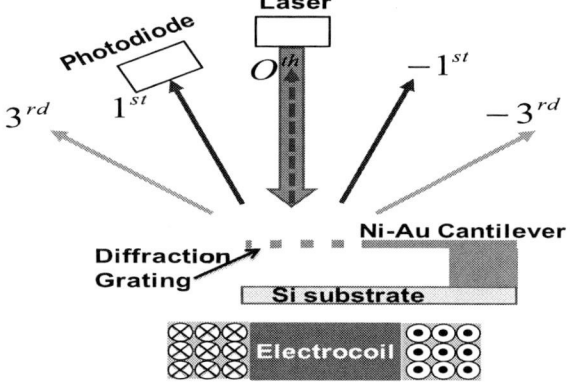

Figure 2. Conceptual drawing of the optical readout and magnetic actuation

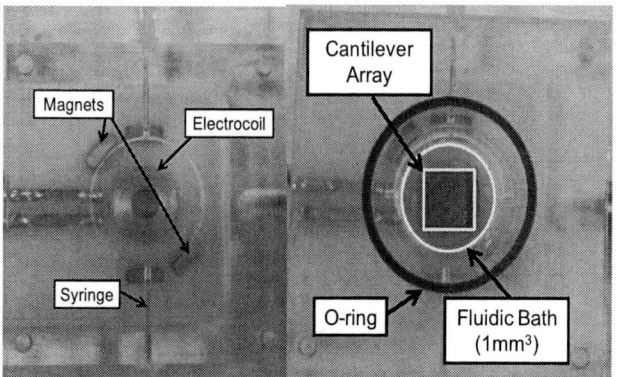

Fig. 4 Integrated electronics and cantilever array inside the fluidic Chamber

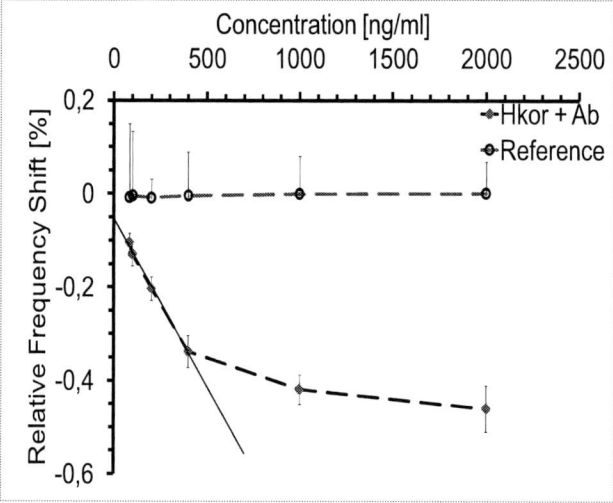

Figure 5. The relative frequency shift due to Hkor Antigen and Antibody and reference measurements.

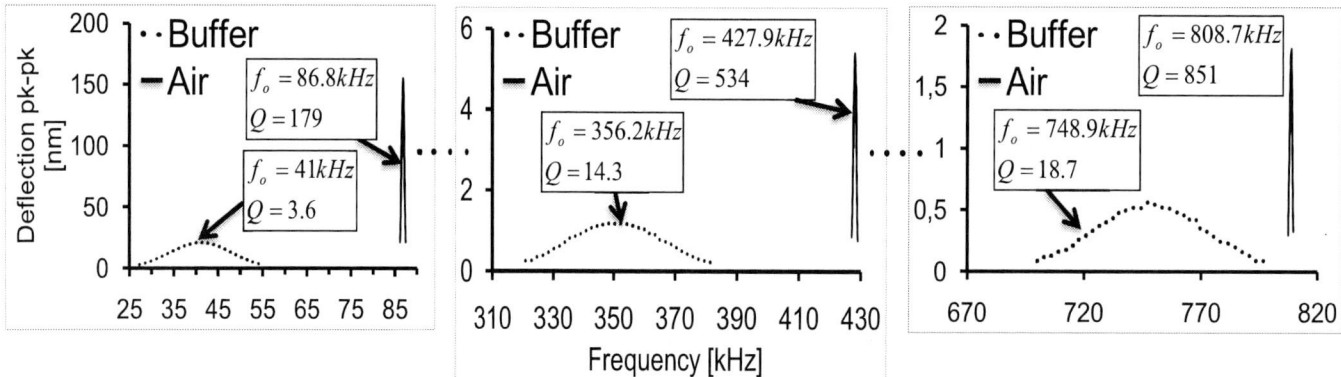

Figure 3: Measured frequency response for 1st, 2nd, and 3rd bending modes in air and in buffer solution

978-1-4244-8926-8/10 $26.00 © 2010 IEEE

Development of an integrated microsystem for the multiplexed detection of protein markers in serum using electrochemical immunosensors

Ciara K. O' Sullivan

Department d'Enginyeria Química, Universitat Rovira i Virgili, Av. Països Catalans, 26, 43007, Tarragona, Spain
Institució Catalana de Recerca i Estudis Avançats, Passeig Lluís Companys 23, 08010 Barcelona, Spain.

Recent advances in the fabrication of microfluidic platforms initiated during the late 90s have facilitated the realisation of micro total analysis systems [1]. The integration of miniaturised fluidic handling and delivery systems with chemical and biochemical sensors provide applied scientists with powerful tools for in-field measurements away from central laboratories [2]. Amongst the various classes of elements able to transduce a chemical or biochemical events into a measurable signal, electrochemical platforms undoubtedly present the most promising advantages. Electrodes of all type, sizes and geometries can easily be integrated within a microfluidic platform and provide excellent sensitivity and versatility in comparison to other transduction techniques based on for example optical or mass sensing [3]. Furthermore, the associated electronics used to drive the electrochemical detection and signal processing can also be easily miniaturised and integrated onto the same platform by carefully designing application specific integrated circuits [4]. We have recently reported a simple and rapid approach for prototype microfluidics and sensor assembly to perform complex protein and genetic electrochemical assays with excellent reproducibility [5]. The microfluidic platform was realized by high precision milling of polycarbonate sheets, which offers flexibility and rapid turn over of the desired designs. Sixteen-electrode sensor arrays were fabricated using photolithographic deposition technologies in order to realize three-electrodes cells comprising of gold counter and working electrodes as well as silver reference electrode. Fluidic chips and electrode arrays were assembled via a laser machined double-sided adhesive gaskets, creating the microchannels necessary for sample and reagent delivery. Surface chemistry methodologies were evaluated in order to achieve the double function of eliminating non-specific binding and optimal spacing of the anchor biocomponents for maximum accessibility to the target proteins. Storage conditions were optimized, demonstrating a long-term stability of the reporter conjugates jointly stored within a single reservoir in the microsystem. The final system has been optimized in terms of incubation times, temperatures and simultaneous, multiplexed detection of the protein markers was achieved in less than 10 minutes with less than ng/mL detection limits. The microsystem has been validated using real patient serum samples and excellent correlation with ELISA results obtained.

Keywords: electrochemical biosensor, microsystem, clinical diagnostics, surface chemistry

978-1-4244-8926-8/10 $26.00 © 2010 IEEE

TUNABLE OPTOFLUIDIC MICRO-IRIS
Philipp Müller, Nils Spengler, Wolfgang Mönch, and Hans Zappe
University of Freiburg, Germany

ABSTRACT

A tunable aperture stop based on optofluidic technology is presented. Using the high absorption of aqueous pigment dispersions, we demonstrate that this approach may be used to define optical stops of high contrast. Our highly flexible design is based on photolithographic patterning of dry film resist and allows control of laminar flow in microfluidic chambers.

INTRODUCTION

A basic element of many optical systems is a variable stop, or iris, which allows the adjustment of important parameters such as the system f/#, the field stop diameter or the overall energy throughput. Most approaches for micro-optical tunable stops have been based on micromechanics, using established micro-machining fabrication technology [1], [2]. While these approaches have enabled highly integrated devices, the functionality is often limited by complicated actuation mechanisms and limited fabrication tolerances. Moreover these approaches generally suffer from a disadvantageous ratio between tunable aperture area and total device size.

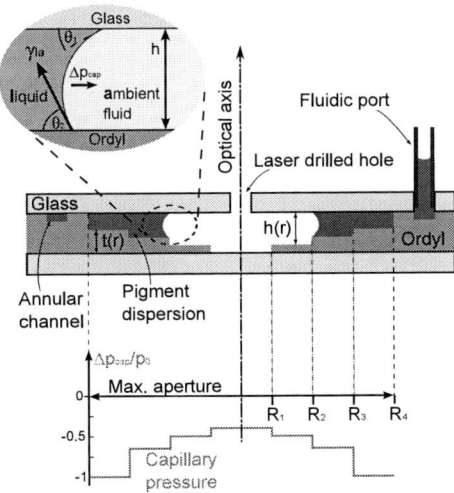

Figure 1: Schematic cross-sectional view of the microfluidic structure. The resist thickness t(r) is varied for circular areas of radius R_i to create different chamber heights h(r), resulting in different capillary pressures Δp_{cap}.

Optofluidic technology, on the other hand, is a promising approach for conceiving ultra-compact micro-optical devices. We present here a new type of miniaturized tunable iris based on optofluidics in which the key properties of liquids (easily reshapeable interfaces, high mobility) may be exploited for creating a perfectly circular stop without the need for any mov-able parts. The flow control in our design is based exclusively on capillary forces.

DESIGN

As illustrated by the schematic cross-sectional view of Figure 1, the capillary forces are generated by a closed fluidic chamber geometry of varying height h(r), which is defined by photolithographic patterning of dry film resist. In equilibrium, the capillary pressure difference across the interface between the liquid and the surrounding ambient fluid (which can be either a gas, or a second liquid) can be approximated by:

$$\Delta p_{cap} \cong -\frac{\gamma_{la}}{h}(\cos \theta_1 + \cos \theta_2). \quad (1)$$

Here, γ_{la} denotes the interfacial tension between liquid and ambient fluid and θ_1, θ_2 are the static contact angles at the upper and lower interfaces limiting the chamber (Fig. 1). By defining the areas of varying height h(r) in circular form (Fig. 3), a stepwise decrease of capillary pressure in radial direction is created (Fig. 1), which forces the liquid meniscus to take a toroidal shape.

FABRICATION

The microfabrication process is shown in Figure 2. The fluidic structure is defined by successive lamina-

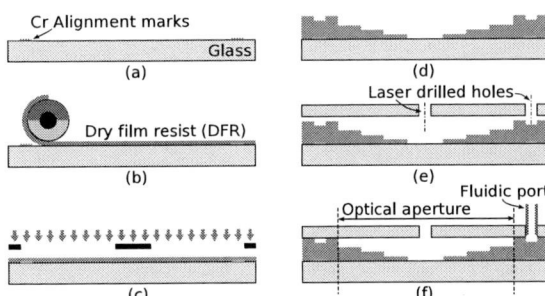

Figure 2: Fabrication process flow for the optofluidic stop.

tion and photolithographic patterning of five layers of 30 μm thick dry film resist (*Ordyl SY300* from *ElgaEurope*). Fluidic ports are drilled using a UV-laser. The microfluidic chambers are formed by full wafer bonding of two 4" glass wafers with the patterned resist in between, followed by chip dicing. Dry film lamination results in very uniform coatings with low thickness variation, essential for a high bond yield. A final tempering step at 200°C greatly improves the transmittance of the resist film.

EXPERIMENTAL RESULTS

To evaluate the fluidic operation, the optofluidic iris was actuated with an aqueous pigment dispersion and air as the second "fluid", using a programmable syringe drive pump. Numerical FEM-simulations of

978-1-4244-8926-8/10 $26.00 © 2010 IEEE

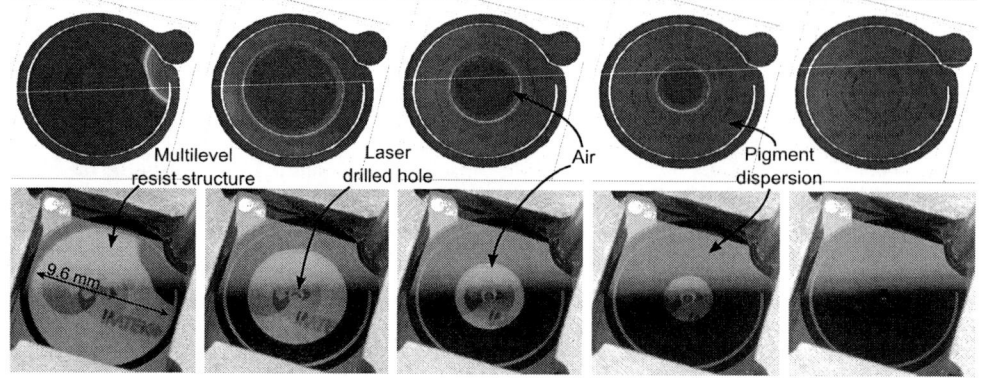

Figure 3: Liquid flow in the optofluidic stop: Comparison of FEM simulations (top) to experimental actuation (below).

the transient flow were done with *COMSOL Multi-physics®3.5a*, based on measured contact angles of the dispersions on glass and Ordyl. From Figure 3, it is clearly visible that the circular steps in the multilevel resist structure act as pinning barriers for the advancing liquid meniscus, as predicted by the capillary model and the FEM simulations. The actuation could be performed reversibly and repeatedly without degradation of stability.

MEASUREMENT RESULTS

Figure 4 provides data from transmission measurements that were performed with the optofluidic stop and a standard mechanical iris diaphragm for comparison. As can be seen, the transmittance of the optoflu-

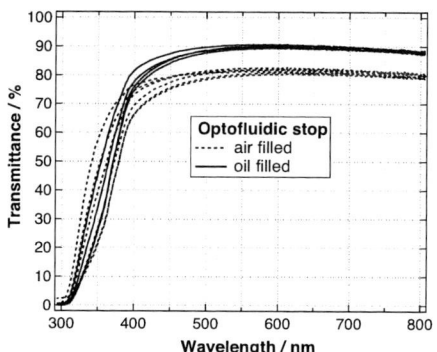

Figure 5: Absolute transmittance spectra of the optofluidic stop measured at different positions within the aperture. The device was first filled with air and then with an index matching oil (n=1.47).

CONCLUSION AND OUTLOOK

A novel design for a miniaturized tunable optical iris has been presented. We demonstrated control of laminar flow exclusively by capillary forces, which may be of interest for other optofluidic MEMS. The device allows tuning a 9.6 mm aperture in five discrete steps, with a maximum transmission of up to 90 % in the visible range.

Currently we are developing a more advanced prototype that will feature two-level-fluidics and an integrated electrowetting actuation to achieve a fully encapsulated device that can be electrically controlled.

REFERENCES

[1] C. H. Kim, K. D. Jung, and W. Kim: *A wafer-level micro mechanical global shutter for a micro camera*. In *Optical MEMS 2009, Sorrento Italy*, 2009.

[2] R. R. A. Syms, H. Zou, J. Stagg, and H. Veladi: *Sliding-blade MEMS iris and variable optical attenuator*. Journal of Micromechanics and Microengineering, 14:1700–1710, 2004.

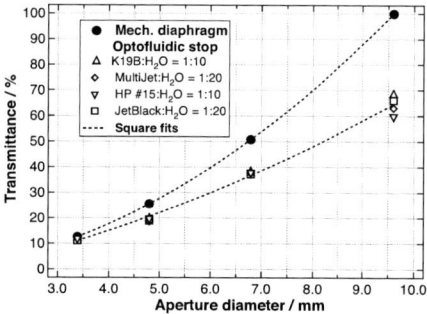

Figure 4: Measured absolute transmission of the optofluidic stop, filled with a number of pigment dispersions and air as the second fluid, in comparison to a mechanical diaphragm for different aperture diameters.

idic stop can be tuned in a manner similar to a mechanical diaphragm. Due to Fresnel reflections at the glass-air interfaces and scattering at the rough resist surface ($R_{rms} \approx 120$ nm), the maximum transmittance is limited to approximately 65%. By using a counter-fluid of higher refractive index, optical losses in the fluidic chamber can be nearly completely eliminated, resulting in high uniform transmission in the visible range, as illustrated in Figure 5.

SINGLE CHAMBER ADAPTIVE MEMBRANE LENS WITH INTEGRATED ACTUATION

Jan Draheim[1], Florian Schneider[1], Tobias Burger[1], Robert Kamberger[1] and Ulrike Wallrabe[1,2]

[1] Department of Microsystems Engineering - IMTEK, University of Freiburg, 79110 Freiburg, Germany

[2] Freiburg Institute for Advanced Studies (FRIAS), University of Freiburg, 79104 Freiburg, Germany

jan.draheim@imtek.uni-freiburg.de

ABSTRACT

In this work we introduce an extremely thin single chamber adaptive fluidic membrane lens with an integrated piezoelectric bending actuator. The height of the system is 1.76 mm and the membrane diameter 10 mm. The lens consists of a ring shaped piezo bending actuator with an optical clear silicone membrane in the center. A refractive power range of 11.3 dpt is achieved at a piezo voltage between ±40 V. The system shows a maximum resolution of 189 lp/mm measured at 50 % contrast and a focal length of 200 mm.

1 INTRODUCTION

In most available autofocus systems the change of the focal length is achieved by modifying the position of lenses which is very elaborate and costly. Adaptive lenses overcome these disadvantages by using different working principles (e.g. electrowetting [1], photo polymers [2]). Fluidic membrane lenses were published with different designs. However the fabrication often requires cleanroom processes (SU-8 lithography, ICP-RIE etching) [3-5], which make these systems costly. Furthermore some of these systems need an external pressure controller to change the focal length [3].

The objective of this project is the design of a thin low cost silicone membrane lens based on a single fluid chamber including a pumping unit. The focal length of the adaptive component changes by applying a voltage to the actuator.

2 DESIGN AND SIMULATION

The lens comprises one silicone chamber with an integrated piezoelectric bending actuator and a glass substrate (Fig. 1). The enclosed volume is filled with an optical transparent fluid. The piezoelectric bending actuator consists of two active layers.

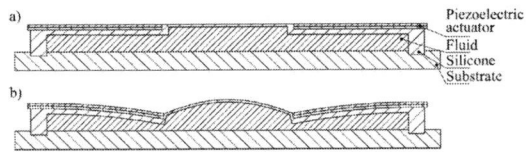

Fig. 1: Design and working principle of the liquid membrane lens. a) $U = 0$ V b) $U > 0$ V.

A ring trench is structured to the glass substrate for the alignment of the lens chamber. In the initial state (Fig. 1a) no voltage is applied to the piezoelectric bending actuator, hence the membrane is flat. If a positive voltage is applied to the piezoelectric bending actuator the inner actuator part displaces downwards. Since the fluid volume inside the chamber is constant the silicone membrane bulges upwards (Fig. 1b) and a plano-convex lens is formed. At a negative voltage the principle works in the opposite direction with a concave shape. The total lens has a height of 1.76 mm and a diameter (incl. glass substrate) of 25 mm. Due to the variability of the inner and outer piezo radii and the lens chamber the system is easily scalable.

The lens is calculated with a mechanical simulation in Comsol Multyphysics. The influence of the lens fluid is modelled by a constant volume domain inside the chamber. Afterwards the refractive power is computed by the analysis of the membrane shape over the aperture.

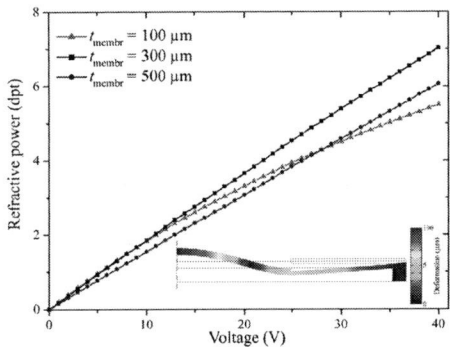

Fig. 2: Simulation of the refractive power as a function of the applied voltage for different membrane thicknesses (t_{membr}). The voltage ranges from 0 V to 40 V (d_{lens} = 10 mm, d_{ap} = 5 mm).

Fig. 2 shows the simulated refractive power as a function of the applied voltage (0 V to 40 V) for membrane thicknesses from 100 μm to 500 μm. The membrane diameter d_{lens} is 10 mm and the analyzed aperture d_{ap} = 5 mm. The largest refractive power range of 7.0 dpt is achieved for a medium thickness of 300 μm. This is caused, on the one hand, by the decrease of the pump volume for increasing membrane thickness, and by the increasing radius of curvature of the membrane for decreasing membrane thickness, on the other hand. Due to the higher membrane stiffness the two thicker membranes (t_{membr} = 300 μm, t_{membr} = 500 μm) depict an almost linear curve progression in the entire voltage regime. The membrane with t_{membr} = 100 μm shows a non linear behaviour because the inflection point is shifted towards the outer edge.

978-1-4244-8926-8/10 $26.00 © 2010 IEEE

3 FABRICATION

The fabrication of the lens comprises three main steps. The first step is the manufacture of the piezoelectric bending actuator (Ekulit GmbH), followed by the casting of the lens chamber (Fig. 3a/b) and the filling process. The home-made two-layer actuator is structured with an UV-laser.

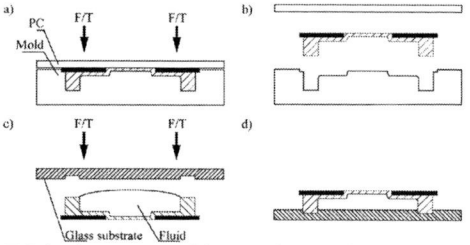

Fig. 3: Fabrication process of the adaptive membrane lens.

After contacting the lower electrode by soldering a 100 µm thick copper wire, the piezo is placed into the aluminum mold which is readily filled with silicone (RTV 615). Subsequently, the mold is sealed with a polycarbonate foil with optical surface quality (R_a = 1.5 ± 0.25 nm). A weight of m = 1.6 kg is placed onto the mold while the silicone is cured at 100 °C for 75 min in an oven (Fig. 3a). After demolding (Fig. 3b) the center and top electrode are contacted by soldering. The lens is then placed top down and filled with an optical fluid. Meanwhile, a glass platelet is fabricated using the same laser. A ring shaped alignment trench with a depth of 150 µm is lasered into the glass. Silicone is dispensed in the trench and degassed.

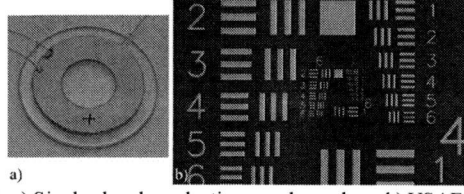

Fig. 4: a) Single chamber adaptive membrane lens. b) USAF 1951 test chart for a lens with t_{membr} = 300 µm at f = 500 mm.

Finally the glass is pressed top down onto the filled cavity and aligned by the trench (Fig. 3c). The compound is then loaded with a weight and cured at 100 °C for 60 min. A picture of the final device is shown in Fig. 4a.

4 SYSTEM PERFORMANCE

The optical system behaviour is measured for lenses with membrane thicknesses of 100 µm and 300 µm.
The resolution of the lens is determined performing a MTF measurement using an USAF 1951 test chart (Fig. 4b). The system with t_{membr} = 100 µm achieves a maximum resolution of 189 lp/mm at 50 % contrast and a focal length of 200 mm exemplarily. Furthermore the refractive power as a function of the applied piezo voltage (Fig. 5) is calculated. The membrane deflection is measured with a laser triangulation sensor for this purpose. The refractive power is calculated by a spherical fit across an aperture of 5 mm. A refractive power of -5.0 dpt to 4.3 dpt for t_{membr} = 100 µm and ±5.6 dpt for t_{membr} = 300 µm is achieved at piezo voltages between ±40 V. Due to initial

membrane stress induced by volume shrinkage during the fabrication the measured refractive power is below the values predicted through the simulations of Fig. 2. The curve shows the typical hysteresis behaviour for piezoelectric materials.

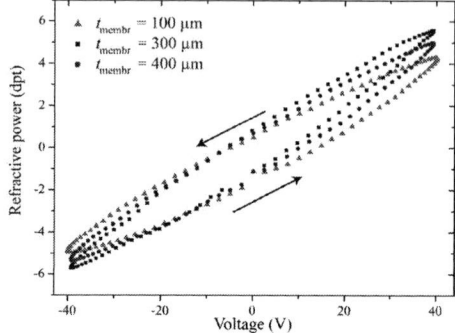

Fig. 5: Refractive power as a function of the piezo voltage for a system with t_{membr} = 100 µm, t_{membr} = 300 µm, t_{membr} = 400 µm (d_{ap} = 5 mm).

Finally the dynamic behavior of the lens with t_{membr} = 300 µm is measured. Fig. 6 shows the refractive power as a function of the time for a step function from -40 V to 40 V.

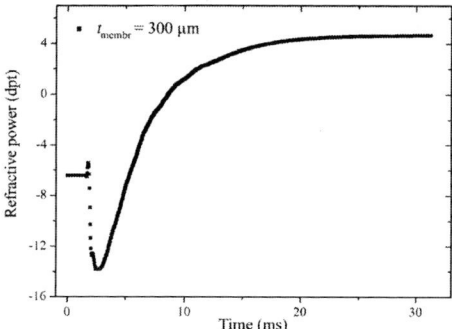

Fig. 6: Resolution of a membrane lens with a membrane thickness of t_{membr} = 100 µm (d_{ap} = 5 mm).

Due to mechanic fluidic interaction a slight increase of the refractive power at the beginning of the step is observed. The curve shows a damped behavior with a response time of 24 ms. The damping is caused by the viscosity of the fluid and the stiffness of the membrane

5 CONCLUSIONS

A simple and flat design for an adaptive lens as well as an easy and lean fabrication process without the need of cleanroom processes is demonstrated. A refractive power of ±5.6 dpt and a response time of 24 ms are obtained by a lens with 300 µm membrane thickness.

6 REFERENCES

[1]   B. Berge et al., *Eur. Phys. J. E*, 3, pp. 159-163, 2000
[2]   S. Xu et al., *Opt.Express*, 17, no. 20, pp. 17590-595, 2009
[3]   A. Werber et al., *Appl. Opt.*, 44, no. 16, pp. 3238-45, 2005
[4]   D.Y. Zhang et al., *Opt. Lett.*, 29, no. 24, pp. 2855-57, 2004
[5]   J. Draheim et al., *J. Micromech. Microeng.*, 19, 2009

EYE-SHAPED COPLANAR VARIABLE LIQUID LENS USING SPHERICAL POLYMER ENCAPSULATION

Jae Yong An[1], Ji Hwan Hur[1], Sung Kil Lee[1] and Jong-Hyun Lee[1,2]
[1]School of Mechatronics, [2]Department of Nanobio Materials and Electronics,
Gwangju Institute of Science and Technology (GIST),
261 Cheomdan-gwagiro (Oryong-dong), Buk-gu, Gwangju, Korea 500-712
Tel +82-62-715-2395, Fax +82-62-715-2384, E-mail jonghyun@gist.ac.kr

Abstract

An eye-shaped coplanar liquid lens was fabricated using spherical polymer encapsulation to enhance the focusing power. For the spherical encapsulation, a thin film polymer layer was deposited by chemical vapor deposition on an outer surface of the top liquid. Polymer thin film (parylene) encapsulation with coplanar structure provides a unique advantage of wafer level packaging of liquid lens and convenient accessibility of the electrical signal. The fabricated variable liquid lens with an initial focal length of 35mm was tuned up to 12mm at V_{rms} 70 V.

Keywords: variable focus, liquid lens, spherical encapsulation, electrowetting, coplanar, polymer

1. INTRODUCTION

Variable focal lens is one of the essential elements for many optical systems for mobile phone camera, endoscope imaging, optical communication and so on. In conventional lens systems, mechanical translation of optical component was required to change the focal length. While they worked well in macro system, miniaturization is limited without micro fabrication and assembly of small elements. Recently, many kinds of variable micro lens were developed using pneumatic [1], thermal [2], electrostatic [3] and electrowetting actuation [4]. Among them, electrowetting is very attractive because of its simple fabrication, fast response and no mechanical moving part. In this paper, electrowetting based variable liquid lens was suggested using spherical polymer encapsulation to enhance the focusing power and realize the wafer level packaging. For easy electrical accessibility, coplanar electrode structure was also employed.

2. DEVICE DESIGN

The schematic view of the device was shown in Fig. 1. It consists of outer parylene layer, two immiscible liquids (oil and KCL mixture), hydrophobic dielectric layer and two ITO (indium tin oxide) electrodes (ground and signal line).

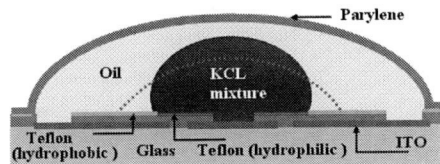

Fig. 1. Schematic view of the eye shaped variable liquid lens.

Droplet of KCL mixture is located at the central Teflon region (hydrophilic) and mineral oil was sat having a contact with the outer ITO region (hydrophilic). When the AC voltage is applied to the electrode, the KCK mixture changes its radius curvature leading to the adjustment of the focal length.

By adopting the spherical polymer encapsulation, high focusing power can be realized at a low operation voltage. Meanwhile, wafer level packaging is preferable to the conventional glass bonding in view of fabrication cost and mass production. Coplanar electrodes were also implemented for easy electrical accessibility.

3. FABRICATION

Fig. 2 shows the fabricate sequence of the variable liquid lens. Firstly, Cr (200 *nm*) was sputtered on an ITO (250 *nm*) coated glass substrate, and align marks and electrode are patterned using photolithography followed by wet etching. Secondly, parylene C (900 *nm*) was deposited for the insulation layer and Teflon AF (150 *nm*) was spin-coated for the hydrophobic surface. Thirdly, double-layer lithography using LOR 30B (Microchem) and RIE (reactive ion etching) were conducted for Teflon and parylene patterning. In order to position the droplet on the central area, the Teflon surface (1 *mm* in diameter) was changed using a double-layer lithography and oxygen plasma (10 *s*, 10 *Watt*). Remained parts of the PR and LOR 30B were removed by acetone and developer, respectively. Then, KCL mixture and oil were dispensed positioning on the optical axis around the hydrophilic area. Finally, parylene C (1.5 *um*) was deposited on the device.

As a surrounding liquid, mineral oil (refractive index 1.467) was used. KCL (0.1 M) and glycerin (KCL: glycerin=1:0.3 weight ratio) were mixed to make the liquid with 1.37 of

978-1-4244-8926-8/10 $26.00 © 2010 IEEE

refractive index. Through the process sequence, a variable liquid lens was successfully fabricated as shown in Fig. 3. The size of the fabricated device is 15x15 mm^2, and the lens diameter is 3.4 mm.

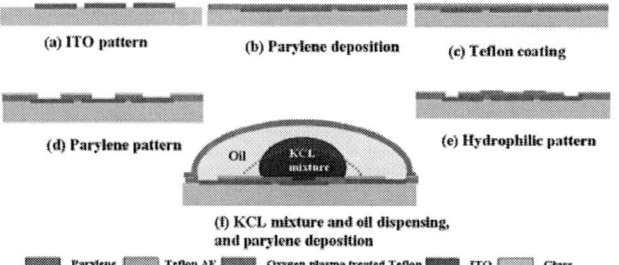

Fig. 2. Fabrication sequences of the variable liquid lens.

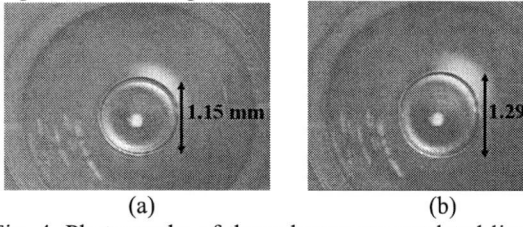

Fig. 3. Photograph of the fabricated variable liquid lens.

4. RESULTS AND DISCUSSIONS

The radius of curvature of KCL mixture was changed with respect to the applied voltage to adjust the focal length as shown in Fig. 4. Due to the hydrophilic surface of Teflon, the droplet of KCL mixture spreading was balanced keeping its position on the optical axis.

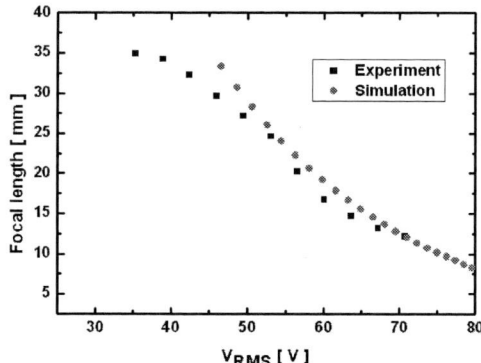

Fig. 4. Photographs of the polymer encapsulated liquid lens with the applied voltage: (a) off state, (b) V_{RMS} 50 V.

The fabricated variable liquid lens was evaluated in terms of driving voltage and focal length. A laser diode (LD), optical fibers and collimator were prepared to direct the collimated light source to the central area of the lens. Focal length was measured using a confocal microscope (Nano focus) by tracking the confocal image of the collimated beam with a translation stage, as shown in Fig. 5.
Measured initial focal length is 35mm with the spherical mineral oil (2280 um in radius of curvature), and KCL

mixture droplet of which estimated radius of curvature is 548 um. The device could be tuned up to 14mm with V_{RMS} 70 V at the driving frequency of 50 kHz. There are no apparent focal length changes below V_{RMS} 35 V because of the static friction between KCL mixture and hydrophilic Teflon region. Positive focusing power was obtainable owing to the spherical encapsulation even though the light passing through the KCL mixture initially diverges through the mineral oil.

The measured focal length was compared with ray tracing simulation. In the initial operation with low voltage, the comparison shows a discrepancy, which might be caused by imperfect hydrophilic surface of the Teflon layer.

Fig. 5. Experimental results of the focal length with respect to the applied voltage at driving frequency of 50 kHz.

5. CONCLUSION

An eye shaped coplanar variable liquid lens was successfully demonstrated using spherical polymer encapsulation. The proposed variable liquid lens can provide high focusing power and easy wafer-level packaging. Coplanar electrodes were also implemented for easy electrical accessibility. The variable liquid lens having a 35 mm of initial focal length could be tuned up to 12 mm with V_{RMS} 70 V at 50 kHz.

ACKNOWLEDGMENT

This work was supported by the World Class University (WCU) program at GIST through a grant provided by the Ministry of Education, Science and Technology (MEST) of Korea (Project No. R31-2008-000-10026-0).

REFERENCES

[1] Peter M. M. et al., Applied Physic Letter, Vol. 88, 041120, 2006
[2] Weisong. W. et al., IEEE Photonics Technology Letters, Vol. 17, No. 12, pp. 2643-2645, 2005
[3] Nguyen. B. et al., Applied Physics Letter, Vol. 93, 124101, 2008
[4] Kuiper S. et al., Applied Physics Letter, Vol. 85, No.7, pp.1128-1130, 2004

Dynamic Response of Dielectric Liquid Microlens

Chih-Cheng Yang[1], Yih-Ching Wang[2] and J. Andrew Yeh[1,2]*

[1]Institute of NanoEngineering and MicroSystems

[2]Department of Power Mechanical Engineering

National Tsing Hua University, Hsinchu, Taiwan

*Phone: 886-3-5715131 ext. 42912, E-mail: jayeh@mx.nthu.edu.tw

Abstract

A dynamic response of liquid microlens using dielectric force was demonstrated using comparison between experimental and simulation results. The theoretical results of the simulation match fairly well with the experiment in liquid microlens with a diameter of 500 μm. An high speed motion camera with a speed of 3000 fps was used to capture the images of dynamic response for 500μm, 2mm and 5mm droplets in diameter, respectively. Results indicate that a viscosity strongly affects dynamic response time for 2mm droplet in diameter, but only weakly impacts on final contact angle. A propagation of wave on the droplet surface was found when $100V_{rms}$ voltage applied on microlens with a diameter of 5mm. The study may provide the basis for development of optical liquid microlens.

Keywords: liquid microlens, dielectric force, dynamic response, contact angle

1 INTRODUCTION

In recent years, tunable liquid microlens has received serious attention research because that this technology can adjust focal length with no mechanical moving parts. Liquid lenses emerge to be adopted in these regimes, such as mobile phones, endoscopic imaging, light valve, optical communication systems [1-3]. The dynamic response of contact angle of droplet interfaces plays an important and interesting physical phenomenon in liquid microlens applications.

However, the physics behind contact angle movement is not entirely understood. A difficulty has been proposed that the fluidic Navier-Stokes equations coupled with electric Laplace equation in multi-physics analysis. In this study, we report that an implemented multi-physics model that captures the effect of dynamic contact angle. The dynamic response of contact angle for liquid microlens was successfully tracked in a COMSOL simulation. This simulation is currently verified against experiments. The predicted model shows a good agreement well with experimental data in a diameter of 500 μm droplet. A dynamic response for viscosity effect was also investigated in liquid lens with a diameter of 2mm.

2. MECHANISM AND SIMULATON

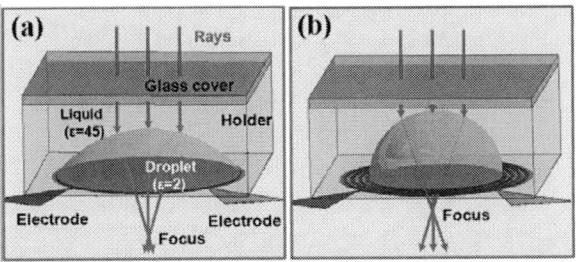

Figure 1: Schematic of liquid microlens sets on concentric electric inside a square sealing chamber (a) no voltage applied condition. (b) Contact angle changes when a voltage applied.

As shown in Figure 1(a), a liquid microlens system consists of a low dielectric constant droplet, a high dielectric constant sealing liquid. The two non-conductive liquids are packaged in a chamber under iso-density. Liquid droplet is set on coplanar concentric electrodes which provide non-uniform electric field to induce dielectric force. The shape and contact angle of liquid droplet are modified by dielectric force and surface tension, when a voltage applied. An adjusting shape and focal length of liquid microlens are clearly translated in Figure 1(b). The mechanical equilibrium between surface tension, electric field and gravity can be described mathematically by the augmented Navier-Stokes equations [4]:

$$\rho\frac{\partial u}{\partial t}+\rho(u\bullet\nabla)u=\nabla\bullet[-pI+\mu(\nabla u+(\nabla u)^T]+F_g+F_{elec}+F_{surface} \quad (1)$$

$$\nabla u=0 \quad (2)$$

where ρ denotes density of the fluid, u is fluid velocity, μ is dynamic viscosity , p is the pressure, I is the identity matrixan and F is a body force term such as gravity, electric and interfacial surface tension. The equation (1) is the momentum balance, and the equation (2) is the equation of continuity. Here a COMSOL simulation package was performed to investigate the dynamic droplet's shape in the electrical equilibrium state. This simulation considered an oil droplet with permittivity ε = 2, viscosity μ = 9×10^{-2} Pa-s and density is 1,024 kg/m³, surrounded by the mixed-alcohol with permittivity ε = 45, viscosity μ = 5×10^{-3} Pa-s. The diameter of the droplet is 500 μm.

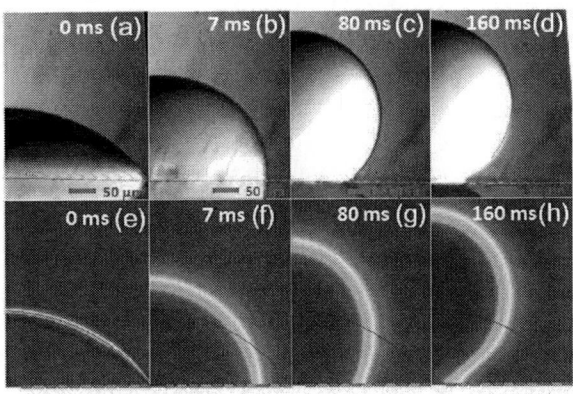

Figure 2: Dynamic response comparison of experiment images and simulation results for liquid microlens with a diameter of 500μm. The images captured by high speed camera with 3000 fps at (a)0ms (b)7ms (c)80ms (d)160ms when applied $70V_{rms}$@1kHz. Snapshots of simulation results show at time step (e)0ms (f)7ms (g)80ms (h)160ms when $70V_{rms}$ was used .

978-1-4244-8926-8/10 $26.00 © 2010 IEEE

Figure 2(a)-(d) show images in a resolution of 1024 by 1024 pixels were captured using an high speed motion camera (Redlake MotionXtra Y4) with a speed of 3000 fps. The drop starts from an initial position with a contact angle of 51°. When a potential of 70 V_{rms} was applied to the concentric electrodes, the unbalanced dielectric force along the droplet surface pushed the low dielectric contact fluid to move inward and raise up. Results of the simulation indicate a change of contact angle of a sessile drop at time steps t = 0, 7, 80 and 160 milliseconds, respectively, as shown in Figure 2(e)-(h). Apparent from figure 2(e)-(h), snapshots of simulation indicate that that the simulation from COMSOL model can be seen as good agreement with the experiment.

3 EXPERIMENT AND DISCUSSION

An high speed motion camera with a speed of 3000 fps was used to capture dynamic response of a liquid microlens in a resolution of 1024 by 1024 pixels. Image software (Sindatek Model 100SB MagicDroplet) was used to analyze the images with curve fitting to calculate contact angle. Figure 3 displays dynamic responses of a microlens with a diameter of 2 mm were captured when 60 V_{rms} voltage was applied. Four different viscosities in droplet were changed at 10 cSt, 50 cSt, 100 cSt and 200 cSt condition, respectively. The experiments are conducted under same initial condition at an equilibrium state, which initial contact angle of droplet is 30 degrees. Figure 3 evidently indicates that the viscosity has a strongly affect dynamic response time when 60 V_{rms} voltage applied. Lower viscosity of droplet has a quick response than higher one for the reason of low friction resistance. Final contact angle reaches the same 71 degree for different viscosities after a transient period.

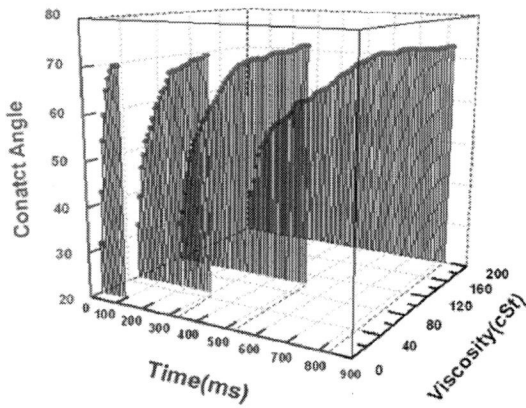

Figure 3: Dynamic responses of a liquid microlens with a diameter of 2 mm were captured when applied 60 V_{rms} voltage. Four different viscosities of droplet were investigated at 10 cSt, 50 cSt, 100 cSt and 200 cSt conditions in liquid microlens.

Figure 4 shows an interesting phenomenon for a wave propagating on the droplet surface between a transient period with 5 mm droplet in diameter. With no electrical actuation, the initial contact angle is 15 degree. The liquid droplet with kinematic viscosity 3.5 cSt surrounded by alcohol-mixture with 4.9 cSt is contained in a sealing box. The image of droplet was captured using high speed camera with a 3000 fps speed when 100V_{rms} @1kHz voltage was applied. Initially, the oil droplet displays the spherical shape as shown in Figure 4(a). Figure 4(b) and 4(c) show the deformation of the droplet causes by high non-uniform electric field near the contact line close to electrode gap. As approaching the base electrodes, the electric force abruptly increases to a tremendous value. The droplet therefore is pushed inward near the base electrodes because of larger dielectric force. Furthermore, the dielectric force locally resulted in the generation of wave propagating on the droplet surface can be observed. And the transient droplet surface can be thought as a result of wave superposition. Figure 4(d) shows the wave generate from two sides of the droplet and superpose on the middle part. The height of the droplet rise sharply as shown in Figure 4(e). Finally, as shown in Figure 4(f), the droplet changes to a new spherical shape by balancing the surface tension and the external electric force.

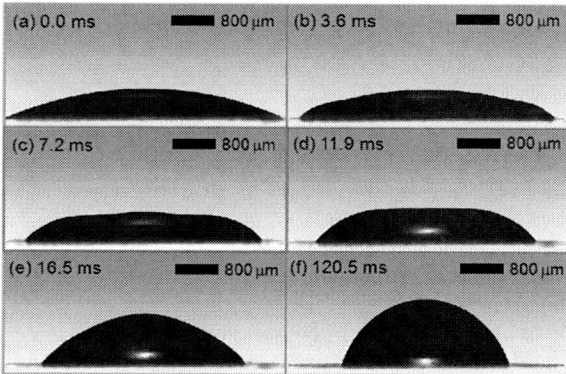

Figure 4: Snapshots of a oil droplet with a diameter of 5mm were captured with a speed of 3000 fps at (a) 0 ms (b) 3.6ms (c) 7.2 ms (d)11.9 ms (e) 16.5 ms (f)120.5 ms when 100V_{rms}@1kHz voltage applied on electrode.

4 CONCLUSION

A dynamic response of dielectric liquid microlens was demonstrated in this study. By successfully including an assembly of the two-phase fluid dynamics and electrically actuated boundary conditions, this well predicted model captures the effect of dynamic contact angle and compares with experimental results in a dielectric liquid microlens with a diameter of 500 μm. Experimental results indicated that a viscosity strongly affect dynamic response time when voltage applied on 2mm droplet in diameter. In contrast, final contact angle keep the same when different viscosities of droplet are used. The wave propagation was captured using an high speed motion camera with a speed of 3000 fps after voltage applied on electrode with a diameter of 5mm droplet. The study of dynamic behavior could provide the basis for development of optical liquid microlens.

REFERENCES

[1] R. A. Hayes and B. J. Feenstra, "Video-speed electronic paper based on electrowetting," Nature **425**, 383-385 (2003).

[2] H. Ren and S. T. Wu, "Tunable-focus liquid microlens array using dielectrophoretic effect," Optics Express **16** (4), 2646-2652 (2008).

[3] C. C. Cheng and J. A. Yeh, "Dielectrically actuated liquid lens," Optics Express **15** (12), 7140-7145 (2007).

[4] O. Lastow and W. Balachandran, "Numerical simulation of electrohydrodynamic atomization," Journal of Electrostatics **64**, 850-859 (2006).

OPTICAL SCANNING WITH MEMS IN-PLANE VIBRATORY GRATINGS AND ITS APPLICATIONS

Guangya Zhou*, Yu Du, Kelvin K.L. Cheo, Hongbin Yu, Fook Siong Chau

Dept. of Mechanical Engineering, National University of Singapore

*Corresponding author: Guangya Zhou, mpezgy@nus.edu.sg

ABSTRACT

MEMS optical scanners are highly desired due to their low-power, high-speed scanning. The in-plane vibratory grating scanner is a development in this area which possesses several unique features. The in-plane scanning minimizes the dynamic deformation, allowing for higher-resolution displays. The dispersive element permits splitting the incoming beam into its constituents for analysis and imaging. Coupling a grating platform to an in-plane moving structure is useful for real-time motion measurement which would otherwise be difficult to analyze. These applications are described including a recent development in the structural design of a double-layer layout which further improves the performance of the grating scanner.

INTRODUCTION

Microelectromechanical systems (MEMS) based micromirror laser scanners [1][2], which utilize out-of-plane deflection of a reflective surface to scan the laser beam, were mostly developed due to their outstanding advantages, such as having a miniaturized device size, a high scanning speed, low power consumption and a low per-unit cost, compared with macro laser scanners. The need to balance between maximizing scanning performance with minimizing the dynamic non-rigid body deformation of the reflective surface and significant aberration to the optical system during high speed scanning [3][4] has been a limiting factor in their development.

MEMS vibratory grating scanners [5]-[7], which utilize in-plane rotation of a diffraction grating, have the potential to scan at a high scanning speed with enhanced mechanical stability and subjected to less dynamic deformation. Multi-wavelength collinear laser scanning, suitable for miniaturized raster-scanning color display applications, was demonstrated [8]. High optical efficiency was also achieved (close to that of a coated scanning mirror) using subwavelength gold-coated diffraction gratings [9] and large optical scan angle using a 2-DOF circular resonator design [10].

DOUBLE-LAYER MEMS GRATING SCANNER

The initial designs had the actuation mechanism and the grating platform in the same plane. Since the rotational angle of the grating platform is inversely proportional to the diameter of the diffraction grating when the maximum allowable deformation of platform suspension flexures is fixed, the aperture size and the optical scan angle cannot be increased simultaneously. This significantly limited both the aperture size and the scanning angle.

Recent double-layer design overcomes this limitation. In this configuration, the driving actuators (Fig.1a) and the grating platform (Fig.1b) are fabricated separately. They are then post-assembled to form a double-layer design (Fig.1c & 1d). The size of the diffraction grating can be increased and the size of the connection platform reduced independently, thus increasing the aperture size and the scanning amplitude simultaneously. Moreover, the structural thickness of the grating platform and the actuation layer can be varied independently. Thinning the grating platform and increasing the actuators can lower the rotational inertia and increase the structural rigidity respectively. This is useful in maintaining the operating frequencies on the back of having a larger rotational inertia when the aperture size increases.

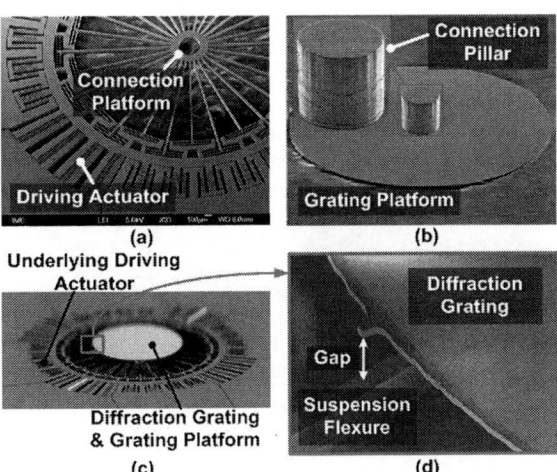

Figure 1: a) The actuation layer with connection platform, b) the bottom of grating platform with connection pillar, c) the assembled double-layer device, d) close-up showing the separated layers of the top grating and the bottom suspension flexures.

A prototype scanner with a 2mm diameter diffraction grating was fabricated. Fig. 2 shows the measured frequency response of the prototype grating scanner near the region of the first resonating mode. The inset shows a photograph of the projected laser scanning trajectory on a projection screen, which is located at a distance of 100 mm from the grating scanner. We were able to achieve scanning at a frequency of 23.391 kHz with an optical scan angle of 33°, resulting a $\theta_{optical}D$ product of 66 deg·mm.

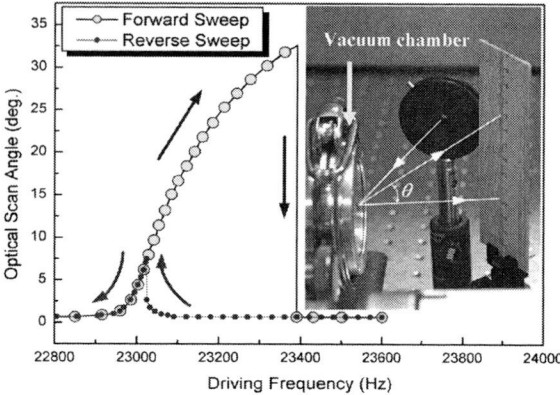

Figure 2: Frequency response of the grating scanner with inset of projected scan line on a projection screen.

APPLICATIONS

The in-plane vibratory grating scanner mechanism opens up several developments. Fundamentally, it is a potential large-scan-amplitude and high-frequency optical scanner. With the double-layer design, high-resolution laser projection displays [11] are possible. The presence of a dispersive grating element can also be utilized to configure the grating scanner as a low-cost, miniature line hyperspectral imager [12]. As shown in Fig.3, we had previously demonstrated the viability to capture the spectral image of two different wavelength laser diodes positioned in a straight line 11mm apart.

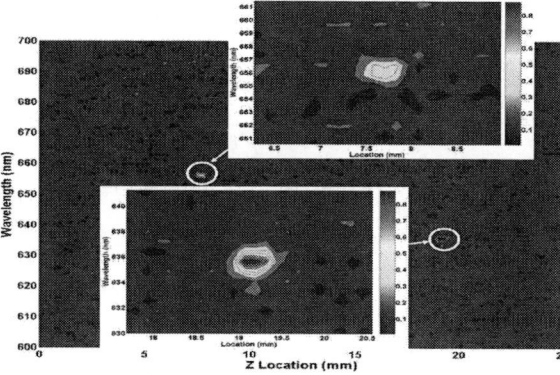

Figure 3: Captured spectral image showing the location and corresponding wavelength of the two laser diode inputs.

Another novel application is in the real-time precision measurement of in-plane motion of micro devices. Similar to the geometrical signal amplification of an AFM tip for out-of-plane deflection, measurement of in-plane motion of microstructures can be augmented using a grating platform. With one end of the grating directly connected to the movable platform under test, while the other end is fixed to the substrate, the in-plane movement of the platform will be translated into the grating's rotation. With a laser beam incident on the grating, the small rotation angle of the grating is magnified into a diffracted linear scan onto a PSD for measurement as in Fig.4.

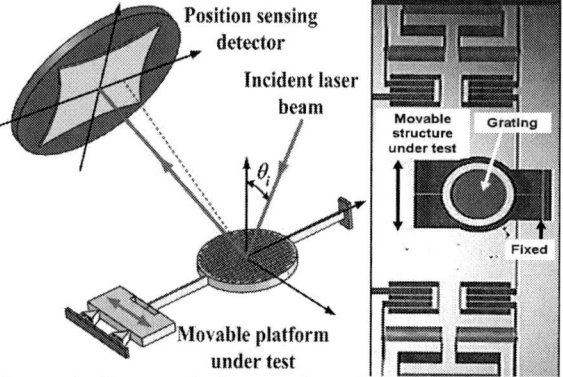

Figure 4: Grating platform configured to measure in-plane displacement of a movable structure.

REFERENCES

[1] P.M. Hagelin and O. Solgaard, *IEEE J. Sel. Topics Quantum Electron.*, Vol.5, No.1, p. 67-74 (1999).

[2] R.S. Muller and K.Y. Lau, *Proceedings of the IEEE*, Vol. 86, pp. 1705-1720 (1998).

[3] R.A. Conant, J.T. Nee, K.Y. Lau and R.S. Muller, *IEEE/LEOS int. Conf. on Optical MEMS*, p. 49-50 (2000).

[4] J.T. Nee, R.A. Conant, R.S. Muller and K.Y. Lau, *IEEE/LEOS int. Conf. on Optical MEMS*, p. 9-10 (2000).

[5] G. Zhou, L. Vj, F.S. Chau, and F.E.H Tay, *IEEE photonic technology letters*, Vol. 16, p. 2293-2295 (2004).

[6] G. Zhou, L. Vj and F.S. Chau, *U. S. Patent 7542188 B2*, 28th, July, 2005.

[7] Y. Du, G. Zhou, K.L. Cheo, Q.X. Zhang, H.H. Feng, B. Yang and F.S. Chau, *Sensors and actuators A*, Vol. 156, No. 1, p. 134–144 (2009).

[8] G. Zhou and F.S. Chau, *IEEE J MEMS*, Vol. 15, No. 6, p.1777-1788 (2006).

[9] G. Zhou, Y. Du, Q.X. Zhang, H.H. Feng and F.S. Chau, *J. Micromech. Microeng.*, Vol. 18, No. 8, p.085013 (2008).

[10] Y. Du, G. Zhou, K.K.L. Cheo, Q.X. Zhang, H.H. Feng and F.S. Chau, *IEEE J MEMS*, Vol. 18, No. 4, p. 892-904 (2009).

[11] H. Urey, D. Wine, and T. Osborn, *Proc. SPIE, MOEMS Miniaturized Systems*, vol. 4178, p. 176-185 (2000).

[12] G. Zhou, K.L. Cheo, Y. Du, F.S. Chau, H.H. Feng, and Q.X. Zhang, *Optics Letters*, Vol. 34, No. 6, p. 764-766 (2009).

A LARGE ROTATIONAL ANGLE MICROMIRROR BASED ON HYPOCYCLOIDAL ELECTROTHERMAL ACTUATORS FOR ENDOSCOPIC IMAGING

Xiaojing Mu[1,2], Yingshun Xu[1], Janak Singh[1], Nanguang Chen[2], Hanhua Feng[1], Guangya Zhou[2], Aibin Yu[1], Chee Wei Tan[1], Kelvin Wei Sheng Chen[1], Fook Siong Chau[2]

[1] Institute of Microelectronics, A*STAR (Agency for Science, Technology and Research), 11 Science Park Road, Singapore Science Park II, Singapore 117685
[2] National University of Singapore, 9 Engineering Drive 1, 117576, Singapore

ABSTRACT

The paper presents a large rotational angle micromirror base on hypocycloidal electrothermal actuators for circumferential endoscopic imaging. The micromirror consists of a double-side Cr/Au coated high reflective mirror plate (1mm by 0.8mm) laterally supported by two hypocycloidal electrothermal actuators on both sides (Fig. 1(a)). In our design, 1μm PVD Al deposited on 2μm single crystal silicon (SCS) forms a bimorph microstructure with the length of 800 μm and the width of 60μm. Four bimorph structures were staggerly connected in parallel to form a hypocycloidal electrothermal actuator. In this configuration, a metal layer was on a silicon backbone in one bimorph structure while the metal layer was deposited below the silicon backbone in adjacent bimorph structures (Fig. 1(b)). Since the radius of curvature of each bimorph structure is the same, the deflection of each structure is the same. Hence the rotational axis keeps still and there is no lateral shifting effect. Simulations via finite element analysis (FEA) show that the mechanical deflection angle of a micromirror significantly increases by using this actuator design. 141.2° was found in the design with fully double-side Al coated actuators (Fig. 2(a, b)) and 68.6° was found in the design with only frontside Al coated actuators (Fig. 2(c, d)). Micromirrors were fabricated by a post-CMOS MEMS process on 8 inches SOI wafers. An optical microscopic image and a scanning electron microscope (SEM) micrograph of a released micromirror are shown in Fig. 3(a) and (b), respectively. However, so far we have not successfully patterned Al layer below the SCS layer as part of the actuator and therefore only micromirrors equipped by frontside Al coated actuators were experimentally characterized (Fig. 4). ~35° mechanical deflection was achieved by 2.6 V DC input voltage (Fig. 5). It has a discrepancy in comparison in comparison with the FEA simulation. -3dB cutoff frequency was found to be about 29 Hz as the large signal frequency response (Fig. 5). Current-voltage relationship of an electrothermal actuator is also shown in Fig. 5. A series of frames from a video of a switching micromirror shows various tilting angles of the micromirror under a sinusoidal drive signal with the amplitude of 2.6 V was still with absence of microstructures with backside Al coated, the concept of achieving large deflection angle by using hypocycloidal electrothermal actuators has been demonstrated. Both FEA simulation and experimental results prove the capability of the Single-axis rotational micromirror device.

References:

[1] D. E. Glumac, et al., *IEEE Trans. Ultrason., Ferroelect., Freq. Contr.,* vol. 45, no. 5, pp. 1145-1150, 1998.

[2] S. T. Todd, et al., *J. Opt. A: Pure Appl. Opt.,* vol. 8, pp. S352-S359, 2006.

[3] P. J. Gilgunn, et al., *J. Microelectromech. Syst.,* vol. 17, no. 1, pp. 103-114, 2008.

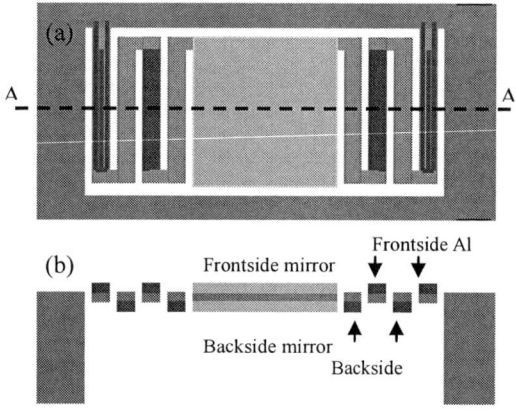

Figure 1. Schematic view of a large rotational angle micromirror with two hypocycloidal electrothermal actuators. (a) Top view and (b) cross-sectional view of the micromirror. A-A' shows the rotational axis.

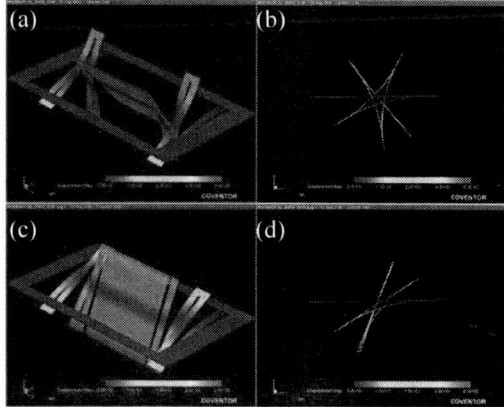

Figure 2. FEA results: (a) and (b) for the micromirror with double-side Al coated actuators; (c) and (d) for the micromirror with frontside Al coated actuators.

Figure 3. SEM micrograph of a released micromirror its insertion is an optical microscopic image and initial tilting of the mirror plate can be easily observed.

Figure 4. Experimental setup for micromirror characterization.

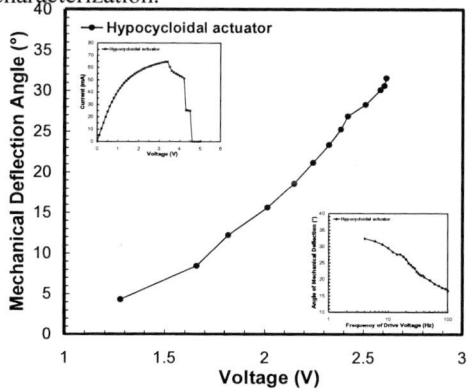

Figure 5. Mechanical deflection angle versus drive voltage. (top) Current-voltage relationship of an electrothermal actuator. (bottom) large signal frequency response of the micromirror

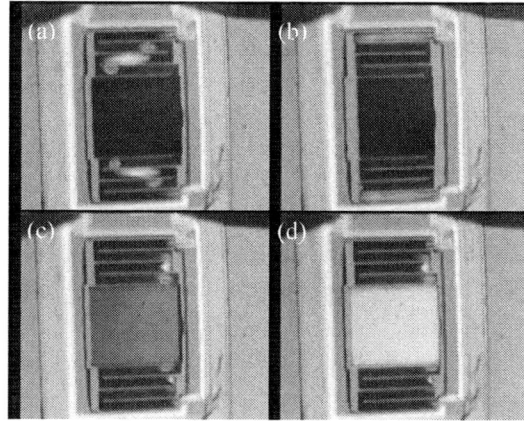

Figure 6. (a-d) A series of frames from a video of a switching micromirror show various tilting angles of the micromirror under a sine drive signal with the amplitude of 2.6 V and the frequency of 4 Hz.

TRANSLATORY MEMS ACTUATOR WITH EXTRAORDINARY LARGE STROKE FOR OPTICAL PATH LENGTH MODULATION

Thilo Sandner, Thomas Grasshoff, and Harald Schenk

Fraunhofer Institute for Photonic Microsystems (IPMS), Grenzstr. 28, D-01109 Dresden

Phone: +49-351-8823-152, Fax: +49-351-8823-266, E-mail: thilo.sandner@ipms.fraunhofer.de

ABSTRACT

A translatory MOEMS actuator with extraordinary large stroke – especially developed for fast optical path length modulation in miniaturized FTIR-spectrometers - is presented for the first time. A precise translational out-of-plane oscillation at 500 Hz with large stroke of up to 1 mm is realized by means of a new suspension design of the comparative large mirror plate with 19.6 mm² aperture using four pantographs. The MOEMS device is driven electrostatically resonant and is manufactured in a CMOS compatible SOI process. Up to ± 500 Hz amplitude has been measured in vacuum of 50 Pa and 90 V driving voltage.

Keywords: Optical SOI-MEMS, translatory micro mirrors, optical path length modulation, Fourier-transform infrared spectrometers, infrared spectrometry

1 INTRODUCTION

Fourier Transform Infrared (FT-IR) spectroscopy is a widely used method to analyze different materials - organic and inorganic. Current FT-IR spectrometers are large, usually static, and are operated by qualified personnel. By using translational MOEMS devices for optical path length modulation instead of conventional highly shock sensitive mirror drives a new class of miniaturized, robust, high speed and cost efficient FTIR-systems can be addressed.

An early approach of such a miniaturized MEMS based FTIR spectrometer has been developed in the past by the Fraunhofer Institute of Microsystem Technology (IPMS) and the Carinthian Tech Research (CTR) allowing dynamic FTIR measurements in the ms-range [1]. It was a combination of classical infrared optics with a translatory 5 kHz MEMS mirror using a folded bending spring mechanism. Due to the limited amplitude of ± 100µm a spectral resolution of 30 cm⁻¹ was realized [1]. In [2] a translational 500 Hz MEMS device with two pantograph mirror suspension – originally designed for larger stroke of 500 µm – was reported but only ± 140µm amplitude could be achieved experimentally in vacuum of 20 Pa due to superimposed parasitic torsional modes. Now, we present an optimized MEMS device which overcomes the previous limitations enabling an extraordinary large stroke of 1 mm.

2 TRANSLATORY MEMS Mirror

The novel translatory MOEMS actuator was specially designed to enable a miniaturized MEMS based FTIR spectrometer with improved system performance of 5 cm⁻¹ spectral resolution (λ =2.5…16µm), SNR > 1000 and fast operation of ≥ 500 scans / sec. Hence, a large mirror aperture of 5 mm, enhanced amplitude of ± 500 µm and a small dynamic deformation of < λ/4 is required.

2.1 Large stroke MEMS design

To realize a large stroke of the mirror plate a pantograph like suspension was chosen. Here, torsional springs are used us deflectable elements instead of bending springs which reduces significantly parasitic mirror deformation due to mechanical stress. The mirror plate is supported symmetrically by four pantograph suspensions (see fig. 1). One single pantograph consists of six torsional springs – two springs arranged on the same axis – and connected by stiff leavers (see fig. 3). The new translatory MEMS actuator consists of four symmetric pantograph suspensions in contrast to two pantographs used for a first MEMS design [2], where only ± 140µm amplitude could be achieved due to parasitic tilt modes [2].

Fig 1: Photograph of translatory MEMS @ 400µm mech. pre-deflected

In Figure 2 and table 1 a comparison of the new four pantographs and the previous two pantograph design is shown for the 1st and 2nd mode of the FEA modal analysis. Whereas a crucial tilting (2nd) mode at 1024 Hz is obvious for the old design with two pantographs an excellent mode separation for precise translation could be realized by using the new translational MEMS design with four pantographs.

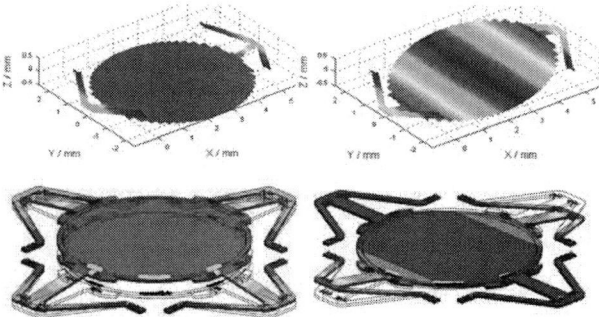

Fig. 2: Comparison of 2 (old design; upper figure) and 4 (new design; bellow) pantograph design in terms of modal analysis results. Shown are the 1st (left) and 2nd mode (right). A bad mode separation is obvious for the old design (see tilting 2nd mode).

Table 1: Parameter of translatory MEMS designs

Parameter	New design	Old design [2]
Design type	4 pantographs	2 pantographs
Mirror diameter	5 mm	3 mm
Amplitude	± 500 μm	± 250 μm
Frequency (1st mode)	500 Hz	500 Hz
Next parasitic (2nd) mode		1022 Hz

The translatory MEMS actuators are driven electro-statically resonant using in-plane vertical comb drives [3] located on each of the 4 pantographs for optimized driving efficiency.

2.2 Fabrication

The translatory MOEMS device are manufactured in a CMOS compatible SOI process [3] using a highly p-doped device layer of 75 μm and vertical open trenches for electrical isolation of out-of-plane comb drive. For electrical characterization and later vacuum packaging the MEMS chips are chip bonded on a ceramic wiring board.

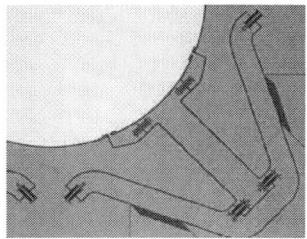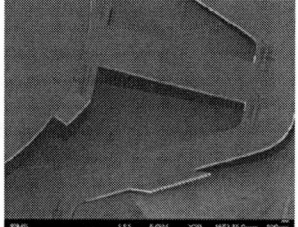

Fig. 3: Photographs of four pantograph suspension; un-deflected (left) and SEM of 400 μm pre-deflected translatory MEMS (right)

2.3 Experimental results

The frequency response behavior of the translatory MEMS devices have been measured at varied driving voltage and pressure by means of a MICHELSON interferometer setup. To vary the ambient pressure the MEMS sample under test was encapsulated in a small vacuum camber. The devices are driven in open loop operation [2] with a pulsed driving voltage of 50 % duty cycle and a pulse frequency twice the mechanical oscillation. Up to ± 500 μm amplitude has been measured in vacuum of 50 Pa and 90 V driving voltage; whereas no oscillation occurs in normal ambient due to squezze film and viscous gas damping.

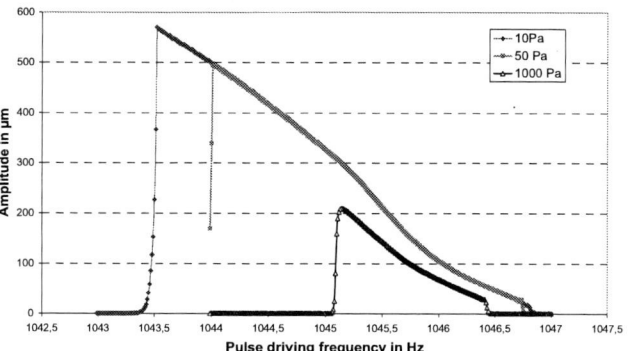

Fig. 4: Frequency response for varied vacuum pressure @ 90V

3 CONCLUSIONS and OUTLOOK

Due to an optimized mechanical design using 4 pantograph suspensions of a 5 mm large mirror plate the new translatory MEMS actuator can provide a precise out-of-plane translation with ± 500 μm amplitude in vacuum of 50 Pa. This enables a completely new family of low cost handheld FTIR analyzers with a spectral resolution of up to 5cm⁻¹, 500 scans/s and SNR > 1000 e.g. applied by individuals for ad-hock inspection of food or environmental parameters. We acknowledge financial support by the European Commission in the context of the FP7 project MEMFIS.

REFERENCES

[1] A. Kenda, et al., "Application of a micromachined translatory actuator to an optical FTIR spectrometer", Proc. SPIE 6186, pp. 78-88, 2006

[2] T. Sandner, et al., "Translatory MEMS actuators for optical path length modulation in miniaturized fourier-transform infrared spectrometers", Journal of micro/nanolithography, MEMS and MOEMS vol. 7 (2), pp. 021006: 1-12, 2008

[3] H. Schenk, et al., "Large Deflection Micromechanical Scanning Mirrors for Linear Scans and Pattern Generation", Journal of Selected Topics of Quantum Electronics vol. 6, pp. 715- 722, 2000

POLYMER-MEMS TORSION MIRROR WITH LARGE ROTATION ANGLE AND LOW DRIVING VOLTAGE

Dzung Viet Dao[1], Satoshi Amaya[2], and Susumu Sugiyama[1]

[1]Ritsumeikan University, Japan, [2]TOWA Corporation, Kyoto, JAPAN

ABSTRACT

This paper presents a novel fabrication of a monolithic PMMA torsional mirror utilizing hot embossing, surface-activated direct bonding, and the elliptical vibration cutting. The robustness and capability of the method are demonstrated through the fabrication of sophisticated PMMA freestanding micro structures. An efficient technique using reinforcement material to protect the PMMA microstructures during release process was proposed. Monolithic PMMA torsional mirror actuated by vertical comb actuator has been fabricated and tested successfully. Since the Young's modulus is 50 times lower than that of Si, the driving voltage of the PMMA actuator should be 7 times lower than that of silicon counterpart.

INTRODUCTION

Recently, PMMA becomes promising material for polymer-MEMS devices because of its several advantages against the silicon material, such as much lower stiffness, transparency, higher coefficient of thermal expansion, better biocompatibility and environmental friendly [1], lower material cost, and easier fabrication by mass-replication technologies, e.g. injection molding, hot embossing [2], and X-ray lithography. The transparency of PMMA is important for optical devices, such as lens, waveguide, etc [3], and optical-based biomedical or microfluidic devices. Small Young's modulus and high thermal expansion of PMMA are the advantages for electrostatic and thermal actuators [4], respectively, because it will require less power than silicon counterparts to produce the same displacement.

Recently, photoresist-based vertical comb actuators were developed by using UV-lithography method [5]. The fabrication process was simple, however, the polymers were limited to photoresist-based polymers, and the device was not monolithic since a glass substrate was used as a carry wafer. In another work, lateral PMMA electrostatic comb actuator was fabricated by hot embossing and O_2 plasma etching [6]. However, heat generation during the O_2 plasma strongly affected the surface of the device. In our recent report, by using elliptical vibration cutting, we could avoid the heat problem, and therefore, successfully developed a lateral PMMA actuator [2]. Basically, our proposed method can be applied for varieties of polymers, and the developments of polymer-MEMS devices are straightforward.

In this paper, the fabrication of a PMMA torsion mirror actuated by electrostatic vertical comb actuator is described to further demonstrate the applicability of the process into polymer-MEMS field.

STRUCTURES OF THE TORSION MIRROR

In order to demonstrate the robustness of the proposed fabrication method for polymer-MEMS devices, a PMMA torsion mirror device, which consists of a mirror plate driven by vertical comb actuators, will be developed. Figure 1 shows the structure of the torsion mirror device. The mirror plate is suspended by two torsion bars. The mirror plate can rotate reciprocally around the torsion bars by electrostatic force generated by the vertical comb-drive actuators located at both sides of the mirror plate. The vertical comb actuators are formed by pushing down the fixed comb fingers to create a vertical distance offset between the fixed and movable comb fingers as shown in figure 2. Dimensions of a finger are $60 \times 5 \times 60 \mu m^3$ ($L \times W \times T$), gap and overlap between two fingers are $5 \mu m$ and $50 \mu m$, respectively, and dimensions of torsion bar are $200 \times 10 \times 60 \mu m^3$. The size of the mirror plate is $1 \times 1 mm^2$. The gap between movable structure layer and the substrate is about $30 \mu m$.

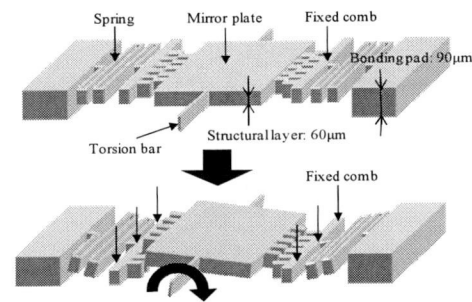

Figure 1. Schematic illustration of the formation of vertical comb actuator. The lower figure shows the vertical comb actuator is created by pushing downward the fixed comb fingers to create a vertical offset between the fixed and movable fingers.

FABRICATION PROCESS

The fabrication process of the PMMA mirror device with vertical comb actuator is shown schematically in figure 2. First, a two-step Si mold was fabricated by Deep-RIE (figure 2a), and then the device structures were formed by hot embossing process (figure 2b). Then, the sample was bonded to another PMMA substrate by surface activation method to create a firm bonding without using intermediate material (figure 2c). A PMMA layer with a thickness of about $40 \mu m$ is always remained after hot embossing process. In order to release the movable structures, this layer must be removed. In this study, this layer was removed by using the elliptical vibration cutting as shown in figure 2e. To protect the PMMA microstructures from fracture and deformation during the mechanically cutting process, reinforcement material was filled into the gaps of embossed PMMA structures (figure 2d). After the cutting process, reinforcement material was self-evaporated in vacuum at room temperature, and the freestanding PMMA microstructures, such as fingers, beams and mirror plate, were successfully released (figure 4). The surface roughness of the mirror was measured to be less than 87nm (Rz). Next, the fixed comb was pushed downward to create the vertical offset between movable and fixed electrodes (figure 2f). Finally, a gold layer was sputtered to the surfaces of the PMMA mirror device to create the electrode for the actuators (figure 2g).

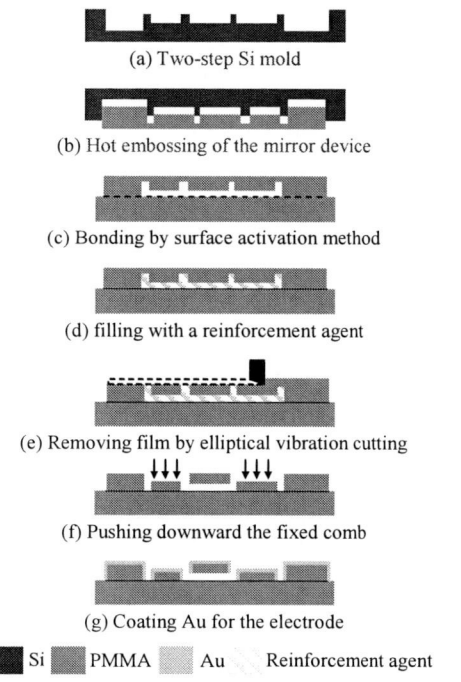

(a) Two-step Si mold

(b) Hot embossing of the mirror device

(c) Bonding by surface activation method

(d) filling with a reinforcement agent

(e) Removing film by elliptical vibration cutting

(f) Pushing downward the fixed comb

(g) Coating Au for the electrode

■ Si　■ PMMA　■ Au　Reinforcement agent

Figure 2.　Fabrication process of the PMMA mirror device.

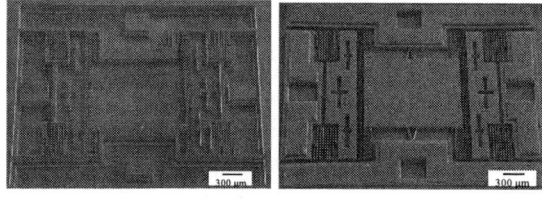

Figure 3.　SEM images of the two-step Si mold (left) and the PMMA mirror structures (right).

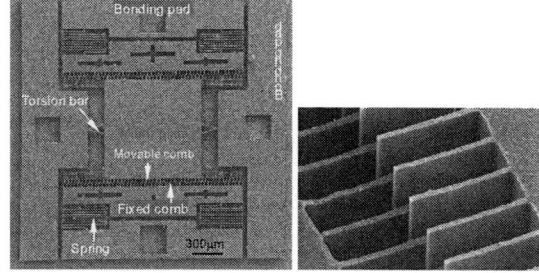

Figure 4.　SEM images of PMMA torsion mirror (left), and vertical comb actuator (right) with vertical offset is 25μm.

OPERATIONAL TEST

The relation between the applied voltage and the mirror rotational angle is a quadratic function with parabolic curve as shown in figure 5. Maximum angle was 6.5° obtained at a driving voltage of 67V. With higher applied voltage, the rotation angle was unchanged. This means that the edge of the mirror had touched to the substrate. If voltage is applied to the actuator at the other side, the total scan angle is 13°.

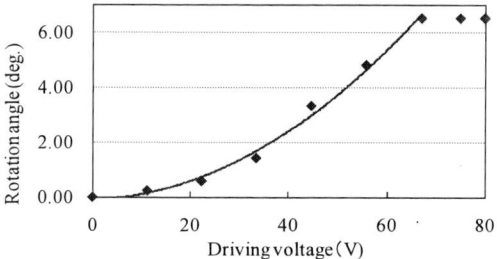

Figure 5.　Relation between the applied voltage and the mirror rotation angle. Resonant frequency was calculated to be 3.5kHz.

CONCLUSIONS

Monolithic PMMA mirror driven by vertical comb actuators has been successfully fabricated from PMMA by hot embossing, surface-activated direct bonding, and elliptical vibration cutting. An efficient protection technique for PMMA freestanding microstructures using reinforcement agent was proposed. The operational test results showed the vertical comb actuator worked properly. The fabrication process was reliable and robust with several advantages, such as simple and low cost process, no alignment is necessary during the bonding, reproduction is straightforward with reusable silicon mold, PMMA actuator requires lower driving voltage than that of Si counterpart, device surface is very smooth, and many types of thermoplastic can be used. The future work will focus on the new type of polymers such as electrical conductive and environment friendly polymers. More efficient releasing processes, such as polishing, are also being considered to further reduce the fabrication cost.

References

[1] Boger A, Bisig A, Bohner M, Heini P, Schneider E., Variation of the mechanical properties of PMMA to suit osteoporotic cancellous bone, *J Biomater Sci Polym Ed.* 2008, 19(9), pp. 1125-42

[2] S. Amaya, D. V. Dao and S. Sugiyama, Development of polymer electrostatic comb-drive actuator using hot embossing and ultraprecision cutting technology, *J. Micro/Nanolith. MEMS MOEMS*, **8**, 043065 (2009)

[3] A. Yeniay, K Renyuan Gao Takayama, A. F. Renfeng Gao Garito, Ultra-low-loss polymer waveguides, *IEEE J. Lightwave Technology* **22**, 154, 2004

[4] B. Chen, S. Liu, J. R. G. Evans, Polymeric Thermal Actuation Using Laminates Based on Polymer–Clay Nanocomposites, *Journal of Applied Polymer Science* **109**, Issue 3, p 1480-1483

[5] J. Chung, Y. Huang, and W. Hsu, Fabrication of polymer-based vertical comb drive using a double-side multiple partial exposure method, *Proceedings of MEMS 2008*, pp 475- 478

[6] Y. Zhao and T. Cui, Fabrication of high-aspect-ratio Polymer-based Electrostatic Comb Drives Using the Hot Embossing Technique, *J. Micromech. Microeng.* **13**, pp 430-435 (2003)

Low cost and large deflection angle polymer MEMS mirror using glass substrate

Osamu Sasaki[*], Takaaki Suzuki[*], Kyouhei Terao[*],
Hidekuni Takao[*] and Fumikazu Oohira[*]

*Department of Intelligent Mechanical Systems Engineering, Faculty of Engineering, Kagawa University
2217-20, Hayashi-cho, Takamatsu, Kagawa, 761-0396, Japan
Tel +81-87-864-2341, Fax +81-87-864-2341, E-mail: s10g509@stmail.eng.kagawa-u.ac.jp

ABSTRACT

In this paper, we propose a polymer MEMS mirror device which has the futures of the large deflection angle and the low cost fabrication. Many conventional MEMS mirror devices have been composed of Si wafer and, the expensive dry etching equipment has been necessary when etching the Si substrate. Then, we propose new composition and the novel fabrication method without the dry etching process by using the inexpensive glass substrate. Also, the torsion bar is composed of a photosensitive polymer that is a low rigid material and can be easily fabricated by the photolithography process. A multilayer wiring process is examined so that the low current actuation and the large deflection angle is attained. The fabricated device showed the large optical deflection angle of more than ±40 degrees at the current of ±16mA when the torsion bar length was 1800μm. The variation of the optical deflection angle was within 0.8 degrees at the 10^6 times repeatability test.

Keywords: MEMS Mirror, Photosensitive Polymer, Low cost, Glass substrate, Large deflection angle, Electromagnetic actuation

1. Introduction

The MEMS mirrors are applied for various fields as the optical deflection devices, such as the laser display, the laser radar, and the optical switch [1]. In recent years, the needs for the high performance and the low cost optical deflection mirror devices are increasing rapidly. For this purpose, various fabrication methods and systems have been proposed aiming at the high performance and the low cost deflection mirror device. For example, the mirror devices with the torsion bar made by the polymer material have been studied for the purpose of the large deflection angle. [2,3] But, many of these MEMS mirror devices are composed of Si or SOI wafer as the substrate, and the expensive dry etching equipment is necessary to fabricate the torsion bar. Therefore, there is a limit for reducing the fabrication cost. So, this study aims to realize the low cost polymer MEMS mirror device without using Si, SOI wafer and dry etching equipment. We propose the electromagnetic actuation type polymer MEMS mirror which is composed of the glass as the substrate and the photosensitive polymer as the torsion bar. We designed the torsion bar by the theoretical analysis, and examined the fabrication methods for the low cost process. We examined the fabrication conditions and fabricated the mirror device. Finally we evaluated the actuation characteristics and also tested the repeatability of the optical deflection angle.

2. Composition of the proposed mirror device

Fig.1 shows the composition image of the proposed mirror device. We aim at the large deflection angle and low cost fabrication method in this device. For this purpose, first, the glass substrate was used as the substrate. The glass substrate is inexpensive also it can be easily wet etched with high rate. As the result, the dry etching processing is unnecessary and the shortening of fabrication time can be expected. Next, the torsion bar is composed of the photosensitive polymer that is a very low rigid material. Then, it is fabricated by the easy photolithography process. The drive principle of the proposed device is the electromagnetic actuation by using the Lorentz force, and multilayer wiring is used to generate the large generative force. As the result, the device can realize the large deflection angle in the non resonance mode.

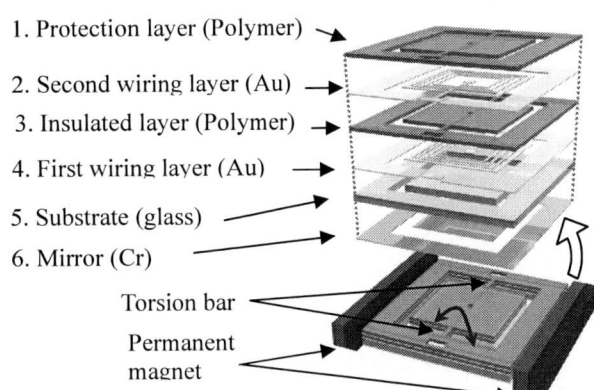

1. Protection layer (Polymer)
2. Second wiring layer (Au)
3. Insulated layer (Polymer)
4. First wiring layer (Au)
5. Substrate (glass)
6. Mirror (Cr)

Torsion bar
Permanent magnet

Fig.1 The composition image of the proposed mirror device

3. Fabrication of the proposed mirror device

Fig.2 shows the fabrication process flow.

(1) Cr layer is deposited by a sputtering on the surface side of the glass substrate, and it is patterned as the mirror part.

(2) Au is sputtered on the back side surface of the glass substrate, and it is patterned as the wiring part by the photolithography and the wet etching.

(3) Next, the photosensitive polyimide is spin coated and patterned as the torsion bar part and insulating layer by the photolithography.

(4) Cr and Au layers are sputtered on the surface of the imidized polyimide, then the wiring part is patterned by the photolithography and the wet etching.

(5) After the unnecessary resist is removed, the photosensitive polyimide is spin coated and patterned on the wiring part and the torsion bar part.

(6) Finally, the unnecessary glass part is wet etched by hydrofluoric acid, and then the MEMS mirror is completed.

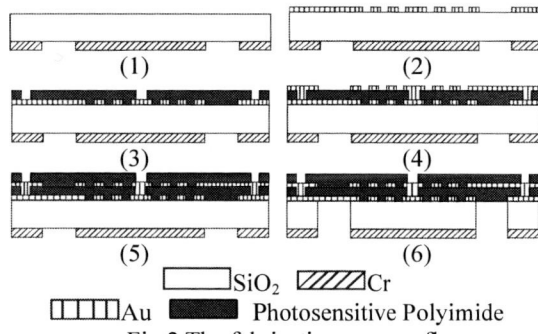

Fig.2 The fabrication process flow

4. Fabrication result of the polymer MEMS mirror

Table.1 shows the design parameters of the proposed polymer MEMS mirror device. Fig.3 (a) to (c) shows the fabricated polymer MEMS mirror and SEM image of the torsion bar. The mirror side has the optical mirror surface. Torsion bar is the moving part and is the thin part of the polymer, but it has no crack or damage, and it has the smooth surface.

Table 1 Parameters of designed mirror device

Mirror area	10mm×10mm	Glass thickness	150μm
Torsion bar length	1500μm / 1800μm	Wiring turn number	10
Torsion bar width	400μm	Wiring width	400μm
Torsion bar thickness	20μm	Wiring thickness	1μm

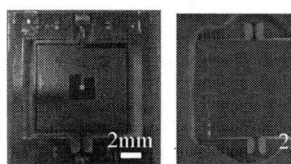
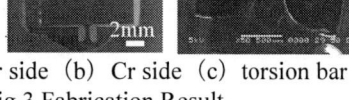

(a) Polymer side (b) Cr side (c) torsion bar
Fig.3 Fabrication Result

5. Characteristics of the electromagnetic actuation type polymer MEMS mirror

Fig.4 shows the actuation characteristics of the fabricated polymer MEMS mirror device. The fabricated device attained the optical deflection angle of ±30 degrees at the current of ±17mA when the torsion bar length was 1500μm, and ±40 degrees at the current of ±16mA when the torsion bar length was 1800μm. Fig.5

shows the measured resonance frequency of the dynamic characteristic. The fabricated device has the resonance frequency of 20.25 Hz. Fig.6 shows the repeatability of the optical deflection angle. The variation was within 0.8 degrees in the measurement range at 10^6, and more repeatability test is undergoing. The warp of the mirror surface is about 30~40μm and it will be reduced by improving the Cr layer, polymer layer and Au layer deposition conditions.

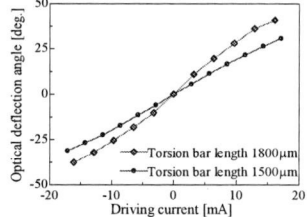

Fig.4 The deflection angle characteristics

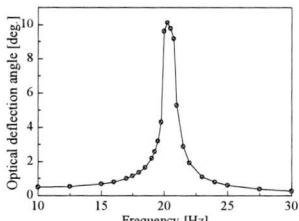

Fig.5 The resonance frequency

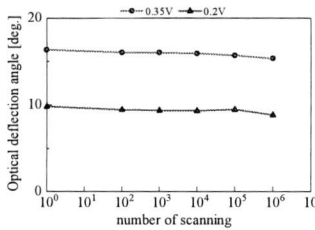

Fig.6 The angle repeatability of the fabricated mirror device

6. Conclusions

We proposed a polymer MEMS mirror device which has the futures of the large deflection angle and the low cost fabrication. For the purpose, we proposed new composition and the novel fabrication method without the dry etching process by using the inexpensive glass substrate and the photosensitive polymer. As the result, we confirmed the possibility to realize the large deflection angle and the low cost polymer MEMS mirror.

Reference

[1] ECO SCAN, Nippon Signal Co. : http://www.ecoscan.jp/

[2] H. Miyajima, K. Murakami and M. Katashiro Olympus Co.: "MEMS ptical Scanners for Microscopes, IEEE Journal of Selected Topics in Quantum Electronics, Vol.10, No.3, MAY/JUNE, 2004, pp.514-527.

[3] N. Yoshida, T. Fujita, K. Maenaka and Y. Takayama: "A Si MEMS Mirror using SU-8 Torsion Beam with Large Reflection Area", Papers of Technical Meeting on Micromachine and Sensor System, IEE Japan, Vol.MSS-05, 2005, No.21-44, pp.109-112 (in Japanese)

978-1-4244-8926-8/10 $26.00 © 2010 IEEE

PHOSPHOR SENSORS USING MECHANOLUMINESCENCE

Chao-Nan Xu

National Institute of Advanced Industrial Science and Technology, Japan

ABSTRACT

Elasticoluminescence (ESL) is a kind of mechanoluminescence (ML). ESL materials are novel functional materials that can convert elastic deformation energy into visible light directly. Utilizing the materials, novel sensing devices and various applications are now under development. These materials can allow direct viewing of stress distribution.

INTRODUCTION

ESL materials are novel multi-functional materials that can convert deformation energy into visible light directly. Utilizing these materials, novel sensing devices and various applications are now under development. These materials can allow direct viewing of stress distribution. The direct visualization and monitoring of stress distribution can be achieved using the materials. It would help to health monitoring structures, such as buildings, bridges and aircrafts, to maintain safe and secure [1] ~ [6].

EXPERIMENTAL

E S L materials of ceramics powders have been developed with certain structure of aluminates [7], silicates [8], and phosphates. The ceramics fine powders have been then to disperse in certain polymer develop the smart paint of ESL. The paint can be coated on various target structures and the emission intensity of which under the application of mechanical stress has been evaluated using the ML measurement system.

RESULTS ANS DISCUSSION

Figure 1 shows the example of developed ESL material of strontium aluminates $SrAl_2O_4:Eu^{2+}$.
Among ESL materials, $SrAl_2O_4: Eu^{2+}$ (SAOE) has the strongest ESL, it shows green emission even under a weak elastic deformation.
The ESL intensity increases with the increase of deformation energy (Fig. 2).
Figure 3 shows the TEM image of nano-particles of SAOE [9]. Each nano-particle of SAOE gave ESL pulse (green line) under the application of mechanical load (yellow line).

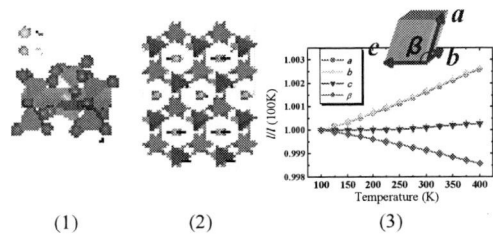

Fig. 1 Structural features of ESL material of SAOE
(1)The luminescent center locates at non-inversion site
(2) A flexible framework structure
(3) Anisotropic thermal expansions

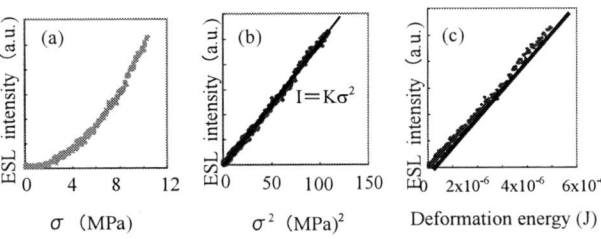

Fig. 2 (a) Relationship between ESL intensity and mechanical stress
(b) Relationship between ESL intensity and square of mechanical stress
(c) Relationship between E S L intensity and deformation energy

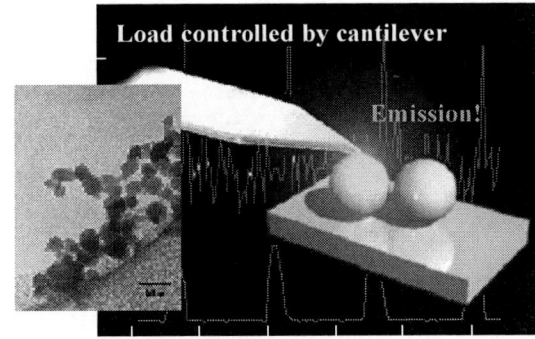

Fig. 3 TEM image of nano-particles and evaluation device for stress-induced luminescence
(load/light emission controlled by cantilever)

Figure 4 illustrates the ESL materials, and the luminescence image in pseudo-color displays the stress distribution on steel plate with a circle hole [10].

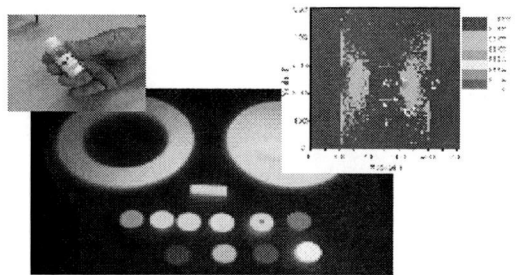

Fig. 4 ESL materials and the luminescence image in pseudo-color displays the stress distribution on steel plate with a circle hole

SUMMARY

We demonstrate that ESL materials developed possess the following features.

· Fine powders of ESL have high luminosity upon stress-stimulation, and the intensity of light is proportional to the deformation energy.
· Paint of ESL can coat various objects and make it possible to direct visualization of stress distribution.
· The light emitted from ESL can be controlled with a wavelength from ultraviolet to near infrared range.
· Each nano particle of these materials is able to work as a stress sensor individually, such as 20 nm force sensor available.

Utilizing these features, dynamical stress distribution can be visualized directly and easily. These materials can realize both wide area and high resolution stress distribution sensing. Various applications are now under developing, such as on-field visualization of stress distribution, novel mechanical sensors, nano-micro stress sensing, product proto-type evaluation, stress-sensitive displays with multi-colors, amusements, etc.

REFERENCES

[1] http://unit.aist.go.jp/msrc/ci/organization/ouryoku. html, National Institute of Advanced Industrial Science and Technology, Measurement Solution Research Center

[2] C.N. Xu, "Sensing Technology with Elasticoluminescence – Visualizing 'Invisible' Defects in Structures" Bull. Ceram. Soc. Japan, Vol. 44, No.2, pp.154−160, 2009

[3] C.N. Xu, Encyclopedia of Smart Materials: John Wiley & Sons, Vol.1, 2002

[4] C.N. Xu, T. Watanabe, M. Akiyama and X. G. Zheng, "Artificial skin to sense mechanical stress by visible light emission" Appl. Phys. Lett., Vol. 74, No.9, pp.1236-1238, 1999

[5] C.N. Xu, T. Watanabe, M. Akiyama and X.G. Zheng, "Direct view of stress distribution by mechanoluminescence" Appl. Phys. Lett., Vol.74, No.17, pp.2414-2416, 1999

[6] C.N. Xu, X.G. Zheng, M. Akiyama, K. Nonaka and T. Watanabe, "Dynamic visualization of stress distribution by mechanoluminescence image" Appl. Phys. Lett., Vol.76, No.2, pp.179-181, 2000

[7] C.N. Xu, H. Yamada, X.S. Wang, X.G. Zheng, "Strong elasticoluminescence from monoclinic-structure $SrAl_2O_4$" Appl. Phys. Lett., Vol.84, No.16, pp.3040-3042, 2004

[8] L. Zhang, H. Yamada, Y. Imai, C.N. Xu, Observation of elasticoluminescence from $CaAl_2Si_2O_8:Eu^{2+}$ and its water resistance behavior" J. Electrochem. Soc., Vol.155, No.3, J63-65, 2008

[9] C.Z. Li, Y. Imai, Y. Adachi, H. Yamada, K. Nishikubo, C.N. Xu, "One-step synthesis of luminescent nanoparticles of complex oxide, strontium aluminate" J. Am. Ceram. Soc., Vol.90, No.7, pp.2273-2275, 2007

[10] C.S. Li, C.N. Xu, Y. Imai, W-X. Wang, L. Zhang, H. Yamada, "Real-time monitoring of dynamic stress concentration by mechanoluminescent sensing film" Appl. Mechanics and Materials, Vol.13-14, pp.247-250, 2008

A NEW FABRICATION TECHNIQUE FOR INTEGRATING SILICA OPTICAL DEVICES AND MEMS

Karen E. Grutter, Anthony M. Yeh, Susant K. Patra, Ming C. Wu
Berkeley Sensor and Actuator Center (BSAC)
University of California, Berkeley, U.S.A.

ABSTRACT

We have developed a novel fabrication process which integrates silicon MEMS actuators with silica optical components. Suspended silica optical waveguides are actuated by a silicon electrostatic comb drive actuator, with a maximum displacement of 8μm at 35V bias.

INTRODUCTION

There are many potential applications of optical tunable-bandwidth filters in optical communications and signal processing. Using Si photonic MEMS technology, a microdisk filter with a tuning range from 3 to 80 GHz has been demonstrated [1]. Better filter performance can be achieved by using silica as the device material. This has been demonstrated, but the tuning was accomplished using bulk micropositioning stages [2]. In this paper, we report on a novel fabrication process that integrates a silicon-based MEMS comb drive actuator with silica optical components, as shown in Figure 1. To the best of our knowledge, this is the first example of this integration.

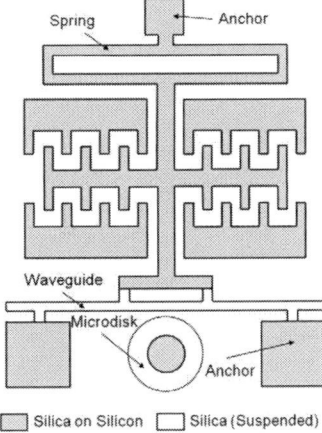

Figure 1: Schematic showing the integration of silicon-based MEMS actuator and silica-only optical components

FABRICATION

Silica is an ideal material for microresonator-based tunable photonic integrated circuits because of its high quality factor ($>10^7$). However, it is an insulator and integration with MEMS actuators is difficult. Here, we use a modified SCREAM fabrication process [3] on an SOI wafer (Figure 2). The optical components are made from a 1μm-thick layer of phosphosilicate glass (PSG) on the SOI. Electrostatic actuation is accomplished through a 25μm-thick silicon layer under the optical components, and electrical isolation is provided by the buried oxide layer. The optical components are protected by photoresist during the deep reactive ion etching (DRIE) step (Figure 2(c)). Upon release (Fig. 2(f)), the silica waveguides and microdisks are completely suspended to avoid substrate leakage. The fabricated device is shown in Figure 3.

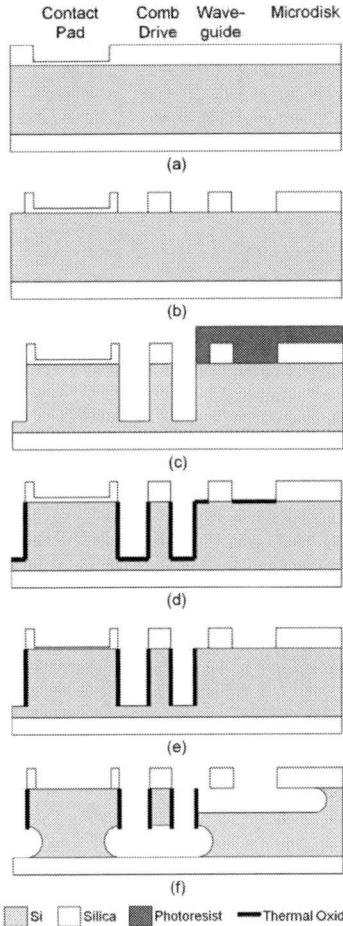

Figure 2: Fabrication process on SOI wafer. (a) Deposit phosphorus-doped silica and partially etch MEMS contact pads. (b) Etch device layer. (c) Deep Si etch with PR protection layer over optical parts. (d) Remove PR, oxidize Si. (e) Remove floor oxide with RIE. (f) XeF$_2$ release. After this step, silica can be reflowed to increase smoothness.

978-1-4244-8926-8/10 $26.00 © 2010 IEEE   33

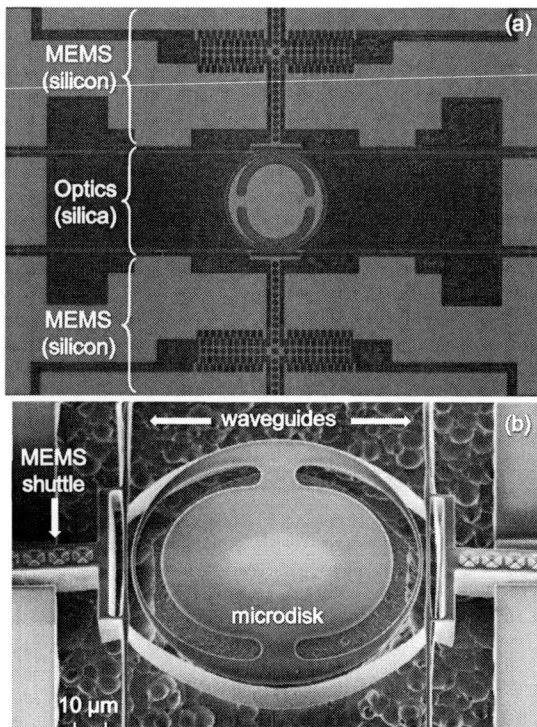

Figure 3: (a) Fabricated device composed of integrated silica optics and silicon MEMS. (b) Scanning electron micrograph of device; silica-only optical components (waveguides, microdisk) are suspended above substrate while connected to released silica-on-silicon MEMS (shuttle). Note: the microdisk includes adiabatic tapering to improve phase matching with the waveguide.

DEVICE CHARACTERIZATION

The suspended optical waveguide has a cross section of 1μm x 1μm. The measured fiber-to-fiber insertion loss of a 1mm-long silica waveguide is less than 13dB, limited mainly by the coupling loss. The MEMS comb drive actuator with 20 comb fingers and 3μm finger spacing has a maximum displacement of 8μm at 35V bias. This range of motion is more than sufficient for our application, in which we predict a maximum required disk-waveguide distance of 3μm in order to achieve a passband of <1GHz.

CONCLUSION

We have demonstrated a new fabrication process which integrates silicon MEMS actuators with silica optical components. Our specific device, a tunable-bandwidth filter, demonstrates the use of an electrostatic comb drive to change the coupling of the filter by moving the silica waveguide with respect to the microdisk. Optical performance will be presented at the conference.

The authors would like to acknowledge support from the National Science Foundation through CIAN NSF ERC as well as the NSF Graduate Research Fellowship Program. The authors would also like to thank Ming-Chun Tien for his guidance in the early stages of this project.

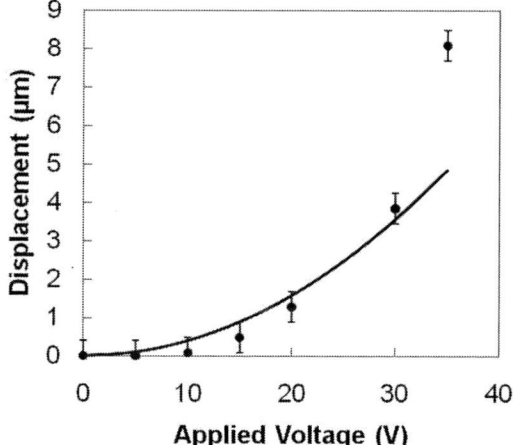

Figure 4: MEMS actuator movement with respect to applied voltage. The solid line is a best fit of the measured displacement. The final displacement does not follow the trend due to nonlinear effects as the actuator approaches maximum displacement.

REFERENCES

[1] J. Yao and M.C. Wu, "Bandwidth-tunable add-drop filters based on micro-electro-mechanical-system actuated silicon microtoroidal resonators," Optics Letters, vol. 34, 2009, pp. 2557-2559.

[2] M. Hossein-Zadeh and K.J. Vahala, "Free ultra-high-Q microtoroid: a tool for designing photonic devices," *Optics Express*, vol. 15, Jan. 2007, pp. 166-175.

[3] K.A. Shaw, Z.L. Zhang, and N.C. MacDonald, "SCREAM I: A single mask, single-crystal silicon, reactive ion etching process for microelectromechanical structures," Sensors and Actuators A: Physical, vol. 40, 1994, pp. 63-70.

FOUR-MASK PROCESS BASED ON SPACER TECHNOLOGY FOR SCALED-DOWN LATERAL NEM ELECTROSTATIC ACTUATORS

Daesung Lee, W. Scott Lee, Subhasish Mitra, Roger T. Howe, and H.-S. Philip Wong
Stanford University, USA

ABSTRACT

This paper presents a four-mask fabrication process of lateral nanoelectromechanical (NEM) electrostatic actuators based on spacer technology. Critical dimensions of the actuators, i.e., the beam width and the gap size between the movable and fixed electrodes can be made smaller than the lithographic resolution by creating nitride spacers on an oxide hardmask followed by selective etching of the oxide hardmask. The combined oxide hardmask, nitride spacer, and another mask (photoresist mask) is then transferred to an underlying polysilicon structural layer to create lateral NEM electrostatic actuators with narrow beam and narrow gap.

INTRODUCTION

For lateral NEM electrostatic actuators, achieving both a low static actuation voltage (<10V) and high resonant frequency (>1MHz) is very challenging without aggressively scaling critical dimensions beyond the lithographic resolution [1, 2]. For this scaling, spacer and sidewall stringer approaches were used in the past [2-4]: Submicron interelectrode gap of electrostatic combdrives was obtained by using a sacrificial thermal oxide spacer [3]. In the spacer FinFET process [4], narrow fins and doubled fin density were obtained by using a sacrificial masking layer and a chemical vapor deposition (CVD) spacer. In the three-mask sidewall stringer process [2], sub-lithographic beam width and gap size were obtained by fabricating electrically isolated TiN beams from sidewall stringers formed on a polysilicon mold. The process, however, relies on a very conformal deposition of the structural layer, limiting the choice of materials.

In our approach, sub-lithographic beam width and gap size are defined at a masking step by creating nitride spacers on an oxide hardmask followed by selective etching of the oxide hardmask. The combined mask is then transferred to an underlying planar structural layer, eliminating the need of a mold layer and conformal deposition of the structural layer. The resulting devices are scaled-down devices fabricated by etching a single planar structural layer, allowing additional processing, e.g., sidewall coating.

FABRICATION AND RESULTS

The four-mask spacer process starts with a 0.8μm thick LTO and 0.35μm thick doped-polysilicon deposition, followed by an RTA at 1075°C. Figure 1 shows top views of how the four masks are applied to form a combined mask, which is then transferred to a structural polysilicon layer.

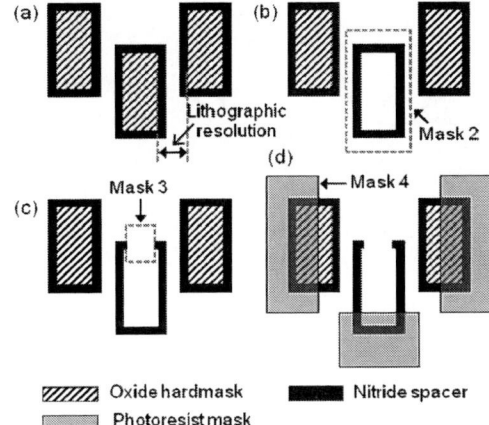

Figure 1. Mask preparation in the proposed four-mask spacer process prior to being transferred to a structural layer (top views).

A 350nm thick LTO is deposited and patterned (Mask 1) using i-line optical lithography and a reactive-ion-etch (RIE) with a minimum feature size of 500nm. Then a 170nm thick low-stress LPCVD nitride is deposited and blanket etched by RIE to form nitride spacer, resulting in a minimum gap of 160nm between spacers from adjacent islands (1a). The oxide hardmask in each island is selectively etched in 6:1 BOE using a photoresist mask (Mask 2) to define regions for narrow beams with the same width as the nitride thickness (1b). The required alignment accuracy at this step is half of the lithographic resolution. In an optional subsequent step, the nitride spacer whose oxide hardmask inside is removed is etched by RIE at some locations (Mask 3) to modify its shape from a closed-boundary to an open-boundary (1c). Next a photoresist mask is patterned (Mask 4) to define regions for anchors and connectors between the narrow beams and moving parts (1d). These features have non-critical

978-1-4244-8926-8/10 $26.00 © 2010 IEEE

dimensions, allowing the use of a thick photoresist mask. The final length of the narrow beams is determined at this step due to the overlapped region between the narrow beams and the photoresist mask. The combined oxide hardmask, nitride spacer, and photoresist mask is transferred to the underlying polysilicon structural layer by RIE. Finally after stripping the photoresist mask, the remaining oxide and nitride masks are removed by RIE and the devices are released in 49% wet HF followed by critical point drying. Note that the three masks (Mask 1, 2, 4) of this process play a similar role as the three of the sidewall stringer process [2].

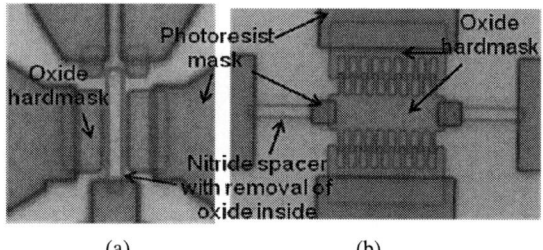

(a)   (b)

Figure 2. Optical micrographs at the completion of the masking step (1d): (a) parallel-plate actuator with a perimeter beam, (b) lateral combdrive with narrow beams and reduced comb gaps.

Figure 2 shows optical micrographs of combined masks (step 1d) where the following features can be observed: the nitride spacer with removal of oxide hardmask inside, reduced gap between the spacers from adjacent islands, oxide hardmask, photoresist mask, overlap region between the oxide hardmask and the photoresist mask, and overlap region between the narrow beams and the photoresist mask.

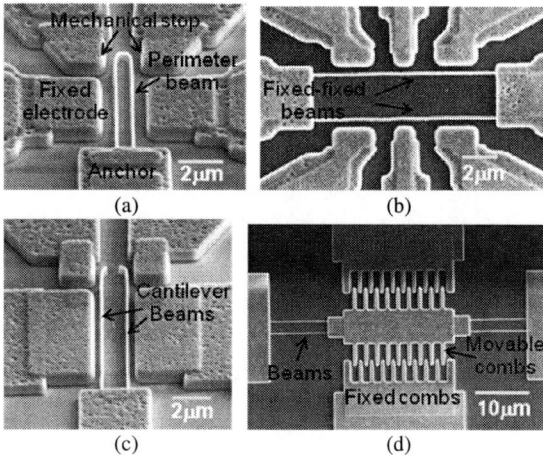

(a)   (b)

(c)   (d)

Figure 3. SEMs of finished NEM electrostatic actuators: parallel-plate actuators with (a) a perimeter (oblique view), (b) two fixed-fixed (top view), (c) two separated cantilever beams (oblique view), (d) lateral combdrive with narrow beams and reduced comb gaps (top view).

Figure 3 shows SEMs of various finished lateral

NEM electrostatic actuators. The beam width is measured to be around 170nm whereas the minimum gap size is measured to be around 160nm. The gap size can be further reduced by adding a step of TEOS deposition followed by a blanket etch after patterning the oxide hardmask. Using 193nm lithography, minimum half-pitch at Mask 1 can be as small as 60nm [5]. Thus, gap sizes of < 10nm can be easily obtained by depositing spacers of only 25nm. Due to non-negligible etch rate of polysilicon during the removal of the remaining masks, regions without oxide hardmask are observed to have reduced thickness, which can be alleviated by using wet etching instead of RIE.

For a similar type of device shown in Figure 3a, the device was actuated by applying a DC driving voltage to the fixed electrode while the beam and mechanical stop was connected to ground and 3V bias, respectively. By monitoring the beam current, the pull-in voltage was measured to be 5V where the beam length is 13μm and the beam-to-electrode and beam-to-stop spacing is 360nm and 260nm, respectively. The estimated resonant frequency of the device is calculated to be around 2MHz. The pull-in voltage of a device in the same design using oxide hardmask only is calculated to be 93V.

CONCLUSIONS

We present a four-mask spacer process that enables minimum beam and gap feature sizes below the lithographic resolution as the sidewall stringer process. This enables a significant reduction in the operating voltage as well as a significant increase in the resonant frequency of the NEM electrostatic actuators. This process uses a similar mask set as the sidewall stringer process, but key advantages include (1) the process does not require conformal deposition of the structural layer, (2) the devices are fabricated from a single planar structural layer, (3) the gap reduction is readily available. This masking technique can be applicable to various CVD masking layers with good etch selectivity and various structural and sacrificial layers. Using 193nm lithography, sub-10nm gaps can be easily obtained. Designs of actuators with decoupled mechanical springs and electrodes can be optimized by adding etch holes and optimizing its dimensions.

ACKNOWLEDGEMENTS: This work was supported by the DARPA MTO program "NEMS" (Contract number: NBCH 1090002, Program manager: Dr. T. Akinwande).

REFERENCES
[1] K. Akarvardar et al., IEDM 2007, pp. 299-302.
[2] D. Lee et al., Proc. IEEE MEMS 2010, pp. 456-459.
[3] T. Hirano et al., JMEMS, vol. 1, no. 1, pp. 52-59, 1992.
[4] Y.-K. Choi et al., Solid-State Electronics, vol. 46, pp.1595-1601, 2002.
[5] S. Natarajan et al., IEDM 2008, pp. 941-943.

NANOSTRUCTURED ORIGAMI™ FOLDING OF PATTERNABLE RESIST FOR 3D LITHOGRAPHY

Se Young Yang, Hyung-ryul Johnny Choi, Martin Deterre, George Barbastathis*

Massachusetts Institute of Technology, Cambridge, USA

*Singapore-MIT Alliance for Research and Technology (SMART), Singapore

ABSTRACT

A new method to fold free standing poly(methyl methacrylate) (PMMA) resist using e-beam exposure is developed and demonstrated. The results prove controllable folding of the patterned PMMA. An explanation of the folding mechanism is proposed based on experimental characterization and theoretical analysis. 3D lithography is achieved by attaching the patterned resist on an adjacent side wall by folding. Patterns are effectively transferred by depositing metal followed by a lift-off process.

INTRODUCTION

The concept of folding nanopatterned resist is inspired by nanostructured origami, the term used to refer Japanese paper art of origami, in which 2D surfaces are folded into 3D volumetric shapes. [1] Nanostructured origami of thin membranes was well demonstrated by previous works of folding [1-5], stacking [6], and alignment [3,5,7]. The focus of this work is to develop a new method to fold patterned PMMA membranes, which would facilitate 3D lithography. The method is to simply expose PMMA with e-beam, varying the acceleration voltage and the dose. A folding mechanism is proposed based on experimental characterization and theoretical analysis. In addition, implementation of PMMA folding for 3D lithography is achieved by transferring the pattern on to a 3D plane, e.g. the side wall. Further potential application to 3D lithography and future works are discussed.

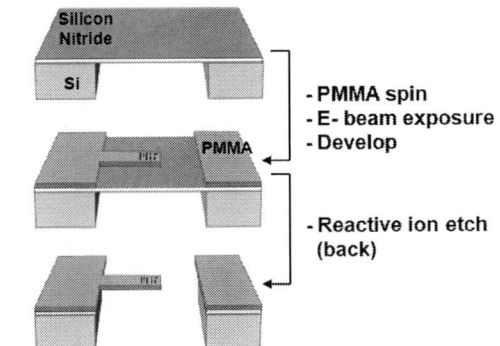

Figure 1. Fabrication process of free standing PMMA conatilever

EXPERIMENT

Sample preparation

The fabrication process for the free standing PMMA cantilever is illustrated schematically in Figure 1. 200nm thick PMMA is spin coated on a silicon nitride membrane which is obtained by KOH etching a silicon wafer coated with LPCVD silicon nitride. The cantilever and the patterns are exposed with a Raith 150 electron-beam lithography system (30 keV, 280 µAs/cm²). After exposure, the films are developed in a 2:1 isopropyl alcohol (IPA): methyl isobutyl ketone (MIBK) mixture for 60 s, followed by a 60 s IPA rinse. Silicon nitride membrane is removed by CF_4 reactive dry ion etching from the back.

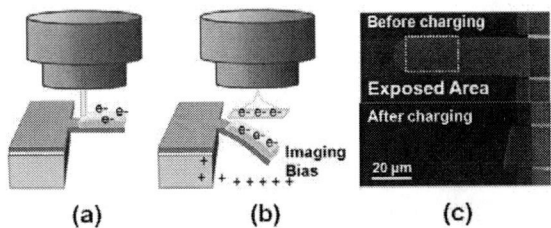

Figure 2. PMMA downward folding with low acceleration voltage using charging effect: (a) Surface charging by e-beam exposure; (b) Cantilever bending due to imaging bias and repulsive force between the scanning beam and the surface charge; (c) SEM image of the actual folding.

Folding

Raith 150 system is used to quantitatively control the dose and the exposure area. Folding of the PMMA cantilever using e-beam exposure is achieved by two different methods depending on the acceleration voltage. At a low acceleration voltage below 2keV, e-beam exposed area charges up significantly. Imaging the exposed surface creates repulsive force between the negatively charged surface and the cantilever beam, resulting in downward folding of the flexible cantilever as in Figure 2. At a high voltage above 10keV, charging becomes negligible, and e-beam exposed area is believed to swell due to de-crosslinking. This results in upward folding, implying that there is a gradient in swelling along the thickness direction (Figure 3). Figure 4 clearly demonstrates that upward folding is controllable by adjusting the dose (100-600 µAs/cm²) which corresponds to adjusting the degree of swelling.

978-1-4244-8926-8/10 $26.00 © 2010 IEEE

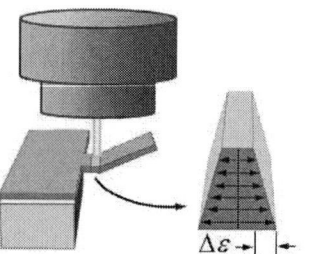

Figure 3. Schematic of PMMA upward folding with high acceleration voltage.

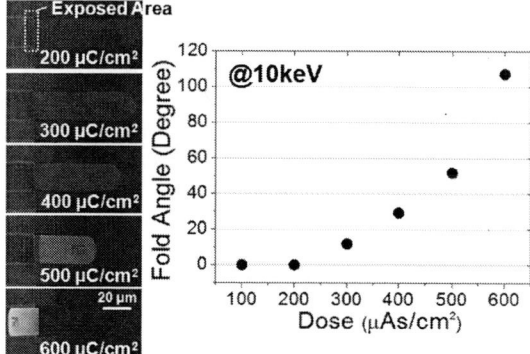

Figure 4. PMMA upward folding at acceleration voltage of 10keV.

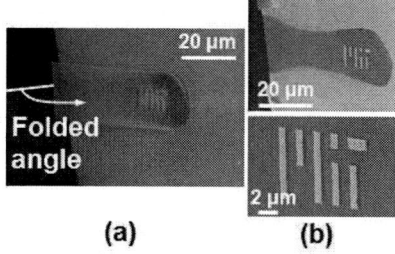

Figure 5. (a) PMMA folded and attached to the side wall due to van der Waals force; (b) Pattern transfer on the side wall with Au e-beam evaporation.

DISCUSSION

Downward folding that utilizes charging effect is limited to single binary actuation. Since charging effect is by its nature hard to control quantitatively, excessive charging is necessary followed by a single imaging actuation. The PMMA is folded and well attached by van der Waals force on to next neighboring surface that lies in the angle of rotation as in Figure 5(a).

Change in modulus for the exposed region has been proven to be negligible [8]. The mechanism for upward folding can be explained by the gradient in swelling along the thickness. Assuming for simplicity that the swelling gradient is linear, we estimate that

the difference in swelling between the top and bottom cantilever surfaces is in the order of 10^{-7} strain for exposure length of 5μm.

To demonstrate 3D patterning, 40nm of Au was e-beam evaporated on the folded PMMA followed by a lift-off process with heated NMP (100°C, 5mins). The 3D lithography result on a side wall is successfully shown in Figure 5(b).

The folded PMMA can be utilized as an etch mask; further implementation of reactive ion etching on the side wall is ongoing. Moreover, multi-step height 3D lithography will be attained by multi-folding the PMMA as illustrated in Figure 6. Further verification of the proposed upward folding mechanism is under investigation.

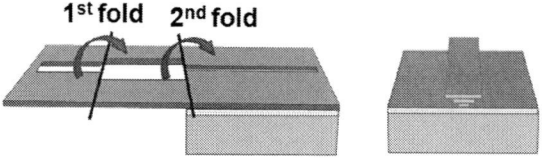

Figure 6. An example of achievable 3D lithography using multi-folding of PMMA

REFERENCES

[1] W. J. Arora et al., "Membrane folding to achieve three-dimensional nanostructures: Nanopatterned silicon nitride folded with stressed chromium hinges," Applied Physics Letters, vol.88, pp.053108, 2006

[2] H. J. In, S. Kumar, Y. Shao-Horn, and G. Barbastathis, Appl. Phy. Lett., vol.88, pp.053104, 2006

[3] A. J. Nichol, P. S. Stellman, W. J. Arora, and G. Barbastathis, "Two-step magnetic self-alignment of folded membranes for 3D nanomanufacturing," Microelectronics Eng., vol.84, pp.1168-1171, 2007

[4] W. J. Arora et al., "Membrane folding by ion implantation induced stress to fabricate three-dimensional nanostructures," Microelec. Eng., vol.84, pp.1454, 2007

[5] H. J. In et al., "Carbon nanotube–based magnetic actuation of origami membranes," Journal of Vacuum Science and Tech. B, vol.26, pp.2509-2512, 2008

[6] A. A. Patel and H. I. Smith, "Membrane stacking: A new approach for three-dimensional nanostructure fabrication," Journal of Vacuum Science and Technology B, vol.25, pp.2662. 2007

[7] A. J. Nichol et al., "Thin membrane self-alignment using nanomagnets for three-dimensional nanomanufacturing," Journal of Vacuum Sci. and Tech. B, vol.24, pp.3128, 2006

[8] S. Eve and J. Mohr, "Effects of UV-irradiation on the thermo-mechanical properties of optical grade poly(methyl methacrylate)," Applied Surface Science, vol.256, pp.2927-2933, 2010

Absorbent liquid immersion angled exposure for 3D Photolithography

Hironori Kubo, Shinya Kumagai, and Minoru Sasaki,

Dept. of Advanced Science and Technology, Toyota Technological Institute,
Hisakata 2-12-1, Tenpaku-ku, Nagoya 468-8511, Japan
E-mail: mnr-sasaki@toyota-ti.ac.jp

Abstract

Photolithography on the sample with vertical side walls is studied. In the angled exposure for patterning side walls or bottoms, the exposure is basically over-dose due to the thinner thickness making the reflection serious for obtaining the defect-free pattern. In addition to the liquid immersion method, the absorbent liquid is introduced. Arbitrary pattern over the trench with aspect ratio of 0.74 is obtained with the better quality than that obtained using the water and the polarization control.

Keywords: 3D photolithography, Vertical side wall, Angled exposure, Absorbent liquid immersion

1. INTRODUCTION

The photolithography can be applied to the planer sample. If the photolithography can be applied to 3D samples, variety of applications is expected. Furthermore, many practical 3D devices have vertical side walls. The photolithography process consists of resist coating and patterning. The spray coating of the resist has reached a level for giving coverage over vertical structures. Another technique for realizing 3D photolithography is an exposure. The normal exposure can not transfer the pattern to the vertical side wall. The angled exposure is indispensable. The reflected light is generated and patterns other surface as shown in Fig.1. This reflection problem is exaggerated by combining with the spatial thickness distribution of the resist. Basically, the film is thinner at the deeper region. Since the exposure dose for the positive resist has to be adjusted for opening the thicker film, the exposure tends to be over-dose for the thinner region.

Previously, our group has reported the liquid immersion angled exposure [1]. Water has the middle value of refractive index (1.33) comparing to those of air (1.00) and photoresist (1.67). The reflection on the resist surface can be reduced. The polarization control against the surface is also confirmed to be effective. Although 30 µm-wide arbitrary patterning is realized for 100 µm-deep trench with 0.5 aspect ratio, the process margin is narrow.

In this study, the absorbent liquid is newly introduced in the angled exposure in place of the pure water. The effect for suppressing the pattern degradation is confirmed.

2. PRINCIPLE

Table 1 shows an example of the film thickness of positive resist (Tokyo Ohka Kogyo, TMMR P-W1000) on trench sample and minimum exposure for opening the pattern in normal exposure. The trench is 135µm wide and 100µm deep (aspect ratio of 0.74 is higher than that in the previous study). On the bottom, film thickness is only ~40% compared to that of top surface. When the absorbent liquid is introduced, the light intensity decreases while UV light propagates from the top to the side wall and the bottom

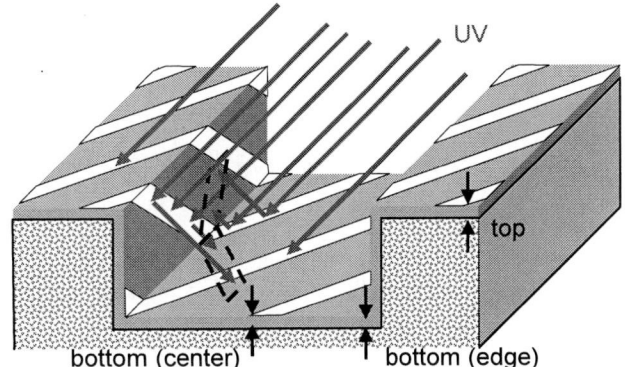

Fig. 1: Schematic drawing for explaining reflections in angled exposure for patterning diagonal line-and-space inside the trench.

Table1: Resist thickness on trench sample and minimum exposure doses for opening the pattern.

surface	top	bottom(center)	bottom(edge)
film thickness [µm]	6.01	2.38	2.17
exposure dose [mJ/cm²]	115±3	50±3	—

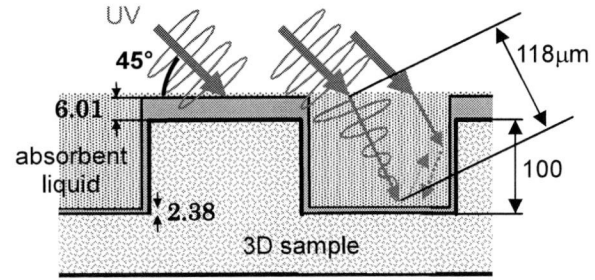

Fig. 2: Schematic drawing for explaining the effect of absorbent liquid in angled exposure.

surface as shown in Fig. 2. Over-dose problem as well as the reflection problem can be decreased.

3. EXPERIMENTS

For the incident angle of 45°, the exposure dose for unit area decreases by the ratio of cos45°. The exposure dose is 300mJ/cm² for opening the pattern at the top surface. If the exposure dose reduces to 30% from the top to the bottom surface, the exposure dose at the bottom surface becomes 90mJ/cm² (63mJ/cm² at angled exposure). As shown in Fig. 2, the UV propagation distance from the top to the bottom surface through the liquid is 118µm. Absorption follows Lambert-Beer's law.

$$I = I_0 e^{-\alpha L} \qquad (1)$$

When L and I/I_0 are 118µm and 0.3, respectively, the absorption coefficient α is 10.2/mm. Such absorbent liquid is prepared by diluting the black ink for the fountain pen (PILOT Corporation). The appropriate absorbent liquid is prepared by diluting the black ink to 12.5 vol. %.

4. RESULTS

Figure 3 shows images of the line-and-space pattern aligned at 45° against the trench length direction as shown in Fig. 1. Conditions are (a) water liquid immersion exposure with p-polarized light, and (b) absorbent liquid immersion exposure without polarization control. On the bottom surface, the pattern quality is nearly same with little defect. On the side wall, Fig. 3(a) shows the clear degradation. This is due to the reflection from the bottom surface to the side wall. The arrow shows the mirroring relation of the pattern. Figure 3(b) does not suffer from such defect. As shown in Fig. 2, the reflection from the bottom surface to the side wall gives the additional propagation distance, enhancing the absorption effect. The obtained pattern in Fig. 3(b) shows the opening width of 70+/–5 µm over the trench while the design width is 70µm.

Figure 4 shows the demonstration pattern obtained with (a) water liquid immersion exposure with p-polarized light and (b) absorbent liquid immersion exposure without polarization control. The typical line width for constructing characters is 30 µm. The pattern can be seen over the top surface, side wall, and the bottom surface. The difference is the separation inside character "A" or "B" on the bottom surface. This separation width in the design is 5 µm. The bottom pattern width is larger due to the over-dose condition. As for the side wall surface, Fig. 4(b) shows some roughness of the resist film.

Absorbent liquid immersion angled exposure is newly proposed. Patterning is possible suppressing the over-dose and reflection problems without polarization light control. The absorption coefficient of liquid can be controlled corresponding to variety of conditions.

This research was supported by a MEXT program for forming strategic research infrastructure for private

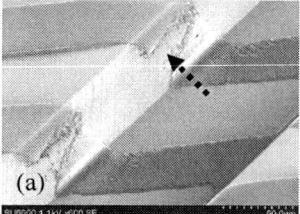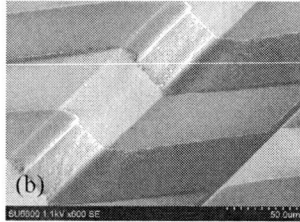

Fig. 3: Diagonal line-and-space pattern over the sample with trench. Exposing conditions are with (a) pure water-resist interface and p-polarization (b) absorbent water-resist interfaces without polarization control.

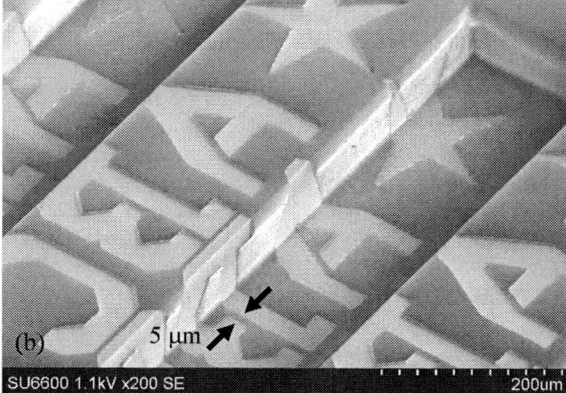

Fig. 4: Demonstration pattern. Exposing conditions are with (a) pure water-resist interface and p-polarization (b) absorbent water-resist interfaces without polarization control.

Universities from 2008, and a Foundation for Technology Promotion of Electronic Circuit Board. Authors thank to Prof. H. Sakaki, Toyota Technological Institute for the experimental collaboration.

REFERENCES

[1] M. Sasaki, T. Hosono, S. Kumagai, Proc. of IEEE/LEOS Int. Conf. Optical MEMS and Nanophotonics, 2009, pp.75-76.

NANOBIODEVICE BASED SINGLE CELL IMAGING FOR CANCER DIAGNOSIS AND *IN VIVO* IMAGING FOR STEM CELL THERAPY

Yoshinobu Baba

Department of Applied Chemistry, School of Engineering, FIRST Research Center for Innovative Nanobiodevice, Nagoya University

ABSTRACT

The lecture will describe new synthetic method for quantum dots (QDs) and their applications to the single molecular real-time imaging of enzymatic reaction, single cancer cell and cancer stem cell imaging, stem cell *in vivo* imaging, imaging of single QD tracking in a single living cell for cancer diagnosis, stem cell therapy, and gene delivery therapy [1-14].

INTRODUCTION

In the cancer diagnosis/therapy and stem cell therapy, it is highly required to develop novel techniques to allow us to diagnosis at the early stage of cancer and in vivo imaging. In this study, we have developed nanobiodevice based techniques for this purposes.

RESULTS AND DISCUSSION

In this research, we developed several types of nano- and micro-fluidic technologies: one for real-time monitoring of interaction and enzymatic reaction between a Qdot-labeled single DNA molecule and an enzyme molecule and another for manipulation of a single cell by balancing laminar flow rate and power of optical force.

Using nano- and micro-microfluidic devices coupled with optical force, we can selectively manipulate and separate single cancer cells and single cancer stem cells by the balancing the nano- and micro-fluidic flow and optical force by laser tweezers. After separation of a cancer cell, we can confirm the separated cell as a single cancer cell through detection of specific sugar chain on the cell surface by using lectin conjugated quantum dot. Quantum dot is applied to the imaging of stem cell and we confirmed that quantum dots could be applied for labeling stem cells without fatal toxicity and serious influence on the ability of differentiation. The quantum dot is utilized for *in vivo* tracing after the transplantation of stem cell to disclose mechanism of therapy for acute liver failure. We developed the cellulose-immobilized carbon nanotube as a novel gene transfection nanomaterial for a plant cell. Quatum dot based technology is applicable to the diagnosis for single cellular level of brain tumor during operation.

ACKNOWLEDGEMENTS

This research is supported by the Japan Society for the Promotion of Science (JSPS) through its "Funding Program for World-Leading Innovative R&D on Science and Technology (FIRST Program)."

REFERENCES

[1] N. Kaji, et al., Nanopillers and Nanoballs for DNA Separation, in *Nanofluidics*, RSC, 2009.

[2] M. Tabuchi, et al.: *Nature Biotech.*, **22**, 337 (2004).

[3] R. Bakalova, et al.: *Nature Biotech.*, **22**, 1360 (2004).

[4] R. Bakalova, et al., *J. Am. Chem. Soc.*, **127**, 11328 (2005).

[5] R. Jose, et al.: *J. Am. Chem. Soc.*, **128**, 629 (2006).

[6] M.R. Mohamadi, et al., *Anal. Chem.*, **79**, 3667 (2007).

[7] K. Hirano, et al., *Anal. Chem.*, **80**, 5197 (2008).

[8] T. Tachi, et al., *Lab on a Chip*, **9**, 966 (2009).

[9] T. Tachi, et al., *Anal. Chem.*, **81**, 3194 (2009).

[10] H. Yukawa, et al., *Cell Transplantation*, **18**, 591 (2009).

[11] F. Dang, et al., *Langmuir*, **25**, 9296 (2009).

[12] F. Dang, et al., *Anal. Chem.*, **81**, 10055 (2009).

[13] Y.S. Park, et al., *ACSNano.*, **4**, 121 (2010).

[14] H. Yukawa, et al., *Biomaterials*, **31**, 4094 (2010)

Pulsed Laser Triggered High Speed Fluorescence Activated Microfluidic Switch

Ting-Hsiang Wu[1], Yue Chen[2], Sung-Young Park[2] Soojung Claire Hur[3], Dino Di Carlo[3] and Eric Pei-Yu Chiou[2]

[1]Department of Electrical Engineering, University of California at Los Angeles (UCLA)

[2]Department of Mechanical and Aerospace Engineering, UCLA

[3]Department of Bioengineering, UCLA

48-121 ENG. IV, 420 Westwood Plaza, Los Angeles, CA 90095-1597, USA

Tel +1-310-825-8620, E-mail: pychiou@seas.ucla.edu

ABSTRACT

We report a high speed fluorescence activated microfluidic switch capable of achieving a switching time of 50 μsec with a detection efficiency of 86.6% and a switching efficiency of 86.5% at a particle flow speed of ~0.7 m/s. The switching mechanism is realized by exciting dynamic vapor bubbles with focused laser pulses in a microfluidic PDMS channel. The explosive bubble expansion generates fast fluid flows which are directed into a neighboring particle channel to switch the particle flow. Fluorescence activated switching of 10 micron polystyrene microspheres in a Y channel has been demonstrated. This ultrafast laser triggered switching mechanism has the potential to advance the sorting speed of the state-of-the-art microscale fluorescent activated cell sorting devices.

INTRODUCTION

Microfluidic cell sorters have several advantages over the conventional fluorescence-activated cell sorters (FACS) such as device miniaturization, integration of multiple functions on one chip and achieving high yields for small sample sizes. Demonstrated switching time range from 2-4 msec with an optical forces [1], 0.5 msec by using dielectrophoresis [2], and 0.1-1 ms by integrating a piezoelectric actuator [3]. The sample flow speeds lay in the range from μm/sec to cm/sec. On the other hand, bubble cavitation has been used for fast microfluidic switching with microsecond switching duration at high flow speed ~1 m/sec. Chen et al. [4] showed that by employing micromachined heaters embedded in silicon based microchannels, an electrically heated bubble is able to deflect a 20 μm particle in 4.9 μsec through a focusing nozzle.

Pulsed laser mirobeams have been used to induce bubble cavitation on the microscale and even nanoscale. When the laser pulse is focused in a liquid medium (e.g. water), the intense optical field generates heated plasma within the focal volume. The heat dissipates into the surrounding liquid and induces cavitation bubbles. In our proposed device, the bubble expansion is used to displace the surrounding liquid, which is channeled into the adjacent particle channel to switch the particle flow. Because of the rapid bubble dynamics, a fast switching time on the order of microseconds can be achieved by controlling the bubble size and the pulsed channel dimensions.

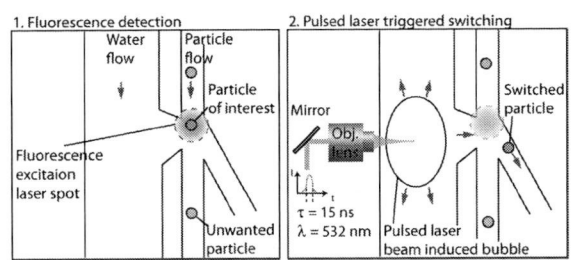

Fig.1 Schematic of the pulsed laser triggered fluorescence activated microfluidic switch.

PRINCIPLE AND DEVICE STRUCTURE

Figure 1 shows the schematic of the pulsed laser triggered fluorescence activated microfluidic switch. The device consists of a main microchannel (or sample channel) with two outlets, collection and waste. Using inertial hydrodynamic focusing, the input particles are focused slightly off the centerline of the main channel and go into the waste channel. Fluorescent particles are detected as they flow through the top of the Y junction. This triggers a laser pulse focused through a high NA objective lens into the pulsed channel running in parallel with the main channel. The focused laser pulse induces optical breakdown of the liquid medium and the subsequent cavitation bubble. As the bubble expands, the surrounding liquid is pushed away and squeezed through the nozzle opening into the main channel. This liquid jet deflects the particle flow into the collection channel.

Rapid switching based on expansion and collapse of the cavitation bubble takes place on the time scale of few μsec to msec, depending on the laser energy and pulsed channel dimensions. Using a nozzle to focus the deflecting liquid jet narrows the switching region to tens of microns near the opening which is essential for achieving high throughput sorting. Combined with the capability of operating at high sample flow speed ~1 m/sec, this device has the potential to achieve high-speed and high-efficiency sorting in a single-layer PDMS microfluidic chip.

EXPERIMENT AND RESULT

Fabrication of the microfluidic chip employed the conventional casting method to produce a single-layer PDMS microchannels bonded to a glass coverslip substrate. The height of the channels was 100 μm. At the switching Y junction, the sample channel had a width of 30 μm and the pulsed channel was 150 μm in

978-1-4244-8926-8/10 $26.00 © 2010 IEEE

width. Channel flows were driven by syringe pumps (KD Scientific). 3D hydrodynamic focusing of the 10 µm polystyrene microspheres into the waste channel was achieved by sheathless inertial focusing [5]. The particle flow was injected through a single inlet. Due to inertial effects, the particles became focused into two parallel streams along the channel wall as they traveled downstream. The channel was split into two where the right particle stream was recycled and the left stream was connected to the switching Y junction. Particle speed was ~0.7 m/s in the main channel.

To induce cavitation bubbles, a Q-switched Nd:YVO$_4$ pulsed laser beam (EKSPLA, Jazz 20) was focused through an objective lens (100x, NA 0.9) into the pulsed channel. The pulse was 15 nsec in pulse width and 532 nm in wavelength with a pulse repetition rate up to 100 kHz. The pulse energy used was 88 µJ. For sample fluorescence excitation, a 10 mW, 488 nm solid state laser was focused into the main channel (spot size ~50 µm in diameter) through a 25×, NA 0.4 objective lens on the other side of the microfluidic chip. The emitted sample fluorescence was collected by the same objective lens and sent into a photomultiplier tube (PMT; Sens-Tech, P30CWAD5). The PMT signal was sampled by a PCI-7831R DAQ card (National Instrument) at a sampling rate of 40 MHz. Fluorescence intensity was obtained by integrating PMT signal every 10 µsec. Codes written in LabView 8.2 were programmed into FPGA logic to perform real-time detection, threshold comparison and timed triggering of the pulsed laser.

Fig. 2 shows the fluorescent signal detection in the system. Fig. 2(a) and 2(b) are the histograms of the background and the fluorescence intensity of the 10 µm green microsphere (Duke Scientific, G1000) respectively. The two populations were clearly separated and a decision threshold to determine whether a particle was detected or not was set at the intensity level $U=0.5U_0$. Each histogram was generated using 4000 events. In Fig 2(c), fluorescence signal over a period of 0.5 sec was plotted. Particles flowing through the detection zone could be distinguished from the background noise based on the predetermined decision threshold.

Fig. 3 demonstrates particle switching at the Y junction triggered by a single laser pulse. Without switching, the particles flew to the waste channel on the left as shown in a time-resolved image (Fig 3(a)) and by fluorescence trace (Fig. 3(b)). A cavitation bubble induced by the focused laser pulse (taken 2 µsec after the laser pulse arrival) pushed the liquid flow into the main channel through a 50 µm wide nozzle to deflect the particle towards the collection channel. The bubble lifetime characterized by the time-resolved images was 50 µsec. Fluorescence activated switching was also confirmed by imaging the fluorescence trace of the switched particle in Fig. 3(b). We further evaluated the detection efficiency and switching efficiency of the

system by sampling events recorded by camera images. Both detection and switching efficiencies can be as high as 90% with means of 86% (Table 1).

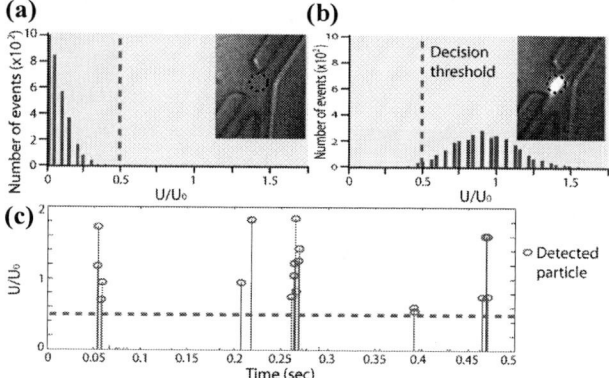

Fig.2 Fluorescence detection. Histograms of fluorescent signals from the (a) background and (b) 10 µm green beads. (c) Time series of the fluorescent signal.

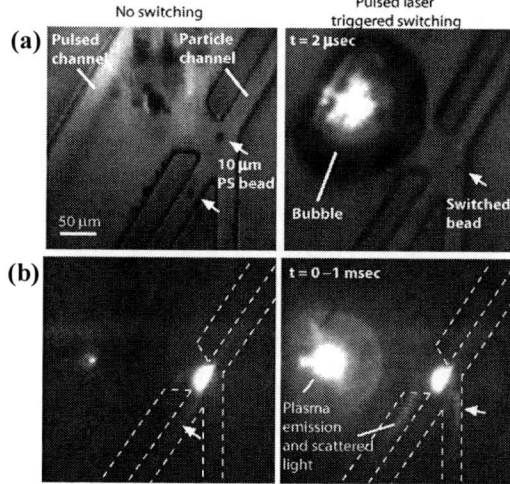

Fig.3 Particle switching triggered by a focused pulsed laser beam. (a) Time-resolved images (b) Fluorescence traces.

Total # sampled	# beads detected	# beads switched	Detection efficiency	Switching efficiency
44	32	28	72.7%	87.5%
54	50	40	92.6%	80.0%
53	50	46	94.3%	92.0%

Table 1. Fluorescence detection and fluorescence activated switching efficiencies of 10 µm beads.

ACKNOLEDGEMENT

This project is supported in parts by the NSF grants NSF ECCS-0901154 and NSF DBI-0852701.

REFERENCE

[1] Wang, M. W. et al., *Nat. Biotechnol.* **23**, 84-87, 2005.
[2] Baret, J. et al., *Lab Chip* **9**, 1850-1858, 2009.
[3] Chen, C. et al., *Biomed Microdevices* **11**, 1223-1231, 2009.
[4] Chen, C. et al., *Sens. Actuators B* **117**, 523-529, 2006.
[5] Hur, S. et al., *Lab Chip* **10**, 274-280, 2010.

A Laser Driven Optofluidic Device for High-Speed and Precise Volume-Controlled Droplet Generation on Demand

Sung-Yong Park, Ting-Hsiang Wu, Yue Chen, and Pei Yu Chiou
Department of Mechanical and Aerospace Engineering, University of California at Los Angeles,
420 Westwood Plaza, Los Angeles, CA 90095-1594, USA
Tel: +1-310-825-9091, E-mail: spark5@ucla.edu

ABSTRACT

We report on an ultra-fast, pulse laser-driven droplet generation (PLDG) mechanism that enables on-demand droplet generation up to 10,000 droplets/sec, and continuous tuning of injected droplet volume ranging from 1pL to 150pL with precise volume control (0.26% volume variation) in a microfluidic device. On-demand droplet generation in PLDG is realized by focusing intense laser pulses into water to induce rapid bubble expansion and perturbing an oil-water interface for droplet formation. Device reliability testing of continuous excitation for 3.6 million laser pulsing cycles is also achieved without any damages.

INTRODUCTION

Droplet-based microfluidic devices have attracted great interests due to the advantages over single-phase continuous-flow microfluidic devices. By isolating aqueous droplets containing biological or biochemical contents in an immiscible oil medium, cross-contamination between droplets is fully eliminated and reagents can be transported without dispersion with rapid mixing. To realize high-throughput and quantitative analyses in these droplet-based microfluidic devices, high-speed droplet generation with precise volume control is important. Passive-type flow-focusing devices are commonly used for continuous droplet generation with highly uniform size at a speed up to 12,000 droplets/sec [1]. However, it is difficult to achieve droplet generation on demand in these passive-type devices.

To achieve on-demand droplet generation, active control valves need to be integrated in microfluidic devices. For example, Zeng et al showed pneumatic microvalve-integrated devices for on-demand droplet generation at a speed up to 100 Hz, and tuning droplet volume by adjusting the on/off valve switching time [2]. However, the volume variation becomes as large as 7.2% at higher generation speed. Hsiung et al showed a microfluidic chip capable of chopping a stream of dispersed phase using a movable PDMS wall structure actuated by an external air pressure. It enables generation of droplets ranging from 10μm to 120μm in diameter at a speed of 20 Hz with volume variation of 9.9% [3]. Most previously demonstrated active-type devices either have a droplet generation speed orders of magnitude slower than the passive-type devices or the droplet volume uniformity is relatively poor at high generation speed [4].

Here, we report on a high-speed, pulse laser-driven droplet generation (PLDG) mechanism enabling on-demand droplet generation up to 10,000 droplets/sec with precise volume control.

DEVICE AND WORKING PRINCIPLE

Figure 1 shows the schematic of a PLDG device and its working principle. The PLDG device consists of two main microchannels (water and oil) connected by a nozzle-shaped opening. A stable water-oil interface is formed at the opening by simply adjusting the flow rates. A pulse laser with 8-ns pulse duration, 100-μJ pulse energy at 532-nm wavelength, and a maximum 100-kHz repetition rate is focused in the middle of the water channel nearby the nozzle opening. An intense laser pulse at the focal point induces water breakdown and triggers a rapidly expanding cavitation bubble to perturb and squeeze the liquid in the water channel into the oil channel, resulting in formation of a droplet. On-demand droplet generation is realized by electrically gating the laser pulses delivered to the excitation channel.

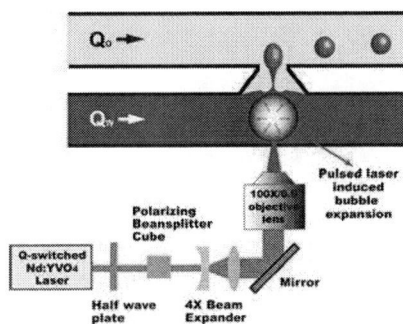

Fig.1 Schematic of a PLDG device and its working principle for on-demand droplet generation.

EXPERIMENTAL DEMONSTRATIONS

Fig. 2 presents the time-resolved images showing on-demand droplet generation induced by a laser pulse. The stable water-oil interface is formed at the nozzle-shaped opening (Fig. 2 (a)). A rapid bubble expansion induced by laser pulsing pushes out a small amount of water into the oil channel (Fig. 2 (b)~(e)). During the bubble collapsing process, the back flow of liquids is created and a neck-shaped connection is formed between a droplet and a main water stream at the opening (Fig. 2 (f)(g)). The process of bubble expansion and collapse causes hydrodynamic instability and breaks

978-1-4244-8926-8/10 $26.00 © 2010 IEEE

this neck connection, resulting in the formation of a droplet in the oil channel (Fig. 2(h)).

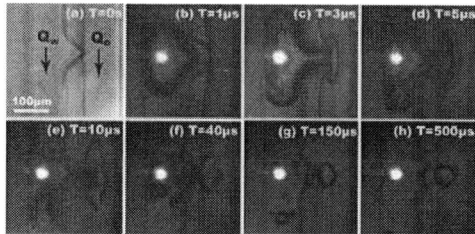

Fig. 2 Time-resolved image series of on-demand droplet generation. A 137-pL droplet is created by a 100-μJ laser pulse at $Q_w=12mL/h$ and $Q_0=0.2mL/h$.

We further analyze injected droplet volume uniformity by taking cross-section images. Six droplets are continuously generated with 1 Hz repetition rate. Their images are taken in gray-scale, which are converted to the logical-scale images. The number of pixels enclosed by a fitting curve are counted and repeated for six droplets, showing a 0.26% volume variation (Fig. 3). Our PLDG mechanism allows continuous tuning of injected droplet volume ranging from 1pL to 150pL either by adjusting the pulse laser energy which controls the bubble size (Fig. 4(a)~(d)) or varying the laser pulsing location that determines the amount of water being pushed into the oil channel (Fig. 4(e)~(g)).

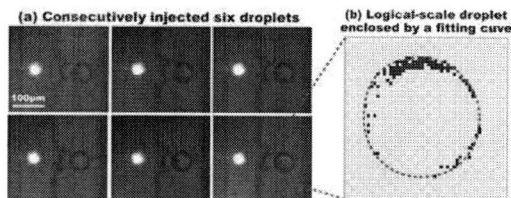

Fig. 3 (a) Gray-scale images of six droplets that are continuously injected with 1 Hz repetition rate. (b) Logical-scale droplets enclosed by an elliptical-shaped fitting curve. The cross sectional areas of the six droplets are estimated to 1128.71 ± 5.88 pixels (0.52% area variation), which corresponds to a 0.26% volume variation. One pixel size is 1.72 μm × 1.72 μm.

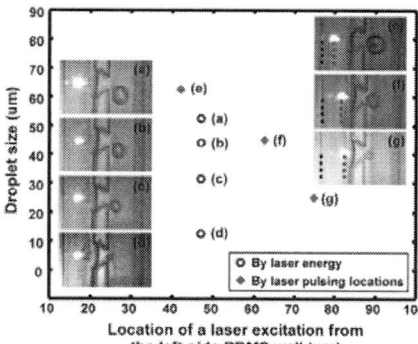

Fig.4 Droplet volume control using PLDG by tuning the laser energy with (a) 100μJ, (b) 90μJ, (c) 80μJ, and (d)

70μJ at a pulsing location 47μm away from the PDMS wall, and by varying the location of a laser excitation (e) 41μm, (f) 62.5μm, and (g) 75μm away from the PDMS wall at a fixed 100-μJ pulsing energy.

Fig. 5 shows continuous droplet generation with various laser pulsing intervals in PLDG. We have achieved droplet generation speed as fast as 10 kHz by using 90-μJ pulse energy at an interval of 100 μs (Fig. 5d). The PLDG device reliability was tested (Fig. 6), showing no damages on the device after 3.6 million times pulsing cycles for 1 hour.

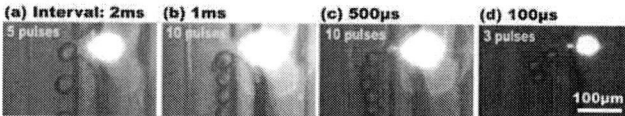

Fig. 5 Continuous droplet generation with various laser pulsing intervals, (a) 2ms, (b) 1ms, (c) 500μs, and (d) 100μs, which control the space between droplets. (d) By using a pulse interval of 100 μs and 90 μJ pulse energy, a maximum droplet generation rate as fast as 10 kHz has been achieved.

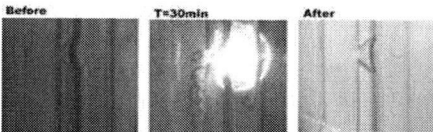

Fig. 6 PLDG device reliability was tested. The laser pulses have been continuously excited to generate droplets with 1 kHz repetition rate for 1 hour. No damages on the device are observed after 3.6 million laser pulsing cycles.

CONCLUSION

We report a pulse laser driven optofluidic device enabling on-demand droplet generation up to 10,000 droplets/sec, and continuous tuning of injected droplet volume ranging from 1pL to 150pL with precise volume control (0.26% volume variation) in a single layer PDMS microfluidic device without any mechanical valves or pumps.

ACKNOLEDGEMENT

This project is supported by the NSF CAREER Award ECCS-0747950, and NSF DBI-0852701.

REFERENCE

[1] L. Yobas, S. Martens, W. Ong, and N. Ranganathan, *Lab Chip*, **6**, 1073-1079, 2006.
[2] S. Zeng, B. Li, X. Su, J. Qin, and B. Lin, *Lab Chip*, **9**, 1340-1343, 2009.
[3] S.-K. Hsiung, C.-T. Chen, and G.-B. Lee, *J. Micromech. Microeng.*, **16**, 2403-2410, 2006.
[4] J. Xu, and D. Attinger, *J. Micromech. Microeng.*, **18**, 065020, 2008.

Reshaping Gold Micro and Nano Structures with Polarization Dependent Photothermal Annealing

Fan Xiao and Pei Yu Chiou

Department of Mechanical and Aerospace Engineering, University of California at Los Angeles, USA
420 Westwood Plaza, Los Angeles, CA 90095-1594, USA
TEL: (310) 8259091, Email: xiaofan@ucla.edu

ABSTRACT

We report a novel polarization dependent photothermal annealing technique for fine-tuning the shape of metallic micro- and nano-structures with flood exposure of nanosecond laser pulses after they are fabricated. This is realized by utilizing the local field enhancement due to collective electron oscillation on metallic nanostructures driven by electromagnetic waves. We have demonstrated that pre-patterned Au squares can be annealed into Au nanospheres; circular disks into elliptical disks; and rectangular Au microwires into Au nanowires with different aspect ratios.

INTRODUCTION

Metallic nanostructures have wide applications in plasmonic devices[1], color filters[2], and biomedical fields such as metallic particle and guided gene delivery[3] and photothermal therapy[4]. Methods for fabricating metallic nanostructures are versatile, ranging from chemical synthesis, electron beam lithography, ion beam lithography, nanoimprint lithography (NIL), and laser induced dewetting processes. Among them, laser induced dewetting process is considered as an economical method for rapid and large area fabrication.

To guide the laser induced dewetting and self-organization processes, two types of methods have been proposed, including pre-patterning the substrates with specific morphologies[5] or spatially shaping the projected light patterns[6]. Multi-beam Interference Lithography (MIL) belongs to the later type and allows low cost and rapid fabrication of two dimensional nanostructures on a continuous thin metallic film. However, MIL is limited to fabricating periodic structures with simple shapes. Xiao et al recently demonstrated a rapid laser printing technology, we call near field photothermal printing (NPTP), capable of fabricating both periodic and non-periodic metallic structures with light shaped by a transparent phase-shifting mask. A laser pulse with uniform spatial light intensity passing through a PDMS phase-shifting mask creates a non-uniform light intensity profile on the underlying gold thin film. The metal film is selectively melted and quickly migrates to cold areas due to surface tension of molten Au. With NPTP, Au nanowires, nanosquares, and nanospheres can be obtained by properly designing the patterns of phase shifting masks and follow-up flood laser exposure.

Here, we study the light polarization dependent nanomorphology evolution process during flood exposure and its applications for fine-tuning the shape of metallic nanostructures. Fig. 1 illustrates the working principle of this mechanism. Laser pulses with uniform light intensity illuminate on these pre-patterned metallic structures on a glass or silicon substrate. Due to the localized collective electron oscillations on these metallic structures, localized surface plasmon waves induce non-uniform field enhancement at different parts of the structures. The kinetic energy of these oscillating electrons quickly converts into lattice heat and selectively melts the high intensity regions. The molten Au migrates to the cold areas due to surface tension before cools down. The field enhancement is polarization dependent. The edges perpendicular to the field polarization direction are enhanced most. Fine-tuning of the shape of metallic nanostructures can be achieved by controlling the pulse energy, numbers, and polarization.

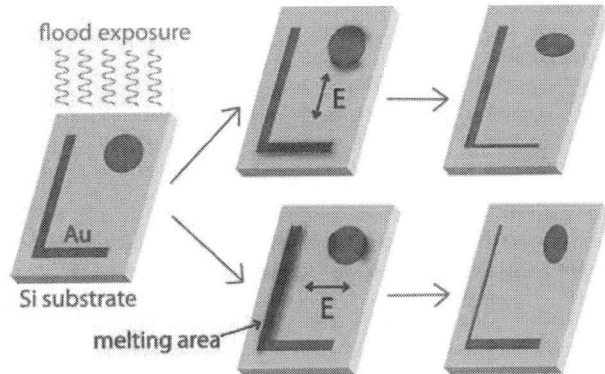

Fig. 1 Schematic of the working principle of polarization dependent photothermal annealing for fine-tuning the shape of metallic micro and nano structures.

SIMULATION AND EXPERIMENTAL RESULTS

Fig. 2 shows the simulated light intensity distribution on a 100nm thick, 2µm wide, and 4µm long metal rectangular disks using the FDTD method (RSoft). Normal incident plane waves with wavelength at 532 nm and E field polarization along the short and long axes of the rectangular disks are used in this simulation. The results indicate that, for both cases, the strong electric energy density occurs at the corners and on the edges perpendicular to the polarization of incident light, predicting stronger photothermal effect in such areas where the melting is initiated.

978-1-4244-8926-8/10 $26.00 © 2010 IEEE

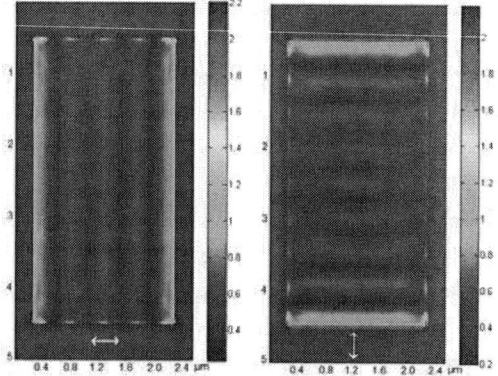

Fig. 2 (a) FDTD simulation showing the electric energy density distribution under light illumination with E field polarization along the short axis (b) E field polarization along the long axis.

In the experiment, a Q-switched Nd:YAG pulse laser with a pulse width of 6 ns and wavelength of 532 nm is applied. 5 nm Ti/100 nm thick microscope Au structures are pre-patterned on Si substrate by photolithography and lift-off approach as show in Fig. 3 (a). By flood exposing the patterns under light pulses with different polarization as indicated in Fig. 3(b) and (c), size reduction occurs in different parts of the initial patterns as predicted by the simulation, which is higher photothermal effect occurring in edges perpendicular to the polarization of light. Fig. 3(d) enlarges the letter "L" in Fig. 3(c) and shows that the initial 2μm wide Au wire evolves into a 400nm nanowire after flood exposure, while the width of the wire parallel to the field polarization remains almost unchanged. Fig. 3(e) and (f) show a circular microdisk evolves into an elliptical disk. With higher pulse energy, the entire microstructures can be melted and quickly evolve into final stable shapes having the lowest surface energy. Fig. (g) and (h) show that an array of Au micro squares are annealed into an array of Au nanospheres using several laser pulses.

CONCLUSION

We have demonstrated a polarization dependent photothermal annealing mechanism capable of fine-tuning the gold micro and nano structures by utilizing the local field enhancement induced by the localized surface plasmon on metallic structures. We have shown that pre-patterned Au micro squares can be annealed into nanospheres; circular disks annealed into elliptical disks; and Au microwires into Au nanowires.

ACKNOWLEDGEMENT

This project is supported in parts by the NSF grants CBET-0853500 and NSF ECCS-0901154.

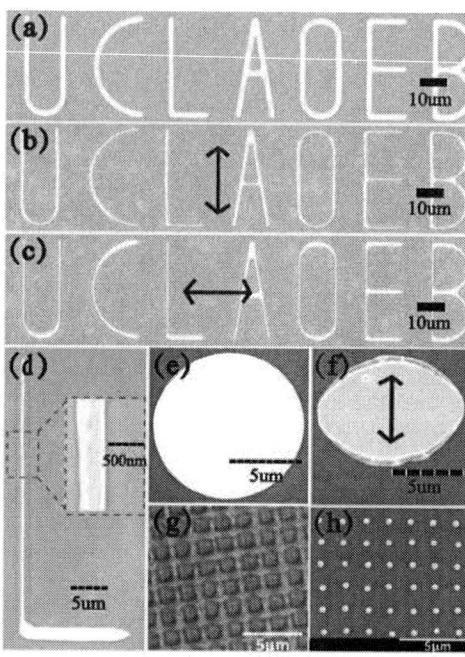

Fig. 3 Experimental results showing laser polarization dependent shape evolution of metallic micro and nanostructures.

REFERENCE

1. M. Righini, A.S. Zelenina, C. Girard, R. Quidant, *Parallel and selective trapping in a patterned plasmonic landscape.* Nature Physics, 2007. **3**(7): p. 477-480.

2. E. Laux, C. Genet, T. Skauli, T.W. Ebbesen, *Plasmonic photon sorters for spectral and polarimetric imaging.* Nature Photonics, 2008. **2**(3): p. 161-164.

3. T.H. Wu, S. Kalim, C. Callahan, M.A. Teitell, P.Y. Chiou, *Image patterned molecular delivery into live cells using gold particle coated substrates.* Opt. Express, 2010. **18**: p. 938-946.

4. J. Kim, S. Park, JE Lee, S.M. Jin, J.H. Lee, I.S. Lee, L. Yang, J.S. Kim, S.K. Kim, M.H. Cho, T. Hyeon, *Designed fabrication of multifunctional magnetic gold nanoshells and their application to magnetic resonance imaging and photothermal therapy.* Angewandte Chemie (International ed. in English), 2006. **45**(46): p. 2138344.

5. Q. Xia, S.Y. Chou, *The fabrication of periodic metal nanodot arrays through pulsed laser melting induced fragmentation of metal nanogratings.* Nanotechnology, 2009. **20**: p. 285310.

6. Y. Kaganovskii, H. Vladomirsky, M. Rosenbluh, *Periodic lines and holes produced in thin Au films by pulsed laser irradiation.* Journal of Applied Physics, 2006. **100**: p. 044317.

NEAR-FIELD PLASMONIC ENHANCEMENT VIA NANOGRATINGS ON HOLLOW PYRAMIDAL APERTURE PROBE TIP

Yuyan Wang, Yu-Yen Huang, Kazunori Hoshino, Ashwini Gopal and Xiaojing Zhang

University of Texas at Austin, Austin, Texas 78712

ABSTRACT

We present the design of hollow near-field scanning microscope (NSOM) probe with nanogratings-on-tip to transport and concentrate localized surface plasmonic polariton (SPP) wave. By adding nano-grooves started from the intensity-maximum locations of lowest transmission mode and with pitch period supporting the metal-air interface SPP mode, the power throughput is increased at over 530 times comparing with single aperture probe with 405nm source and 100nm diameter aperture size. Two types of nanograting probe designs are chosen for fabrication and the power enhancement comparison is examined by probing the near-field fluorescent intensity of excited uniform quantum dots (QDs) layer via micro-contact printing method.

INTRODUCTION

Periodic nano-structuring metal surface is often used for modulating SPP to confine and deliver electromagnetic energy into sub-wavelength spatial volume [1]. Grating structures exploring plasmonic resonance and diffraction have been studied to enhance the transmission of apertureless probe [2]. Aluminum coated fiber tip with periodical surface rings was also introduced for near-field lithography, where the period matches the interference of incident beam and the SPP along probe internal surface [3]. Here, we present a novel concentric grating design on the pyramidal tip, with its pitch period matching the SPP wavelength between probe coating metal and existing medium. Specifically, the first inner groove is positioned such that it is on the incident field intensity maximum of the lowest mode to further increase the power coupling. Theoretical analyses show that such a design provides significant enhancement in average power throughput comparing to that with standard single aperture probe of 100nm diameter aperture within near-field region.

NANOPROBE DESIGN AND SIMULATION

The probe tip structure with nanogratings is shown in Figure 1a. The grating is integrated on the *200nm* Ag coated hollow SiO₂ pyramidal tip [4-5]. The nanogratings are fabricated with focused ion beam (FIB) milling method with pattern shown in Figure 1b.

Figure 2a shows the design principles and design variations of nanogratings. For a single aperture plasmonic probe, most of the incident light is reflected before cut-off. The specific field distribution inside the

probe and along SiO₂-Ag interface does not change with grating design or tip-sample distance according to perturbation theory. With 405nm laser illumination, we design and fabricate nanograting (design A) with period of *177nm* to support air-Ag interface SPP mode, the dominant mode for interface formed with electron beam evaporated Ag layer on dielectrics [6]. Nanograting design B satisfies the constructive interference condition between incident light and inner SiO₂-Ag SPP mode at the tip apex with a period of 225nm. The first inner gratings of A and B are milled at the lowest incident mode intensity maximum (which has the *shortest path* to the tip apex for SPP to travel) for less energy loss and larger power coupling to the aperture. Nanograting C and D are designed to match incident light intensity maxima and minima with various period.

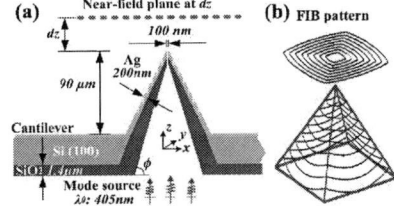

Figure 1. The plasmonic scanning aperture probe with nanograting. (a) Schematic of probe tip structure. (b) nanograting designed on the tip, including the probe pyramidal side-wall shape compensation.

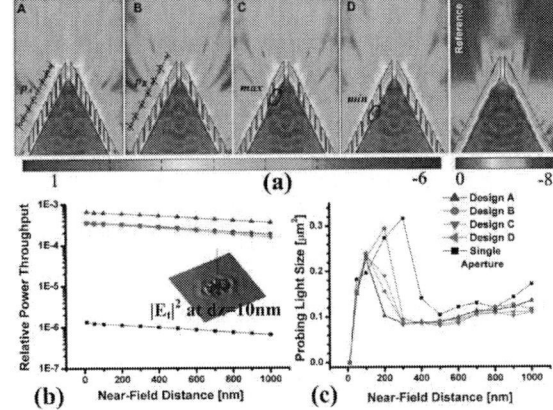

Figure 2. Nanograting design principles and simulation with 405nm laser incidence and 100nm diameter aperture (a) Electric field intensity distribution (in logarithm) along x-z section for 4 nanograting designs with single aperture probe as reference. (b) The relative power throughput at near-field for probes described in (a). The insertion is the probing light electric field intensity distribution at 10nm near-field plane for design A probe. (c) The cross-section area (i.e. beam size) of the probing lights at their

half-maximum. The average size of nanograting A is similar to reference probe and nanograting B is 1.2X to the reference.

The three-dimension SPP distribution (Figure 2) at the probe tip is simulated using commercial finite difference time domain (FDTD) software [7]. A linear polarized p-wave mode source is applied on the model. The center aperture shown in Figure 1c and 2 has a diameter of *100nm*, which is smaller than $\lambda_0/2n_{SiO2}$ (λ_0 is the source wavelength, n_{SiO2} is refractive index of SiO_2), the cut-off width for the lowest TE_{10} and its degenerated mode TE_{01}. This is the key to ensure that only the localized evanescent wave contributes to the probing light.

EXPERIMENT AND RESULT

Figure 3 shows the SEM images of fabricated nanogratings-on-tip structure and cantilever probe overview.

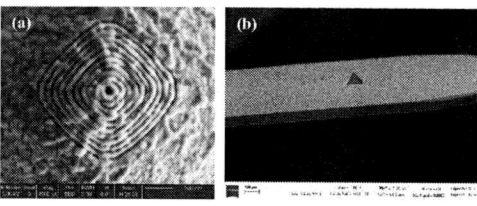

Figure 3. The SEM images of (a) nanogratings-on-tip probe, scale bar is 500nm, and (b) cantilever probe overview the scale bar is 1μm.

Figure 4a shows the experimental setup integrated with customized near-field scanning microscope system. The nanograting probe power throughput is measured by probing the near-field fluorescent intensity of excited uniform QDs layer with varied tip-sample distance. The QDs sample was deposited on a glass slide through micro-contact printing method [8].

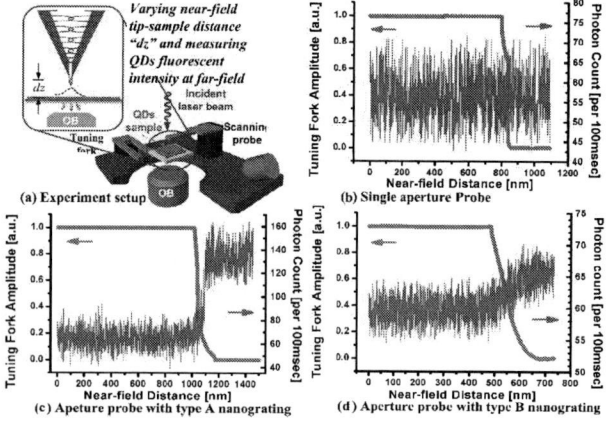

Figure 4. (a) Power enhancement measurement setup. (b)Avalanche photodetector captured fluorescent photons from a single aperture probe with varied tip-sample distance. The tuning fork resonate amplitude is reducing during contact. The near-field intensity from a single aperture is buried by NSOM system noise. (c) and (d) are the measurements from nanograting A and B probes.

We fabricated single aperture probes and probes with nanograting A and B. Their center aperture size is milled via FIB with the same parameter to form a 100nm hole through Ag layer. The emitted photons from the QDs were captured with varied tip-sample distance are shown in Figure 4b-4d. For a single aperture probe, the energy of localized SPP is too weak to suppress the background noise. The nanograting A probe excites fluorescent energy that is twice as large as nanograting B probe, which fits well with simulations (see Figure 2b). These preliminary results demonstrate the effectiveness of near-field energy collecting and coupling by adding properly designed nanogratings.

CONCLUSIONS

In summary, we fabricated a novel plasmonic hollow pyramidal probe with integrated nanograting structure. The grating design was optimized for near-field enhancement based on numerical simulations. Using single aperture probe as a reference, the enhanced power throughput was verified by measuring the near-field excited QDs using two types of nanograting probes. The design A provides twice the power enhancement comparing with the design B, which agrees with the simulation result.

ACKNOWLEDGEMENTS

The financial support for this work by the National Science Foundation (CMMI-0826366) is acknowledged. The work was carried out at the Microelectronics Research Center (MRC) and Nano Science and Technology (NST) Center, the University of Texas at Austin.

REFERENCES

1. H. Raether, "Surface plasmons on smooth and rough surfaces and on gratings," Springer, Berlin, Heidelberg 1988.
2. C. Ropers, et al., "Grating-coupling of surface plasmons onto metallic tips: A nanoconfined light source," Nano Lett., vol. 7, pp. 2784-2788, 2007.
3. Y. Wang, et al., "Plasmonic nearfield scanning probe with high transmission," Nano Letters, vol. 8, pp. 3041-3045, 2008.
4. Y. Wang, et al., "Plasmonic Nanoprobe Integrated with Near-field Scanning Microscope," IEEE Optical MEMS and Nanophotonics, Clearwater Beach, Florida, USA, Aug. 17-20, 2009.
5. Y. Huang, et al., "Fabrication and scanning control of nanoprobe for NSOM applications," SPIE Photonics West, San Francisco, CA USA, Jan. 23-28, 2010.
6. Z. Shi, et al., "Surface-plasmon polaritons on metal-dielectric nanocomposite films," Opt lett., vol. 34, pp. 3535-3537, 2009.
7. Lumerical Solutions, Inc, www.lumerical.com.
8. A. Gopal, et al., "Multi-color colloidal quantum dot based light emitting diodes (QD-LED) micropatterned on silicon hole transporting layers," Nanotech., vol. 20, 235201, 2009.

SURFACE OPTOMECHANICS: MECHANICAL WHISPERING-GALLERY MODES IN MICROSPHERES

John Zehnpfennig, Matthew Tomes, Tal Carmon

The University of Michigan, Ann Arbor, USA

ABSTRACT

We analyze circumferentially circulating mechanical whispering-gallery modes [WGM] resonating in a micron scaled silica sphere. We recently showed that such modes can be excited optically [1]. A variety of modes are calculated in which the deformation is polar, radial or azimuthal.

Additionally, we calculate Rayleigh WGMs for which points on the surface follow a circular path.

INTRODUCTION

Light and sound can travel circumferentially around a dielectric sphere while trapped via total internal reflection [2-5]. These modes are called WGM following their discovery at the St. Peter's Cathedral. Such optical- and mechanical-modes can exchange energy via the interplay between electrostrictive radiation pressure causing deformation and the change of the optical pathlength by this deformation. One can also think about this mechanism as if the optical electro-magnetic wave is writing a train of almost perfectly separated virtual electrodes that collectively excite the mechanical mode.

Here we calculate thmode at a variety of mechanical WGM are possible in photonic-MEMS: longitudinal, shear-polar and -radial, and Rayleigh modes[4]. Further, each of these four families has members with more than one maximum along the radial, polar, or both of these directions, transverse with propagation.

THEORY

We calculate acoustic waves in optical WGM resonators composed of silica glass [6] that are excited by light as we demonstrated in [1]. The acoustical wavelength is half of the exciting optical wavelength as required to conserve momentum [6].

This implies that the frequency f of the acoustical wave can be estimated using the time it takes sound to cross half optical wavelength

$$f \cong \frac{2V n_{eff}}{\lambda}$$

where V is the velocity of sound, n_{eff} is the dielectric medium index of refraction, and 1 is the optical vaccuum wavelength. For example, l≈1.5μm pump light with silica sphere (refractive index ~1.46) a mechanical longitudinal WGM of 11 GHz is excited [1] in accordance with the 5.5km/s velocity for the corresponding longitudinal wave.

As different families of acoustical waves propagate at different velocities within the media, their mechanical frequencies will be different (table 1). Rayleigh waves are expected to propagate at ~3.4 km/s in SiO₂ and are hence expected to oscillate at ~6GHz if excited with the same 1.5μm pump source.

In order to get the accurate value for the frequency and shape of the mechanical WGMs, we numerically solve the exact tensorial stress-strain equation where the only assumption is discretization in space.

RESULTS

Modal structure: different types of modes

While all resonances are propagating azimuthally along the equator, the mode structure can vary. Fig. 1 describes modes with deformation in (a) azimuthal (b) polar (c) radial and (d) radial-azimuthal (Rayleigh-type) directions. These modes have different frequencies due to the different velocities of the corresponding waves. Additionally, the shear-radial and -polar modes have different frequencies because their deformation is different in respect to the air-SiO₂ interface: perpendicular and parallel respectively.

Table 1 Mechanical Waves and Whispering Gallery Resonator Modes [7-9].

V_i represents the velocity of sound of mechanical wave family i. The medium properties are: E elasticity, ν Poisson ratio, and ρ density.

Bulk		60μm SiO₂ Sphere		
Wave	Velocity [m/s]	Mode	Deformation	Velocity [m/s]
Longitudinal	$V_L = \sqrt{\dfrac{E(1-\nu)}{\rho(2\nu^2 + \nu - 1)}} = 5848$	Longitudinal	East-West	5987
Shear	$V_S = \sqrt{\dfrac{E}{2\rho(\nu+1)}} = 3687$	Shear Polar	North-South	3815
		Shear Radial	In-Out	3710
Rayleigh	$V_R = \dfrac{V_s(0.87 + 1.12\nu)}{(1+\nu)} = 3400$	Rayleigh	In-out and East-West	3663

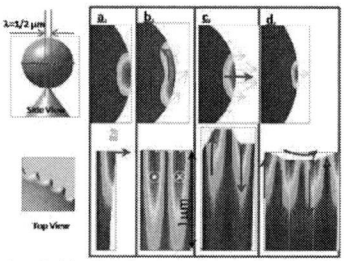

Figure 1: Strain field distribution for a. longitudinal, b. shear polar, c. shear radial, and c. Rayleigh modes. Sphere is 60 micron in diameter and the mode has 734 acoustical wavelengths along the diameter. This corresponds to excitation at $\lambda \approx 1.5\mu m$.

Modal structure: high-order modes

Each mode family described in Fig 1 has members with more than one maximum along the direction transverse with propagation. We called these modes high-order modes and we depict some of these higher order modes in figure 2.

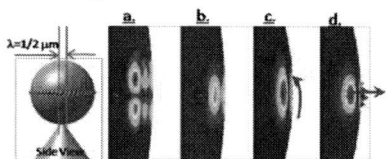

Figure 2: Examples of high-order modes. Shown are: a. (2,3,734) longitudinal b. (1,2,734) longitudinal, c. (1,2,734) shear polar, and d. (1,2,734) shear radial. Arrows indicate deformation direction. (i, j, k) is the (polar, radial, azimuthal) order of the mode.

Modal spectroscopy: acoustic frequency versus sphere size

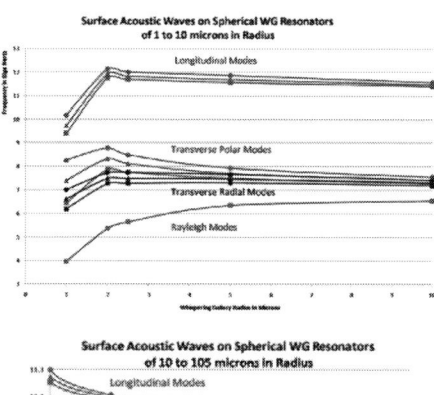

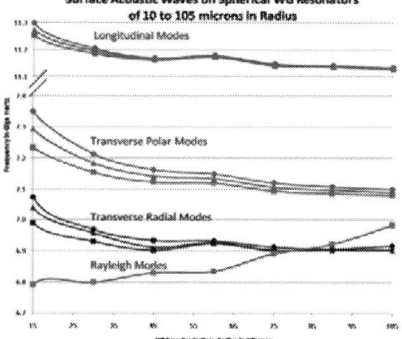

Figure 3: Mode frequency versus sphere size. High order modes of various families are described by lines of the same color.

In figure 3 we calculate the modal frequencies for a typical experiment in which modes are excited with the same $\lambda \approx 1.5$um laser.

We found crossing between modes as the radius changes, similar with what was reported for the corresponding optical WGM [10].

CONCLUSIONS

We calculate a variety of mechanical WGMs that can be optically excited at telecom IR in a µm-scale sphere to provide mechanical resonances from 4 to 12 GHz.

REFERENCES

[1] M. Tomes and T. Carmon, "Photonic micro-electromechanical systems vibrating at X-band (11-GHz) rates," Physical Review Letters, vol. 102, p. 113601, 2009.

[2] K. Vahala, "Optical microcavities," Nature, vol. 424, pp. 839-846, 2003.

[3] R. K. Chang and A. J. Campillo, Optical processes in microcavities. Singapore: World Scientific, 1996.

[4] L. Rayleigh, "On waves propagated along the plane surface of an elastic solid," Proceedings of the London Mathematical Society, vol. 1, p. 4, 1885.

[5] I. Grudinin, A. Matsko, and L. Maleki, "Brillouin Lasing with a CaF_ {2} Whispering Gallery Mode Resonator," Physical Review Letters, vol. 102, p. 43902, 2009.

[6] D. Armani, T. Kippenberg, S. Spillane, and K. Vahala, "Ultra-high-Q toroid microcavity on a chip," Nature, vol. 421, pp. 925-928, 2003.

[7] L. Kinsler, A. Frey, A. Coppens, and J. Sanders, "Fundamentals of acoustics," 1999.

[8] D. S. Ballantine and Knovel, Acoustic wave sensors: theory, design, and physico-chemical applications. San Diego: Academic Press, 1997.

[9] Pham Chi Vinh and P. G. Malischewsky, "Improved Approximations of the Rayleigh Wave Velocity," Journal of Thermoplastic Composite Materials, vol. 21, pp. 337-352, July 1, 2008 2008.

[10] T. Carmon, H. G. L. Schwefel, L. Yang, M. Oxborrow, A. D. Stone, and K. J. Vahala, "Static Envelope Patterns in Composite Resonances Generated by Level Crossing in Optical Toroidal Microcavities," Physical Review Letters, vol. 100, p. 103905 2008.

Silicon Integrated Electronic-Photonic ICs

Patrick Lo and Dim-Lee Kwong
Institute of Microelectronics/A*STAR
11 Science Park Road, Science Park II, Singapore, 117685

ABSTRACT

Copper interconnect is unlikely to be the ultimate solution to support the growing functionalities of next generation microprocessor due to its information latency and power consumption. Converging EPIC on a single chip platform to enable functional diversification emerges as one promising approach to be realized by taking the advantage of low energy and huge data capacity of optical interconnects. In this presentation, we present an overview on the current status of this critical technology development and provide an outlook and strategies for the integration of Si micro- and nano-photonics to meet the bandwidth and energy requirements of data communication in future technology nodes.

Continuous scaling of CMOS transistor over the past decades has led to significant enhancement in the speed performance of electronic IC. As the demand for on-chip functionality continues to grow, it is now widely recognized that electronics technology will increasingly require integration with photonics to keep up with the Moore's Law. Although the exploitation of Cu-wires surrounded by lower permittivity dielectrics had been effective in the past generations in minimizing signal delay, there is little doubt if such option will still be the ultimate solution. The development of a cost-effective and manufacture-able solution addressing these challenges will be a critical, e.g., the optical interconnects realized by a CMOS-compatible platform.

By leveraging on the high index contrast between the Si core and SiO_2-cladding, small foot-print routing of optical signals using photonic micro-waveguide with low propagation loss can be made possible. Furthermore, to harness the increased signal carrying capacity, a wavelength division multiplexing (WDM) could be exploited to provide multiple channels transmission for parallel data processing. Using an optical filter made of ring resonator or AWG, a selected wavelength can be filtered into the photonic waveguide data bus. While Si-based modulator exhibits weaker electro-optical coefficients, a free carrier plasma dispersion effect can be exploited to enable effective optical modulation to convert the electrical signal to optical one. The modulated optical signals are then channeled through the photonic waveguide data bus where multiple wavelengths signals can be delivered on the same channel without suffering interference from one another. At the receiving end, signal of a specific wavelength channel is filtered and fed into a high performance photodetector (PD) for enabling efficient optical-to-electrical encoding. The encoded electrical signal will then pass through the CMOS circuitry for data processing (Fig. 1). Needless to say, the development of both active and passive photonics building blocks is crucial to bring OEIC reality.

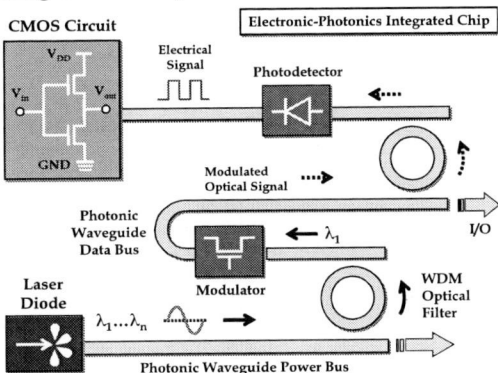

Fig. 1. *Schematic illustration of optical interconnects technology realized using Si photonics platform and its convergence with electronics integrated circuit for high performance functional diversification.*

Recent advancement shows that Ge is attracting growing interest to enable the realization of highly efficient PD in the near-infrared regime. However, technological challenges exist for the epitaxy growth of high quality Ge on Si due to the large lattice mismatch (~4.2%). Two major issues arise: (1) high densities of threading dislocations and (2) rough surface. Both present much concerns for the generation of high leakage current compromising the efficiency of a PD. Unlike the conventional approaches of employing thick SiGe buffer layer, we have developed a low temperature pseudo-graded SiGe buffer approach to relieve the mismatch stress, thereby suppresses the formation of dislocation defects substantially. This allows the elimination of high temperature post-epitaxy anneal that is commonly used for defects reduction, which is attractive for CMOS-compatibility with a low thermal budget (<700°C).

Various Ge-based PDs have been developed in our study (e.g., PiN, Waveguided, APD, and Phototransistors).

Meanwhile, Si-based modulators can be monolithically fabricated by exploiting the free carrier plasma dispersion effect. For wideband operation across a range of wavelengths, the MZI design is typically utilized. By inserting p-n junction phase-shifters in the Si waveguides, the carrier density can be controlled to enable optical intensity modulation at the output. Operation of the phase-shifters in depletion mode results in excellent modulator speed characteristics. We have developed a carrier depletion type modulator with good phase-shifting efficiency ($V_\pi L_\pi \sim 2.56$ Vcm). For 10 Gbps operation, an E.R. of 6.1 dB was achieved for a modulator with a 2-mm-long phase-shifter, which was single-ended driven with a RF signal of 5 V_{pp}. The free carrier density in the rib waveguide phase-shifters is modulated by varying the depletion width of the p-n junction.

Meanwhile, passive photonics components form an important integral for EPIC implementation. Passive device library developed in this work includes nano-tip, filter, directional coupler, splitter, ring resonator, AWG and most importantly Si-waveguides for routing optical signals on the photonics integrated circuit. With availability of HIC system, miniaturization of photonic circuit becomes feasible due to the possibility of reducing the bend radius <5 μm. Using typical CMOS process, negligible waveguide transmission loss of ~0.5 dB/cm has been routinely obtained. Meanwhile, WDMs based on (a) 32-channels AWG composed of rib SOI waveguides and (b) ring resonator are developed to provide multiple channels for parallel data processing.

Furthermore, based on the basic photonics building blocks developed, we have successfully developed a monolithic technology strategy to integrate both electronic and photonic integrated circuits simultaneously on common SOI platform. In particular, carrier depletion type Si modulators and waveguided Ge-PD have been monolithically fabricated using a process integration flow that is compatible with SOI CMOS process technology, both in terms of thermal budget and fabrication feasibility. In such monolithic integration scheme, an 'electronic-first/photonic-last' integration was proposed and demonstrated. In this approach, all high temperature processes (e.g., >900°C) needed to form CMOS logic circuit are performed first prior to the fabrication of photonic components and circuit.

Ultimately, what limits the integration of optical and electronic circuits most is the mismatch in their respective length scales. This is true as electronics devices continue to scale in advanced CMOS technology node, the foot-print of existing photonics devices which are typically in the μm-scale seems to impose much constraint to enable the realization of ultra-compact EPIC.

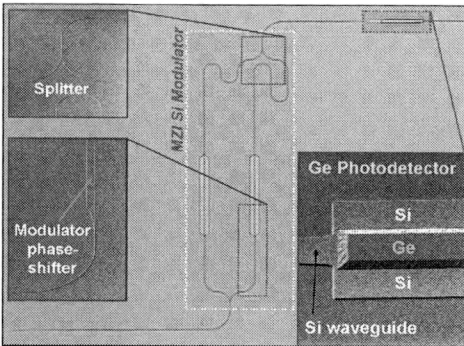

Fig. 2 Monolithically integrated Si optical modulator and Ge photodetector.

In parallel to the research effort to monolithically integrate photonic circuits into existing electronics circuits, one effective strategy could be using the advanced 3-D through-silicon-via (TSV) scheme to stack the photonics circuit over electronics heterogeneously. By this method, the individual circuit could be fully optimized for their own respective performance without having to consider the inevitable trade-off during the whole integration.

In another consideration, miniaturization of photonic devices into the nanoscale regime could propel the development of EPIC towards much small form factor, low power and high speed integration, but such development does not come without fundamental challenges. The exploitation of surface plasmon based nano-photonics may offer a solution to alleviate this problem as the feature of SP enables the possibility of localization and guiding of light in sub-wavelength structures. In our recent developed nano-photodetector, inter-digitated metallic electrodes acting as 1-D rectangular gratings to enhance the TM-mode responsivity. As projected, the application of plasmonics would enable the next wave of development effort in nanophotonics to realize ultra low power and high data rate optical interconnect solutions in the nanoscale regime including low-loss waveguides, ultra-compact modulator, especially if realized by CMOS-compatible technology.

In conclusion, we will present the current status of Si-photonics development, along with several options for strategy to realize the photonics and electronics circuit integration, monolithically and heterogeneously.

978-1-4244-8926-8/10 $26.00 © 2010 IEEE

HIGH POWER THZ PHOTOCONDUCTIVE ANTENNA USING LOCALIZED SURFACE PLASMON RESONANCE

Sang-Gil Park[1], Yongje Choi[1], Minwoo Yi[1], Jun-Hyuk Choi[2], Kyung-Hwan Jin[1], Jong-Chul Ye[1],
Jaewook Ahn[1], and Ki-Hun Jeong[1]

[1]KAIST, Korea
[2]Korea Institute of Machinery & Materials, Korea

ABSTRACT

This work presents a plasmon-enhanced photo-conductive antenna for high power THz emission using the localized surface plasmon resonance of hierarchically patterned silver nanoislands. The silver nanoislands on a photoconductive region enhance the electrical amplitude of pump laser beam and the scattering of the pump beam so that the optical power transmitted toward the substrate is amplified. As a result, a high transition rate of the photo-excited carrier was obtained. The plasmon-enhanced antenna shows 1.6-fold increase of peak-to-peak amplitude of the emitted THz wave compared to the photo-conductive antenna without silver nanoislands.

INTRODUCTION

As developing reliable THz sources, there are a number of interests in THz region and its various applications because THz time-domain spectrum has an ability of identifying intermolecular vibration spectra that FT-NIR could not offer [1]. However, a poor transmittance to polar molecule, in particular water, is major hurdle to bio-medical applications.

A photoconductive antenna (PCA) using Ti:Sapphire oscillator is the most popular THz generator. The amplitude of terahertz wave is proportional to the time derivative of photocurrent. Recently, Ti:Sapphire amplifier system is reported for high power THz emission [2]; however cost efficiency is low.

Herein we demonstrate plasmon-enhanced PCA (P-PCA) of which the optical coupling efficiency is increased by localized surface plasmon resonance (LSPR) on Ag nanoislands (Fig. 1). Electron cloud in nanoislands hierarchically patterned on a photo-conductive region collectively oscillated by incident light induces a electric dipole, and then concentrates light in local field. Its resonance frequency depends on the size and the shape of nanoparticle.

FABRICATION PROCESS

Figure 2 illustrates the fabrication steps of plasmon enhanced photoconductive antenna. Initially, the silver thin film was formed by E-beam evaporation. And the film was patterned by photolithography followed by chemical etching. Then, the antenna electrodes (Cr/Au) were patterned by lift-off (Fig 3a,b). Before the deposition of electrode metal, the silver film was laterally etched to prevent electrical contacts between electrodes and silver nanoparticles. Finally, the nanoislands of which diameters are widely distributed 10nm to 200nm were formed by thermal annealing (Fig. 3c,d).

MEASUREMENT OF OPTICAL EXTINCTION

The scattering intensity of nanoislands is directly measured by using a spectrometer with a dark field objective lens. The resonance peak is observed around 500nm in all samples. (Fig. 4) As the annealing time increases, the scattering intensity is more enhanced due to morphological isolation from a film to nanoparticles.

MEASUREMENT OF TIME-DOMAIN THZ WAVEFORM

The detection of THz emission from the fabricated antenna is based on conventional electro-optic sensing technique [3]. Peak-to-peak amplitude is 1.6-fold enhanced then PCA without nanoislands. Moreover, the waveform of P-PCA is not different from that of conventional PCA and consequently bandwidth was conserved. (Fig. 5)

SUMMARY & CONCLUSION

Plasmon-enhanced PCA with Ti:Sapphire oscillator is proposed for high power emission. 1.6 times THz emission compared to non-plasmonic structured PCA was achieved. Further enhancement will be possible by modulating optical property of nanoislands to their optical pumping wavelength.

REFERENCES

[1] R. Woodward, V. Wallace, R. Pye, B. Cole, D. Arnone, H. Linfield, M. pepper, "Terahertz Pulse Imaging of ex vivo Basal Cell Carcinoma", J. Invest. Dermatol., 2003.

[2] T. Hattori, K. Egawa, S. Ookuma, T. Itatani, "Intense Terahertz Pulse from Large-Aperture Antenna with Interdigitated Electrodes", Jpn. J. Appl. Phys., 2006.

[3] D. Wu et al., "Free-space electro-optic sampling of terahertz beams", Applied Physics Letters, 1995.

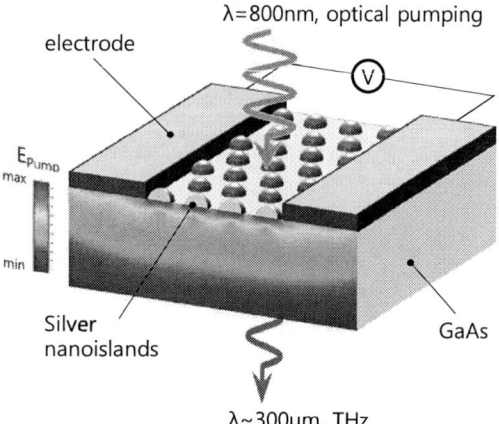

Figure 1. Schematic of surface plasmon enhanced photoconductive antenna with silver nanoislands on photoconductive region

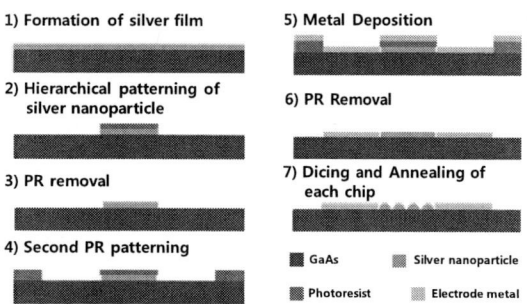

Figure 2. Fabrication process of THz P-PCA. Thin Ag patterns and thick Au electrodes were defined by conventional two-mask process. Thin silver layer was transformed to silver nanoislands by thermal annealing process.

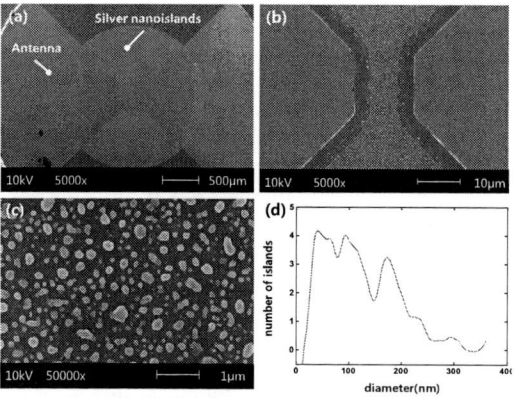

Figure 3. (a) SEM image of bow-tie P-PCA. (b) Photo-conductive region (15μm gap width). (c) Silver nanoislands. (d) Variation of nanoislands diameter.

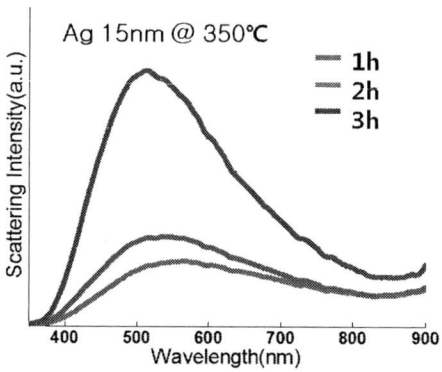

Figure 4. Scattering intensity of Ag nanoislands depending on annealing time. Scattering intensity at optical pumping wavelength (800nm) is enhanced with increasing annealing time.

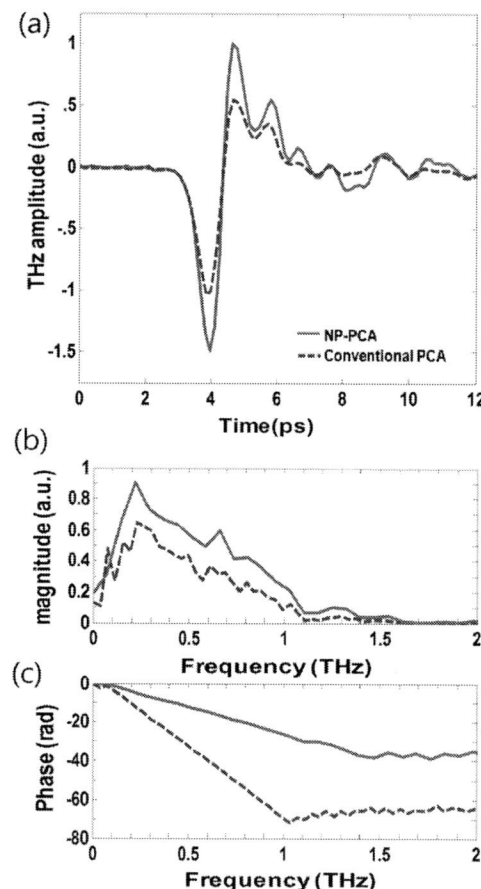

Figure 5. (a) Time-domain THz waveform by conventional electro-optic sensing technique. (b) Magnitude and (c) phase of THz waveform on frequency domain. 1.6 times-enhanced THz peak-to-peak amplitude was detected without loss of frequency bandwidth.

CONTROL OF SOLID-STATE LASERS USING AN INTRA-CAVITY MEMS MICRO-MIRROR

W. Lubeigt[1], J. Gomes[2], G. Brown[2], A. Kelly[1], V. Savitski[1], D. Uttamchandani[2] and D. Burns[1]

1. Institute of Photonics, University of Strathclyde, UK

2. Centre for Microsystems and Photonics, University of Strathclyde, UK

ABSTRACT

Low-cost scanning MEMS micro-mirrors were incorporated within the cavities of Nd:based laser systems to control the output beam characteristics. Successful Q-switching was obtained from these solid-state intra-cavity MEMS lasers resulting in pulse durations of 220ns and peak powers of 13.2W at a wavelength of 1064nm.

INTRODUCTION

Low-cost, intra-cavity scanning MEMS micro-mirrors were used in order to investigate the prospect of enhanced performance and functionality from optically pumped solid-state lasers. For instance, boresight correction, novel routes to active Q-switching and multi-beam laser configurations using either a single MEMS micro-mirror or 2-D MEMS micro-mirror arrays are realistic prospects.

Recently, Fabert *et al.* demonstrated the use of a cantilever-type MEMS micro-mirror as an active Q-switching element in a fibre-based laser system obtaining pulse durations as short as 20ns [1]. In their work, the dimensions of the largest MEMS devices $(150 \times 350 \mu m^2)$ and the maximum reflectivity (R=80%) restricted the possible resonator geometry and, more importantly, significantly limited the overall laser efficiency.

In this paper, we present the use MEMS micro-mirrors incorporated within the optical cavity of diode-pumped solid-state lasers, and discuss the limitations faced by using present MEMS devices.

MEMS MICRO-MIRRORS

The scanning silicon micro-mirrors were fabricated using the SOIMUMPs silicon-on-insulator foundry process from MEMSCAP [2]. Micro-mirrors with optical surfaces having dimensions ranging from 0.5mm x 0.5mm to 3mm x 3mm were fabricated. The micro-mirrors were designed with either one or two-axis actuation enabling angular positional control. Two types of micro-mirrors have been fabricated:

electro-thermal actuated, and electro-static actuated devices. The current-driven, electro-thermal micro-mirror, see figure 1, produces a wide range of angular adjustment, whereas the resonant voltage-driven, electro-static variant, shown in figure 2 provides rapid x-y scanning (10s of kHz) but has no appreciable DC response. A multi-layer dielectric coating with a measured reflectivity of >99% at 1064nm was applied to both types of micro-mirrors enabling their use as low-loss intra-cavity mirrors. Their maximum scanning angle was measured as +/- 10°. Multiple MEMS devices can be located on the same chip as shown in figure 3.

Fig. 1. Image of a 1mm electro-thermal micro-mirror

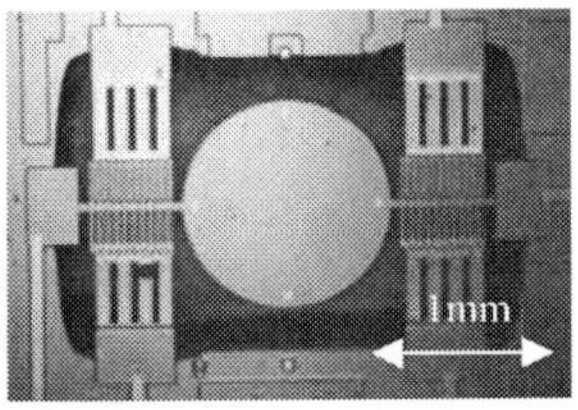

Fig. 2. Image of a 1mm electro-static micro-mirror

Fig. 3. Image of a PCB mounted chip containing 5 thermal micro-mirrors

EXPERIMENTAL SET-UP AND RESULTS

A simple 2-mirror laser cavity, depicted in figure 4, was assembled around a side-pumped Nd:YLF rod and featured an electro-thermal micro-mirror as an end-mirror. In this way, a stable CW output of 20mW at a wavelength of 1064 nm was observed – significant concave deformation of the micro-mirror surface (measured to have a radius of curvature (ROC) of about -80mm) via heating due to residual absorption by the silicon-based chip, limited the

978-1-4244-8926-8/10 $26.00 © 2010 IEEE

output power. Figure 5 illustrates the evolution of the optical power of the laser output as the pump power is increased linearly over time. Initially, CW laser oscillation can be observed; with further increase of the intra-cavity laser power, surface distortion due to heating dominates the laser dynamics leading to the rapid modulation of the laser output, significantly limiting the laser performance. This heating effect presents a major challenge to practical implementation of silicon MEMS devices in lasers; however, above the optical wavelength of ~1.2μm (the absorption edge of silicon) this effect should, in principle, be eliminated. Interestingly, however, this opto-thermally induced surface deformation has the potential to provide new prospects for optically-controlled lasers [3] e.g. optical control of the mirror curvature and new modulation routes.

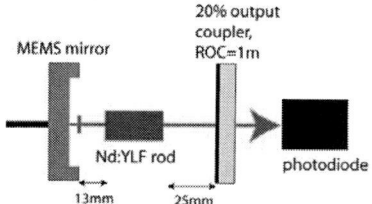

Fig. 4. Schematic of the Nd:YLF cavity featuring the electro-thermal MEMS mirror

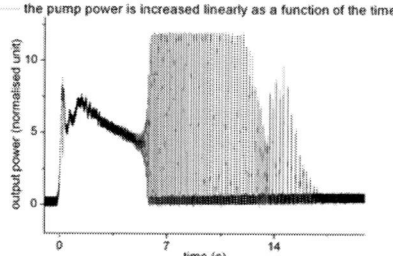

Fig. 5. Output power of the Nd:YLF laser as a function of time (during which the pump power is linearly increased)

In a second experiment, a 3-mirror Nd:YLF laser cavity incorporating an electro-static actuated MEMS micro-mirror was assembled, see figure 6. This system was carefully aligned to ensure that the radius of the fundamental laser mode was reduced to 100μm at the electro-static MEMS mirror surface. In this way, the effects of the thermally-induced surface deformation on the cavity stability were more localised and found to be reduced, while the overall stability of the laser system was increased. With no voltage applied to the micro-mirror, stable CW oscillation was reached resulting in an output power of 200mW. With the maximum drive voltage applied to the actuator (200V), the micro-mirror induced successful Q-switched operation, resulting in pulse

durations of 220ns at a repetition rate of around 10kHz and an average output power of 30mW (see figure 7).

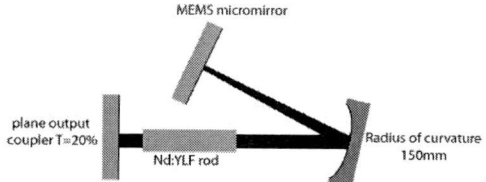

Fig. 6. Schematic of the Nd:YLF laser cavity featuring the electro-static MEMS

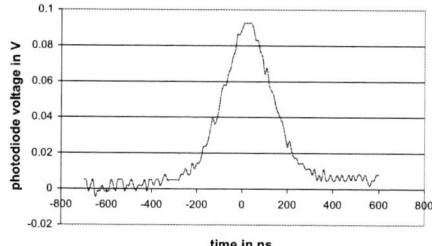

Fig.7. 220ns pulse operating at 10.1kHz

CONCLUSION

We have demonstrated the use of electro-thermal and electro-static MEMS micro-mirrors as flexible, low-cost, intra-cavity optical components in solid-state laser systems. Successful active Q-switching of a Nd:YLF laser has been described. The thermal effects of the micro-mirror and their mitigation have also been described. Results from the application of MEMS micro-mirrors in solid-state lasers operating at 1.5μm (where bulk silicon has low optical absorption) will be described.

REFERENCES

[1] M. Fabert, A. Desfarges-Berthelemot, V. Kermène, A. Crunteanu, D. Bouyge and P. Blondy, "Ytterbium-doped fibre laser Q-switched by a cantilever-type micro-mirror," Opt. Express, vol. 16, pp. 22064-22071 (2008).
[2] MEMSCAP Inc, Parc des fontaines, Bernin, 38926 Crolles, France, www.memscap.com
[3] W. Lubeigt, G. Valentine, and D. Burns, "Enhancement of laser performance using an intracavity deformable membrane mirror," Opt. Express, vol. 16, pp. 10943-10955 (2008)

ACKNOWLEDGEMENT

The work reported was partly funded by the EMRS DTC established by the UK MoD and run by a consortium of SELEX Sensors and Airborne Systems, Thales UK and Roke Manor Research.

LINEAR MEMS MICROMIRROR ARRAY
FOR UV-NIR FEMTOSECOND PULSE SHAPING

Stefan M. Weber[1], Jérôme Extermann[1], Wilfried Noell[2], Fabio Jutzi[2], Sébastien Lani[2],
Denis Kiselev[1], Luigi Bonacina[1], Nico F. de Rooij[2] and Jean-Pierre Wolf[1]

[1] GAP Biophotonics, Université de Genève, Rue de l'École-de-Médecine 20, 1211 Genève 4
[2] EPFL/STI/IMT-NE/SAMLAB, Rue Jaquet-Droz 1, CH-2002 Neuchâtel, Switzerland

ABSTRACT

We report our progress and the first optical application
on the high-aspect ratio micromirror array for UV-NIR
femtosecond (fs) broadband pulse shaping [1]. It is a
bulk-micromachined device, capable of individually
addressing 100 mirrors with a stroke of up to 3 µm
using vertical comb drives in a novel, symmetrical
double-spring design. The device was successfully
implemented in a fs pulse shaper setup at λ_0=795 nm.

INTRODUCTION

MEMS micromirror arrays for femtosecond pulse
shaping have special requirements, such as high reflectivity, flatness, duty cycle, fill factors and low diffractive effects. The advantages over liquid crystals
and acousto-optical modulators are higher damage
thresholds, speeds, and wide-band applicability. Having a versatile device capable of manipulating the
phase of a fs-pulse in the UV could greatly expand the
field of phase-coherent spectroscopic techniques on organic and bio-molecules with large absorption bands in
the DUV and UV [2].

CONCEPT / FABRICATION

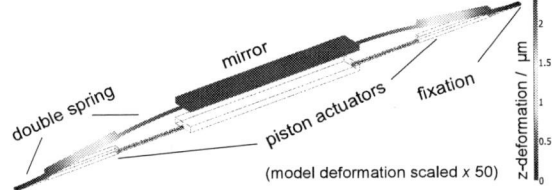

(model deformation scaled x 50)

*Figure 1: Concept/FEM simulation of a symmetrical,
vertically actuated mirror. The springs absorb most of the
stress. Thus the mirror is only deformed e.g. by 3.2 nm when
actuated upwards by 2.35 µm.*

The (120 x 1000) µm² mirrors themselves are fixed on
both sides by two pairs of springs and piston actuators
(Fig. 1). The high aspect-ratio has its origins in the optical configuration, where each mirror addresses a certain wavelength range. This reduces the laser power
density while only 1D diffraction is created. The gaps
between the mirrors are 3 µm wide, resulting in a relatively high fill factor of 97.6%. For the actuation, symmetrical vertical comb drives are used (Fig. 2) as they

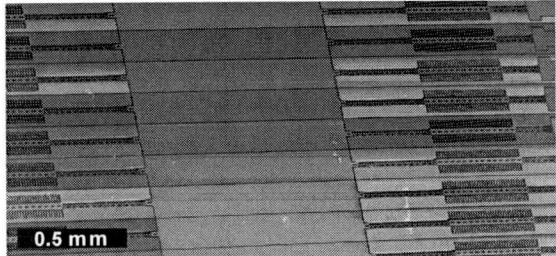

*Figure 2: SEM image of the fabricated device. Mirrors, first
spring, vertical comb drive actuators, second spring,
fixation (center to right).*

have the advantage of providing a continuous force in
the out-of-plane direction using relatively low electrical power. A novel dual-spring concept was applied,
which absorbs much of the residual stress and converts
into into vertical displacement while decreasing the
overall stiffness.

The device was fabricated by timed DRIE from a
35 µm device layer SOI wafer. Release holes in the
springs and a backside opening below the mirrors allow an HF release without mirror release holes. The
surface deformation of individual TiAl coated mirrors
was about 15 nm PTV, similar to [3]. The device was
packaged onto a high-density interconnect PCB board
using fine-pitch gold wire bonding [1].

CHARACTERIZATION / CALIBRATION

The characterization was carried out using a Veeco/
Wyko NT1100 DMEMS optical profiler. Due to the
fabrication process and packaging on materials with
non-negligible roughness, the device presented here
had an overall vertical deformation of 1.7 µm (a radius
of curvature of about 10 m). For the curvature compensation and viable open-loop operation, each individual mirror had to be calibrated. The target displacements even for NIR wavelengths are far from the maximal achievable with the given layer thicknesses.
Therefore, a second degree polynomial is sufficient to
describe the stroke as $z(U)=p_1U^2+p_3,$ where U is the
symmetrically applied voltage, p_1 a mirror coefficient
and p_3 the residual curvature at 0 V (Fig. 3A).

To demonstrate the feasibility to overcome the chips
positive curvature, the device was actuated with a negative, curvature shaped, parabolic pattern using 64 different channels from a home-built, high-voltage DA
converter (Figs. 3B and 4).

978-1-4244-8926-8/10 $26.00 © 2010 IEEE

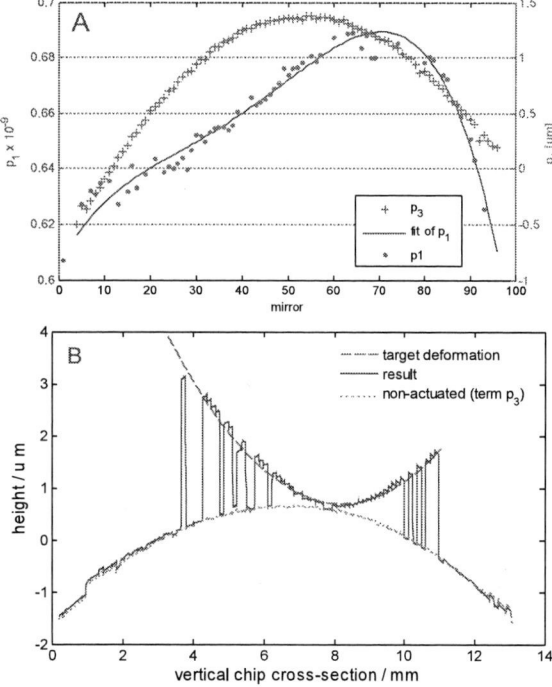

Figure **3A**: chip calibration, terms p_1 and p_3 measured for each mirror. Fig. **3B**: comparison between target and measured deformation for a parabolic pattern, where 64 mirrors were actuated up to 70 V (also shown in Fig. 4).

OPTICAL MEASUREMENTS

The optical characterization in a fs laser setup was performed in a 4f pulse shaper configuration [4] at λ_0=795 nm central wavelength using a Femtosource Synergy fs-oscillator (bandwidth of 60 nm, transform-limited pulse duration ~20 fs). It consists of cylindrical lenses of 15 and 6 cm focal length in a perpendicular configuration and a 600 lines/mm grating, with the micromirror array arranged horizontally in the Fourier plane. The detection was carried out by XFROG [4] in a 10 µm thick BBO (Beta barium borate) SHG (second-harmonic generation) crystal to avoid phase-matching restrictions. The spatial profile after the setup measured with a Newport LPB1 beam profiler is displayed in Fig. 5. The measurements confirmed the necessary surface flatness and open-loop shaping capabilities for this wavelength.

CONCLUSION / OUTLOOK

We demonstrated a working prototype of a 100 mirror, double-spring micromirror array. The first UV applications will encompass open and closed-loop applications for differentiating bio-molecules indistinguishable by standard spectroscopic methods [2].

ACKNOWLEDGEMENTS

We acknowledge the CSEM micro-fabrication facilities, Altatec AG, P. Brühlmeier, and the FNS (CIBA II: 200020-124689/1) for financial support.

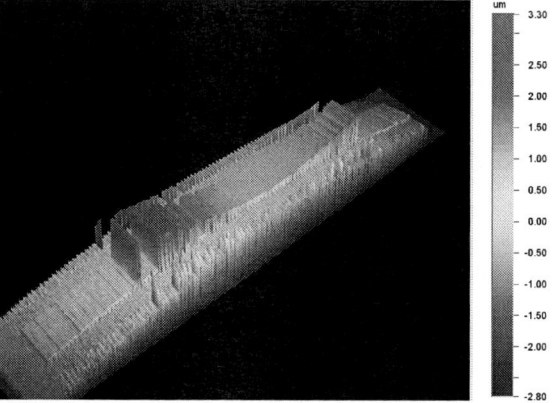

*Figure **4** (color online): Wyko DMEMS optical profiler measurement of the central part of the chip, displaying all 100 mirrors demonstrating the ability to compensate the innate chip curvature by applying a parabolic deformation pattern. The maximal achieved piston with 70 V was 3 µm.*

*Figure **5**: Spatial beam profile, left: reference mirror, right: after the 4f setup including the micromirror array (normalized). The diffractive effects of the devive were found to be very small.*

REFERENCES

[1] S. M. Weber et al., High aspect ratio micromirror array with two degrees of freedom for femtosecond pulse shaping, *Proc. SPIE* **7594**, 75940J (2010)
[2] M. Roth et al, Quantum control of tightly competitive product channels, PRL **102**, 253001 (2009)
[3] S. M. Weber et al., Linear micromirror array for broadband femtosecond pulse shaping in phase and amplitude, *Proc. SPIE*, **7208**, 720805 (2009)
[4] A. M. Weiner et al., Femtosecond pulse shaping using spatial light modulators, Rev. Sci. Instr. **71**, 1929 (2000)
[5] S. Linden et al., XFROG-a new method for amplitude and phase characterization of weak ultrashort pulses, Physica status solidi (b), **206**, 119, (1999)

TUNABLE OPTICAL DIFFUSERS FOR HIGH-POWER LASER APPLICATIONS BASED ON MAGNETICALLY ACTUATED MEMBRANES

Jonathan Masson[1], Andreas Bich[2], Wilfried Noell[1], Reinhard Voelkel[2],
Kenneth J. Weible[2], Nico F. de Rooij[1]

[1] Ecole Polytechnique Fédérale de Lausanne (EPFL), Switzerland
[2] SUSS MicroOptics SA, Switzerland

ABSTRACT

A dynamic laser beam shaper based on MEMS technology is presented. A magnetically actuated deformable single crystal silicon micromembrane is deformed in resonance to diffuse and homogenize laser beams. The large aperture mirror shows line generation with angles up to 1° and line smoothing capabilities. High power density handling is demonstrated up to 140 W/cm^2.

INTRODUCTION

Deformable mirrors are well established technologies used for several years [1] for beam shaping, aberration correction, projection display and adaptive optics. Multiple actuators are used to deform the mirrors and complex driving electronics control precisely the deformations. Deformable mirrors can also be employed for laser beam diffuser applications. For this particular field, simple and rugged devices are needed to satisfy the industrial requirements such as low price devices, high power handling and handy optical component.

In this paper, a diffuser for high power laser is presented. The MEMS-based device focus on small 1D diffusion angles for high power laser applications as required in laser machining, silicon annealing high power illumination and lithography systems. The novelty of this diffuser resides in its dynamic and tunable diffusing angle. The diffuser is based on a magnetically actuated deformable reflecting membrane. Single crystal silicon is used for the membrane material because of its high heat conductivity and optical flatness in rest position.

MAGNETIC ACTUATION

The actuation scheme of the diffuser membrane is based on a single actuating electrical current source and mechanical resonance frequencies. The driving of the system is therefore reduced and simplified to a single AC actuator signal provided by an AC power supply and an external permanent magnet. Mechanical resonance frequencies of the membrane are excited in order to produce a range of shapes using one single actuator.

The membrane is square, free standing and clamped at two opposite extremities (figure 1). The two other sides are free to move because two slits release the membrane from the frame. The backside of the membrane has local mechanical reinforcements in form of integrated and monolithic beams (figure 2a). Specific resonance modes are preferred by the actuation and frequency response. An AC current flows in the membrane from one anchor to the other. The permanent magnet is located about 2 mm under the chip. A vertical Lorentz force is generated on the entire membrane. An upward or a downward force is produced alternatively depending on the current flow direction. When the driving frequency matches a mechanical mode frequency, the membrane is in resonance and the largest amplitudes are reached.

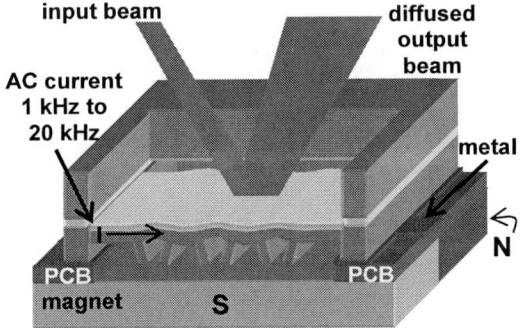

Figure 1 Schematic of the diffuser. An AC current flows in the membrane and a magnet is situated under the PCB. A laser beam is reflected and diffused by the deformed mirror.

FABRICATION

The devices are fabricated from SOI wafers having a device layer of 35 µm, a handle layer of 350 µm and buried oxide (BOX) layer of 2 µm thick. The resistivity of the device layer is 0.02 Ohm·cm. First, the backside of the handle layer is etched using deep reactive ion etching (DRIE) to release the backside of the device layer which is the mirror surface. A timed etch process is used on the front side to provide two silicon levels. First, the two slits are etched by a shallow DRIE using a photoresist mask. Second, a silicon dioxide mask is used to pattern the beams and

978-1-4244-8926-8/10 $26.00 © 2010 IEEE

to thin the membrane to a thickness of 5 μm. The chips are released from the frame by vapor HF etching, which also removes the oxide on the mirror surface. Finally, the mirror is coated with aluminum or gold thin film. Mirrors were fabricated with sizes of 5x5 mm^2, 10x10 mm^2 and 15x15 mm^2. The chips are then assembled on a PCB. A permanent magnet (cubic, 12 mm side, NdFeB, remanence of 1.37 T) is positioned under the PCB (figure 2b).

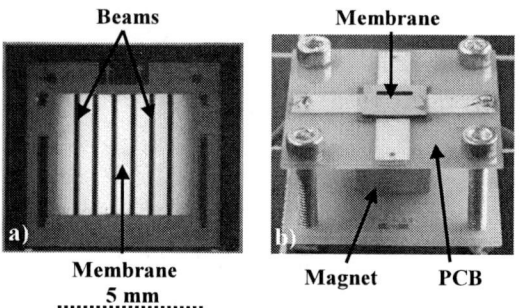

Figure 2 a) Backside of an uncoated membrane with the stiffening beams. b) Assembled chip on a PCB with magnet.

RESULTS

The mode shapes of the membrane were characterized using a laser Doppler velocimeter (LDV). The membrane is connected to a function generator which provides a current up to 60 mA. The first mode at 1.160 kHz could be excited in air. To excite higher order modes, measurements had to be made in a vacuum chamber at 50 mbar. The large size of the membrane introduces air damping and limits the operation at atmospheric pressure for the higher modes. Mode shapes at frequencies ranging from 500 Hz to 20 kHz could be characterized in vacuum. The modes have sine like shapes with half a period for the 1st mode to several periods for higher modes.

For optical characterization a laser beam is expanded and projected on the diffuser. The reflected pattern is observed on a screen. When the diffuser is not actuated it acts like a mirror and the image of the square mirror is reflected. Figure 3 shows a photograph of the reflections of the expanded laser beam when the membrane is at rest (figure 3a) and actuated (figure 3b). The first mode is excited in vacuum at 1.532 kHz and actuated with a current of 40 mA. A line is generated without expansion of the beam in the opposite dimension because of the 1D deformation of the mirror. For the first mode the maximum mechanical tilt angle was determined to be 0.5° for an optical diffusing angle of 1°. The diffusing angle can be tuned within this range by changing the current intensity.

The dynamic diffuser was combined with a static flat

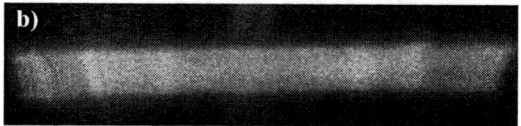

Figure 3 Photographs of the reflection on a screen of a diffused laser beam when the membrane is a) not actuated and b) actuated at 1.532 kHz.

top 1D diffuser having a fixed diffusing angle of 15° [2]. In the far field, the static diffuser generates interference patterns for instance bright and dark lines (figure 4a). When actuated, the dynamic diffuser smooths out the interferences (figure 4b).

Figure 4 Photographs of the reflection on a screen of a diffused laser beam when a static and a dynamic diffuser are combined. a) The static diffuser alone shows interference pattern but b) when the dynamic diffuser is actuated the line is smoothed.

The power handling of the mirror was tested using a supercontinuum laser source Koheras SuperK Extreme. The source was aimed at the center of the membrane for 20 minutes with an optical power density of 140 W/cm^2. After this time, no damage could be observed on the mirror.

CONCLUSION

We show a dynamic and tunable diffuser made of a single crystal silicon layer. Optical coating, high power handling package and control electronics will be addressed for future work.

REFERENCES

[1] L. J. Hornbeck, "128x128 Deformable mirror device" IEEE Transaction on Electron Devices, vol. ED-30 No. 5, pp. 539-545, 1983

[2] R. Bitterli et. al., "Refractive statistical concave 1D diffusers for laser beam shaping" Proc. of SPIE, Laser Beam Shaping IX, San Diego, CA, USA 7062, Aug. 11, 2008, pp. 70620P

PLASMONICS FOR ULTRASENSITIVE BIOMOLECULAR NANOSPECTROSCOPY

Hatice Altug[1,2,3]*, Ahmet A. Yanik[1,2], Ronen Adato[1], Serap Aksu[3], Alp Artar[1], Min Huang[1]

[1]Electrical and Computer Engineering, [2]Photonics Center, [3]Materials Science and Engineering,
Boston University, Boston, MA, 02215
Contact: altug@bu.edu

ABSTRACT

Plasmonics, by localizing light to the sub-wavelength volumes and dramatically enhancing local fields, is enabling myriad of exciting possibilities in bio-detection field [1]. In this talk, first I will demonstrate an ultra-sensitive surface-enhanced infrared spectroscopy enabling direct detection of molecular specific signatures of proteins from monolayer thick films. Our method exploits engineering of diffractive couplings among plasmonic nano-antenna arrays. I will then present a low-cost fabrication method for high-throughput fabrication of these engineered antenna arrays. Finally, I will show an integrated nanoplasmonic-nanofluidic sensor platform leading to targeted analyte delivery and dramatically improved sensor response time.

SUMMARY

Ultrasensitive Infrared Vibrational Nanospectroscopy

Infrared absorption spectroscopy is an important tool for functional studies of bio-molecules. The method enables direct access to the vibrational fingerprints of the biomolecular structure at the mid-infrared spectral region (~3-20μm) [2]. Sensitivity limitations, however, hinders the applicability of the technique to single molecule/monolayer studies. By engineering diffractive couplings among nano-antenna arrays, we will show demonstration of an ultra-sensitive plasmonic vibrational spectroscopy technique with zepto-mole level sensitivities [3]. Engineered arrays support collective plasmonic resonances leading to stronger near-field intensities than what is achievable with individual antennas [4]. Using this approach, we achieve up to 100,000 fold enhancements of the amide-I and II backbone signatures of proteins and obtain direct detection of absorption signals from 145 proteins per antenna.

High-Throughput Fabrication of Engineered NanoAntenna Arrays with Nanostencil Lithography

In addition, we will demonstrate high-throughput fabrication of these tailored infrared plasmonic nanorod antenna arrays using nanostencil lithography (NSL) [5]. NSL, a shadow-mask patterning technique, relies on direct deposition of materials through a pre-patterned stencil mask. We show that optical responses of the engineered antennaarrays fabricated by NSL are comparable to that of the arrays fabricated by electron beam lithography.

978-1-4244-8926-8/10 $26.00 © 2010 IEEE

More importantly, we show that nanostencil masks can be reused multiple times to create series of nanoantenna arrays leading to identical optical responses. This fabrication approach, by enabling the reusability of the stencil and also offering flexibility on the substrate choice and nanopattern design, could facilitate wide-spread use of plasmonics.

Integrated Nanoplasmonic-Nanofluidic Sensor for Targeted Analyte Delivery

Finally, we will also show merging of plasmonics and nanofluidics on nanohole array platform to overcome mass transport limitation [6]. Performances of surface biosensors are often limited by the analyte delivery rate to the sensing surface instead of sensors intrinsic detection capabilities. In a microfluidic channel, diffusive analyte transport to the biosensor surface severely limits the sensor performance [7]. At low concentrations, this limitation causes impractically long detection times. Using our novel platform, we demonstrate how nanoholes can be harnessed both to manipulate light and to transport liquid for targeted analyte delivery. We will present our results on 14-fold increase in mass transport rate constant.

REFERENCES:

[1] A. Artar, A. Ali Yanik and H. Altug "Fabry-Perot Nanocavities in Multi-layered Plasmonic Crystals for Enhanced Biosensing", Appl. Phys. Lett., Vol. 95, 051105 (2009).

[2] A. A. Yanik, X. Wang, S. Erramilli, M.K. Hong, H. Altug, "Extraordinary Mid-infrared Transmission of Rectangular Coaxial Aperture" App. Phys. Lett., Vol 93, 081104 (2008).

[3] R. Adato, A. A. Yanik, J. J. Amsden, D. L. Kaplan, F. G. Omenetto, M. K. Hong, S. Erramilli and H. Altug, "Ultra-sensitive Vibrational Spectroscopy of Protein Monolayers with Plasmonic Nanoantenna Arrays", Proc. Natl. Acad. Sci. U.S.A. 106, 19227 (2009).

[4] R. Adato, A. A. Yanik, C-H Wu, G. Shvets and H. Altug, "Radiative Engineering of Plasmon Lifetimes in Embedded Nanoantenna Arrays," Optics Express Vol. 18, pp. 4526-4537(2010).

[5] S. Aksu, A. Yanik, R. Adato, A. Artar, M. Huang, H. Altug, "High-throughput Nanofabrication of Plasmonic Infrared NanoAntenna Arrays for Vibrational Nanospectroscopy", Nano Letters, ASAP Article (June 2010). DOI: 10.1021/nl101042a

[6] A. A. Yanik, M. Huang, A. Artar, T. Chang, H. Altug, "Integrated Nanoplasmonics-Nanofluidics Biosensor with Targeted Delivery of Analyte", Appl. Phys. Lett., Vol 96, 021101 (2010).

[7] M. Huang, A. A. Yanik, T. Chang, H. Altug, "Sub-wavelength Nanofluidics in Photonic Crystal Sensors" Optics Express, Vol. 17, pp.24224-24233 (2009).

A Photoelectrophyscial Capacitor with Direct Solar Energy Harvesting and Storage Capability

Chi-Wei Lo, Chensha Li, and Hongrui Jiang

University of Wisconsin-Madison, USA

ABSTRACT

Solar energy harvesting and storage are important topics of renewable energy. Current solid-state photovoltaic cells and conventional photoelectrochemical cells cannot harvest and directly store the converted energy within one single structure. We report on a photocapacitor that can convert *and* store solar energy based on a bio-inspired mechanism. The photocapacitor converts the solar energy through photovoltaic effect, and stores the converted energy by maintaining the concentration difference across a membrane upon light irradiation. Results have shown that the device can be charged photovoltaically and hold the voltage of 0.47V with a capacity of 40.63 mC/cm^2 for more than 24 hours.

INTRODUCTION

Conversion and storage of solar energy becomes more and more important [1]. Current solid-state photovoltaic cells and conventional photoelectrochemical cells are not capable of directly storing the converted energy, which has to be facilitated by connecting to external storing devices [2-4], thus increasing the complexity and lowering the overall efficiency of the system. Photovoltaic devices that stores solar energy directly are thus highly desirable [1]. Devices that can harvest and store solar energy simultaneously with single and simple structures is still being sought.

We here report a new kind of photoelectrophysical capacitor which can convert and store solar energy in electrophysical form and has a simple structure design based on bio-inspired method (device picture shown in Fig. 1). The photoelectrophysical capacitor consists of three components: anode (silicon), cathode (carbon), and poly(vinylidene fluoride-*co*-hexafluoropropylene) (PVDF) separator configured as shown in Fig. 2. When silicon is exposed to light, electron-hole pairs are generated through photovoltaic process, thus the energy conversion. However, unlike conventional photoelectrochemical cells and other three-electrode-based photoelectrolytic cells, we introduced an ion conductive PVDF membrane between the anode and the cathode into our photovoltaically rechargeable two-electrode capacitor. The PVDF separator controls the ion diffusion and stabilizes the charges on both electrodes, thus enabling the direct storage capability.

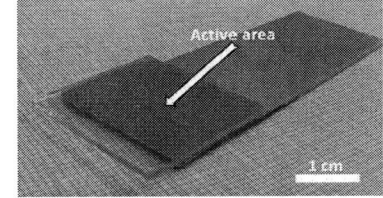

Figure 1. Photo of the photocapacitor device. The active area is 4 cm^2.

EXPERIMENTAL

The detailed fabrication process is shown in Fig 2. The charging characterization of the photocapacitor in open circuit was done by exposing the device to the illumination power of 100 mW/cm^2 (equivalent to 1 Sun). The maximum open-circuit voltage reached 0.47V (Fig. 4). As the photocapacitor was removed from the light source, the voltage was held around 0.5V in dark for more than 24 hours with little drop as shown in Fig. 5, proving the storage capability. We also investigated the device performance by discharging at short-circuit condition when this device was fully charged. Cyclic test of the device demonstrated that it could be charged and discharged for many cycles without observed degradation. Although PVDF is a well known pyroelectric material, pyroelectric effect of the PVDF separator would produce voltages on the order of 1μV under the experimental conditions, thus being negligible.

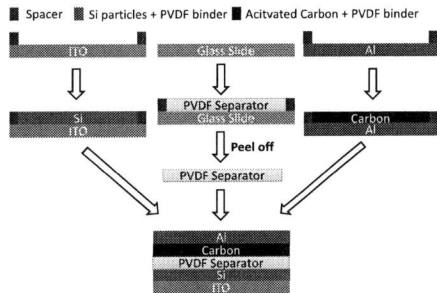

Figure 2. Fabrication process of the photocapacitor. The electrodes are prepared by mixing Si particles (as anode) and activated carbon (as cathode), respectively, with PVDF (the binder; 85:15 mass ratio in organic solution). The two pastes are then cast onto the current collectors: indium tin oxide (ITO) and gold. The electrodes are heated at 75°C for 12 hours to remove the solvent. The structure is then soaked in 0.5M lithium hexafluorophosphate (LiPF$_6$) electrolyte solution followed by final assembly with a PVDF (135 μm thick) layer as separator.

RESULTS AND DISSCUSSION

The underlying mechanism of the photocapacitor is as follows. The incident photons excite the electrons within the semiconducting layer containing Si particles. The holes are injected into the ITO electrode. The electrons attract Li$^+$ ions to the cathode electrode creating concentration gradient across the device (Fig 3). The photocapacitor is charged in this process until a saturated electric potential difference is reached. When the solar irradiance is removed, the diffusional force of the ions and electric field are counter-balanced and maintain a stable electrical double layer across the two electrodes (Fig 3).

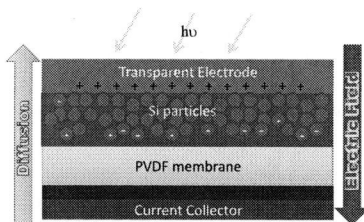

Figure 3. The charging and storage mechanism of the photocapacitor. The energy storage mechanism of the photocapacitor fits the model proposed by Goldman-Hodgkin-Katz equation (1) [5].

$$E_m = \frac{RT}{zF} \ln \frac{P_{Li^+}\left[Li^+\right]_o + P_{PF_6^-}\left[PF_6^-\right]_i}{P_{Li^+}\left[Li^+\right]_i + P_{PF_6^-}\left[PF_6^-\right]_o} \quad (1)$$

E_m is the membrane potential (V), P_{ion} is the permeability for that ion (in meters per second), $[ion]_o$ is the extracellular concentration of that ion, $[ion]_i$ is the intracellular concentration of that ion (in moles per cubic meter), R is the ideal gas constant (8.314 JK^{-1}mol^{-1}), T is the temperature in kelvins, and F is Faraday's constant (96485.34 C mol^{-1})

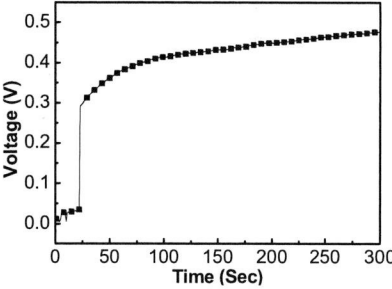

Figure 4. Charging characteristics of the photocapacitor. Open-circuit voltage of the photocapacitor device vs. time during charging under an illumination level of 1 sun. The voltage reaches 0.47V after 300s.

SUMMARY

In summary, we reported on a photocapacitor that can directly convert and store the solar energy based on a bio-inspired method. The device can be charged to 0.47V under the illumination of 1 Sun. The density of stored electrical energy is about 40.63 mC/cm^2. The photocapacitor can hold the stored energy for over 24 hours in a dark environment, and exhibits many cycles of charging and discharging without degradation. This photoelectrophysical capacitor can potentially be used as power sources for a wide range of applications such as portable electronics and miniaturized systems.

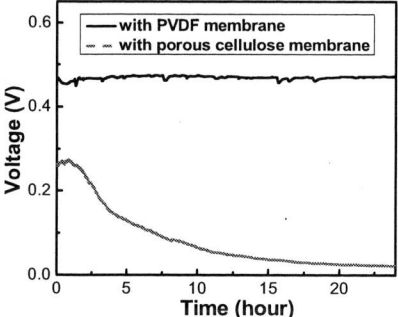

Figure 5. Storage capability of the photocapacitor device in the dark (open-circuited). The maximum charged voltage stays more than 24 hours with little drop. The red line represents a device with porous cellulose membrane which shows much less capability of storing converted energy.

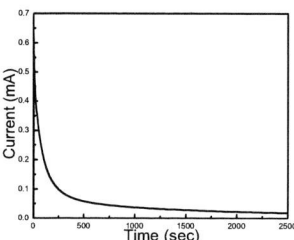

Figure 6. Discharging of the photocapacitor device through short-circuit discharge. The total charge storage capacity of the photocapacitor device is estimated to be 40.63 mC/cm^2, based on the integration of the discharging curve.

ACKNOWLEDGEMENT

This work was partly supported by the US National Science Foundation (ECCS 0702095) and partly by the Wisconsin Institutes for Discovery (WID).

REFRENCES

[1] P. Dutta, *Nature*, vol. 358, pp. 621-621, 1992.
[2] S. Licht, *Nature*, vol. 330, pp. 148-151, 1987.
[3] B. Oregan and M. Gratzel, *Nature*, vol. 353, pp. 737-740, 1991.
[4] T. Miyasaka and T. N. Murakami, *Applied Physics Letters*, vol. 85, pp. 3932-3934, 2004.
[5] D. E. Goldman, *J. Gen. Physiol.*, vol. 27, pp. 37-60, 1943.

REMOTE SWITCHING OF CELLULAR ACTIVITY USING LIGHT THROUGH QUANTUM DOTS

Katherine Lugo[1], Xiaoyu Miao[1,3], Fred Rieke[2] and Lih Y. Lin[1]

University of Washington, Seattle, USA

Department of Electrical Engineering[1], Department of Physiology and Biophysics[2], Current affiliation: Sandia National Laboratories[3]

ABSTRACT

We report integration of CdTe quantum dot (QD) film with LnCap (prostate cancer) cell and CdSe QD probes with cortical neurons for control of cellular activity. We demonstrate the remote switching of cellular activity by exciting QDs with light. Changes in membrane potential and ionic currents are recorded using the patch-clamp method. Upon excitation, the cell shows activation of ion channels and hyperpolarization of the cell membrane.

INTRODUCTION

The study of electrical signaling and communication among neurons has benefited from tools capable of exciting neurons in a spatially and temporally controllable manner. Using light to remotely control the events of cell signaling has been an attractive approach due to its non-invasiveness and flexibility in probing different locations [1,2]. Quantum dots offer unique optoelectronic properties such as spectrally sharp emission peaks when excited and high quantum efficiency. Furthermore, their surface chemistry can be modified for selective attachment to biological particles. QDs have been integrated with cells and used in various applications such as fluorescent bio-labels for targeted imaging. We investigate a new direction: using QDs to control cellular activities and cell signaling through light excitation of ion channels. In this paper we report the integration of CdTe QD film/LnCap cell and CdSe QD probe/neuron cell, in which photo-excitation of QDs caused activation of ion channels and hyperpolarization of the cell membrane.

MODEL OF QD-CELL INTEGRATED SYSTEM

Ion channels in the cell membrane control ion flux and can be turned on and off by changes in membrane potential. These voltage-gated ion channels will open and permit Na^+, K^+ or Cl^- ions to enter or leave the cell given sufficient membrane depolarization. Depolarization refers to the situation when a cell membrane potential is higher than its resting potential. Here we propose to depolarize or hyperpolarize the cell membrane with photo-excited QDs, placed close to cell membrane, as shown in Figure 1. The excited QD experiences electron-hole separation, which exhibits an electric dipole moment

[3] and perturbs the cell membrane potential through dipole induced electric field. The perturbation can result in opening of voltage-gated ion channels, causing further depolarization of the cell and generating an action potential from this pre-synaptic cell; or hyperpolarization can occur if negative ions such as Cl^- enter the cell or positive ions such as K^+ leave the cell.

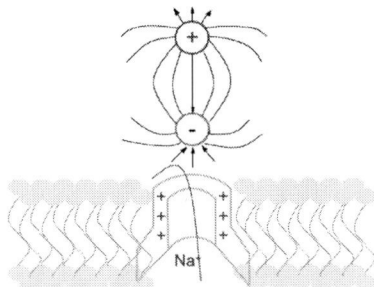

Figure 1. Interaction of a QD with a cell membrane.

QD-CELL INTEGRATION

For integrating QDs with cells two approaches have been used. First, we cultured cells on a CdTe QD film fabricated using electrostatic layer-by-layer self-assembly [4]. Figure 2 shows the LnCap cell cultured on the QD film and the clamped cell for the membrane potential measurement. Second, glass micropipettes used for patch-clamp recording are directly coated with CdSe QDs in hexane solution. Figure 3 shows the fluorescence image of a QD-coated micropipette under mercury-xenon light excitation. In the patch-clamp experiment, one probe coated with CdSe QDs is placed on the cell for light excitation while another one without QDs is used for electric recording.

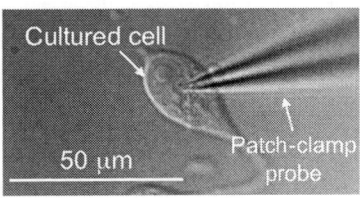

Figure 2. Image of the LnCap cell cultured on the CdTe QD film.

978-1-4244-8926-8/10 $26.00 © 2010 IEEE

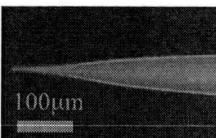

Figure 3. Fluorescence image of a micropipette coated with CdSe QDs

The recorded result shown in Figure 4 was obtained by manually turning on and off a mercury-xenon light source band-pass filtered to 430 nm wavelength illuminating the prostate cancer cell from above. The result shows membrane potential change, indicating activation of voltage-gated ion channels through QDs upon light excitation. The cell membrane is hyperpolarized or reaches a more negative membrane potential upon illumination. With a 740 nm light source (longer than the QD absorption cutoff wavelength), no cellular response was observed.

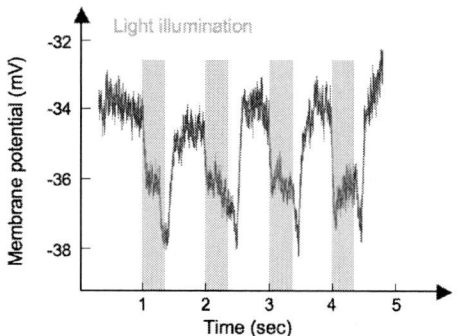

Figure 4. Patch-clamp recording showing effect of CdTe QDs excitation on membrane potential.

Cultured cortical neurons were used to study and record the behavior of ionic current when integrating the CdSe QD-coated probe with the cell. Figure 5 shows the single continuously recorded section of data for 25 seconds using a voltage-clamped cell. In this case the cell voltage was held fixed, figure 5 plots the current required to cancel the current produced by QD activation. Thus QD activation reduced the inward (negative) current. With the voltage allowed to change, the effect of such a reduction in inward current would be to hyperpolarize the cell. Figure 5 (a) shows no change in ionic current when the light (λ = 550 nm) is off and when the light is turned on hyperpolarization occurs in the cell. Figure 5(b) displays a comparison of the current behavior when the CdSe QDs are illuminated with an excitable wavelength (λ = 550 nm) and a non-excitable wavelength (λ = 720 nm). The magnitude of the ionic current decreases when the 550nm wavelength light is applied to the CdSe QDs. On the other hand, there is no change in the ionic current when the QDs are illuminated with the 720 nm wavelength light (longer than the QD absorption cutoff wavelength).

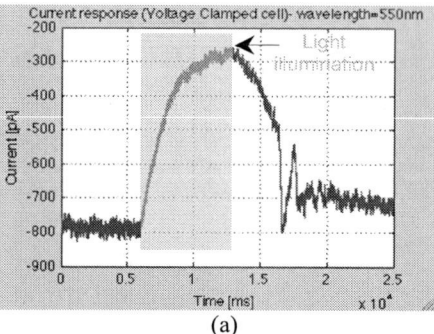

(a)

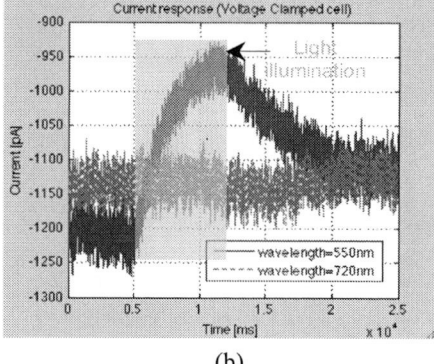

(b)

Figure 5. Patch-clamp recording of CdSe QD-cell system on ionic currents. Light excitation:
(a) λ = 550 nm and (b) λ = 550 nm and 720 nm.

CONCLUSION

Using CdTe and CdSe QDs on cultured cells, we have demonstrated activation of voltage-gated ion channels through light. Hyperpolarization has been achieved and observed with patch-clamp recording as a result primarily of activating K^+ channels. The fields produced by the photo-generated dipoles from the QDs perturb the cell membrane potential, which can potentially generate action potentials that govern communication and signaling among cells. This non-invasive method for switching cell activity can target specific cells through protein binding by changing the QD surface chemistry.

REFERENCES

[1] F. Zhang, A. M. Aravanis, A. Adamantidis, L. Lecea and K. Deisseroth, "Circuit-breakers: optical technologies for probing neural signals and systems," Nature Reviews/Neuros, vol.8, pp.577-581, 2007.

[2] P. Gorostiza and E. Y. Isacoff, "Optical switches for remote and noninvasive control of cell signaling," Science, vol. 322, pp. 395-399, Oct. 17, 2008.

[3] A. P. Alivisatos, "Perspectives on the physical chemistry of semiconductor nanocrystals," J. Phys. Chem., vol. 100, pp. 13226-13239, 1996.

[4] C.-C. Tu and L. Y. Lin, "High efficiency photodetectors fabricated by electrostatic layer-by-Layer self-assembly of CdTe quantum dots," Applied Physics Letters, vol. 93, p. 163107, 2008.

Mechanically tunable coupled photonic crystal nanocavities

Xiongyeu Chew[1], Guangya Zhou*[1], Fook Siong Chau[1], Jie Deng[2], Xiaosong Tang[2], Yee Chong Loke[2]

[1]Micro/Nano Systems Initiatives, National University of Singapore, Singapore
[2]Institute of Materials Research and Engineering, Singapore
*Corresponding Author; E-mail: mpezgy@nus.edu.sg

ABSTRACT

We report a MEMS-integrated on-chip tunable resonator consisting of two coupled one-dimensional (1D) Photonic Crystal nanocavities capable of achieving a relatively large tuning range with no significant deterioration of quality factor and sufficiently large transmissions.

INTRODUCTION

Recently much attention was emphasized on miniaturizing tunable photonic devices capable of manipulating and altering flow of light. One of the prominent designs is the photonic crystal (PhC) nanoresonator which has shown to be a promising building block for dense integration with photonic integration circuits and highly miniaturized low power sensors. It is highly desirable that an optical nanoresonator possess a high quality (Q) factor with small mode volumes [1-2]. PhC nanocavities are ideal candidates for generating such nanoresonators, especially the 1D PhC nanocavities which can be easily integrated with other silicon photonic components. It was recently demonstrated that even 1D PhC cavities can demonstrate high-Q, small mode volumes and high transmission properties. Mechanical tuning of optical resonances utilizing dielectric tips has previously been proposed [3-4]. It was reported that to achieve large tuning effects, it is often desirable to localize the tuning mechanism on the resonators. The strength of the tuning mechanism would be proportional to the effective polarizability of the tip, which in turn is proportional to the tip volume and dielectric constant. In order to achieve large shifts in resonant frequencies, large dielectric constant material with sufficiently large volume is required. Conversely a larger effective polarizability will induce large trade-offs in the Q-factor [3-4].

In this work, we demonstrate an on-chip integrated photonic MEMS device achieving large optical resonance tuning by utilizing coupled interactions of dual identical 1D PhC high transmission nanocavities. This proposed mechanism may be easily integrated with other silicon photonic components / building blocks for future dense integration of photonic integrated circuits such as optical filters, optical add/drop de-multiplexers or even highly sensitive displacement sensors. By utilizing the coupled-cavity approach [5-7]; we can achieve near-lossless and large tuning range resonances while using a simple monolithic fabrication technique. Coupling strength of the PhC nanocavities can be then accurately and stably controlled using mico/nano-electro-mechanical-systems (MEMS/NEMS) actuators [7-8] that are patterned and fabricated together with the PhC structures. Since the perturbing also has an excellent confinement property, we demonstrate that this tuning mechanism offers a lower loss and larger tuning range compared to dielectric tip tuning methods.

DESIGN & FABRICATION

We first design the air-bridge suspended 1D PhC nanocavities having a series of linearly tapered holes for excellent mode matching between the waveguide and Bloch modes. One air-bridge PhC nanocavity is suspended on air for a certain length and connected to a simple silicon-on-insulator (SOI) wire waveguide of width 460 nm and thickness 340 nm. The other nanocavity is attached to a MEMS comb-drive actuator for dynamic tuning. A 1-μm thick layer of buried oxide is necessary beneath the silicon waveguides to ensure sufficient isolation from the silicon substrate. This MEMS tunable coupled-cavity resonator is illustrated in Fig.1 where each nanocavity is formed by using two sets of Bragg-stack mirrors formed by 4 periodic holes with a lattice constant of a = 370nm and radius of $r = 0.246a$. Aperiodic in-tapering holes are introduced at the nanocavity to achieve a gradual confinement of the resonance mode.

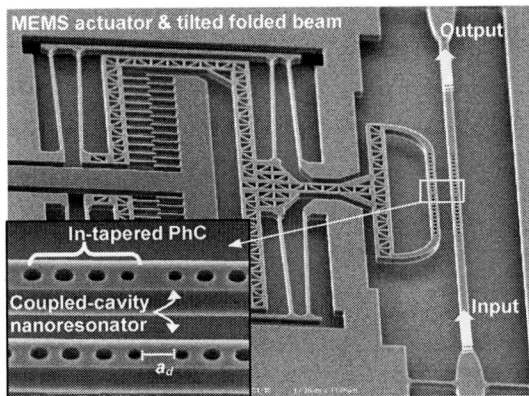

Fig. 1: SEM of the fabricated device based on the proposed tunable PhC coupled-cavity resonator design

978-1-4244-8926-8/10 $26.00 © 2010 IEEE

Such aperiodic tapering has the tendency to reduce radiative losses leading to resonators with higher Q-factor and improved transmission. We implemented a gradual tapering of diameter ranging from $0.18a$ to $0.24a$. We've also designed and optimized a 2-hole aperiodic out-tapering to further improve the modal mismatch to that of the air-bridge waveguide mode, thus further improving the transmission of the resonator. A nanocavity length of $a_d = 1.01a$ yielded a resonance wavelength $\lambda_o = 1.604\mu m$.

In fabrication of the device, we utilized electron-beam lithography to perform patterning on the 340 nm thick silicon layer. The 1μm buried oxide is used as sacrificial oxide for the suspended waveguide and MEMS structures. SF_6 ICP plasma was utilized to etch the submicron holes and waveguides. In the end the sacrificial oxide are removed by simply using HF vapor system.

EXPERIMENTAL RESULTS

The testing of the fabricated device was characterized using a tunable laser (Ando AQ4321) as the light source and coupled to an optical spectrum analyzer (Ando AQ6317). We end-fired a TE polarized light ranging from a single mode fiber. As shown in Fig. 2, the resonance frequency of the resonator, when the MEMS-driven perturbing nanocavity is 1 μm away from the waveguide, falls at 1.604 μm. The experimentally measured full-width half maximum is 0.4 nm, which corresponds to an experimental waveguide-loaded Q-factor of approximately 4000. The modest Q-factor was expected due to present fabrication errors. Further improvements in fabrication techniques and waveguide loadings are on-going to further increase the Q and transmission properties. As voltage is pumped into the MEMS structures, actuators narrowed the gap between the coupled resonators, from meso-field coupling to near-field coupling. A resonant wavelength shift is shown. Fig. 3 illustrates the splitting and shifting of the resonant frequencies where actuation voltages are increased from 0V to 20V. The coupled-cavity resonator achieved a rather equal, weak splitting of 3.14 nm with no significant degradation in the Q-factor. Further increasing the voltages from 20V to 23V, the coupled-cavity resonator achieved strong coupling and large resonance splitting is observed. Again, no significant degradation in Q is observed as light is strongly confined within the nanoscale air-slot region and dielectric region forming the odd and even mode respectively. We experimentally demonstrate a 14 nm splitting in resonance by increasing voltages up to 23V which corresponds to 100 nm gap between both nanocavities.

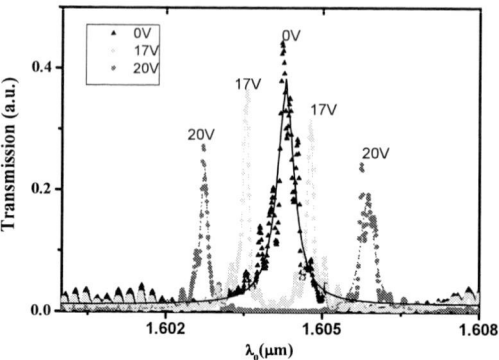

Fig. 2: Normalized spectrum of the device under varying voltages

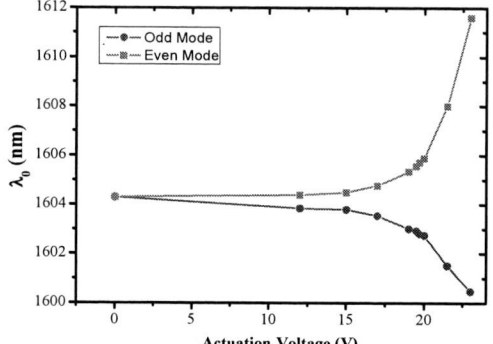

Fig. 3: Relationship of wavelength shift of symmetrical and anti-symmetrical modes at different actuating voltage.

Theoretically, such MEMS actuation mechanism should allow us to achieve a near-field gap of 50 nm, which corresponds to a 20 nm wavelength splitting according to FDTD simulations. We thus demonstrate experimentally an on-chip tunable MEMS photonic resonator with simple monolithic fabrication that is capable of achieving a large wavelength tuning range with high-Q resonances for future dynamic tunable integrated on-chip photonic applications and sensing.

ACKNOWLEDGEMENT

Financial support by the MOE Research grant R-265-000-306-112 is gratefully acknowledged. Fabrication support from SNFC, A*STAR is much appreciated.

REFERENCES

[1] B. Song, S. Noda, T. Asano, Y. Akahane, Nature Mat. **4,** 207, 2005
[2] F. Intonti et. al. Phys. Rev. B **78**, 041401, 2008
[3] L. Lalouat et. al., Phys. Rev. B **76**, 041102, 2007
[4] P. Deotare, M. McCutcheon, I. Frank. M. Khan, M. Loncar, Appl. Phys. Lett **95**, 031102, 2009
[5] A.Yariv et. al., Opt. Lett. **24**, 711, 1999
[6] P. Velha. et. al., Appl. Phys. Lett **89**, 171121, 2006
[7] K. Hane et. al., IEEE trans. Ind. Electronics **56**, 4, 2009
[8] G. Zhou et. al., J. Micromech. Microeng. **13**, 2, 2003

978-1-4244-8926-8/10 $26.00 © 2010 IEEE

RING-RESONATOR REFLECTOR WITH A WAVEGUIDE CROSSING

Wei Shi, Raha Vafaei, Miguel Ángel Guillén Torres,
Nicolas A. F. Jaeger, and Lukas Chrostowski*
University of British Columbia, Vancouver, BC, Canada
* Email: lukasc@ece.ubc.ca

ABSTRACT

We demonstrate the design and performance of a silicon-on-insulator ring-resonator reflector with a low-loss, low-crosstalk waveguide crossing. The device is simulated using the transfer-matrix method and a 2D finite-difference mode solver. It functions as a reflective-type notch filter and can be used for optical communications or thermal, biochemical, or other sensors. An extinction ratio of over 25 dB is observed experimentally.

INTRODUCTION

Waveguide ring-resonator selective reflectors are of great interest for many applications such as tunable lasers, reflective filters, and sensors [1], [2]. They have the advantages of easy fabrication, wide tuning ranges, and easy monolithic integration with other photonic devices. Whereas most of the designs of ring reflectors are reflective-type band-pass filters, H. Sun et al. [3] recently demonstrated a reflective-type notch filter as a promising biochemical sensor. Low loss, low crosstalk waveguide crossings enable more flexible routing and hence more complex photonic circuits. In this work, a new ring-resonator reflector is designed and demonstrated. Using a well designed waveguide crossing [4], it can achieve a high extinction ratio and, potentially, be used as a reflective notch filter in optical communications or as a thermal, biochemical, or other sensor.

DESIGN AND SIMULATION

Transfer-Matrix Analysis
As shown in Fig. 1, the reflection is achieved by "twisting" the ring-resonator using a waveguide crossing and two directional couplers. Assuming that the optical signal is input from the left port, i.e., incoming signal $E_{in} = a_1$ and $c_4 = 0$, we can calculate the reflected signal $E_r = d_1$ and through signal $E_t = b_4$ using the transfer-matrix method [2]. The function of a directional coupler can be described

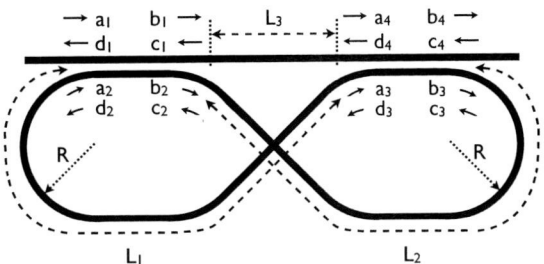

Fig. 1. Racetrack-ring reflector geometry with the transfer-matrix elements: a $(b, c, d)_i$ is the incident / output electrical field in the corresponding direction.

by a transfer matrix:

$$C = \frac{1}{i\kappa} \begin{bmatrix} -\tau & 1 & 0 & 0 \\ -T_c & \tau & 0 & 0 \\ 0 & 0 & -\tau & 1 \\ 0 & 0 & -T_c & \tau \end{bmatrix} \quad (1)$$

where κ and τ are the coupling coefficients and

$$T_c = \kappa^2 + \tau^2 \quad (2)$$

is the total power transfer coefficient. The relation between the incident / reflective port and the transmissive port of the bus waveguide is given by:

$$\begin{pmatrix} a_1 \\ b_1 \\ c_1 \\ d_1 \end{pmatrix} = C_{12} P_{23} C_{34} \begin{pmatrix} a_4 \\ b_4 \\ c_4 \\ d_4 \end{pmatrix} \quad (3)$$

where C_{12} and C_{34} are the transfer matrices of the couplers. P_{23} is the transfer matrix of the two optical paths, L_1 and L_2, and is given by:

$$P_{23} = \begin{bmatrix} 0 & 0 & 0 & P_{L1} \\ 0 & 0 & P_{L2}^{-1} & 0 \\ 0 & P_{L2} & 0 & 0 \\ P_{L1}^{-1} & 0 & 0 & 0 \end{bmatrix} \quad (4)$$

where

$$P_{L1} = te^{-(i\beta+\alpha)L_1} \quad (5)$$

$$P_{L2} = te^{-(i\beta+\alpha)L_2} \quad (6)$$

in which β and α are the waveguide propagation constant and loss coefficient, respectively, and t is the transmission coefficient of the waveguide crossing.

Parameters

We use the typical waveguide parameters for 500 nm wide, 220 nm high, silicon-on-insulator (SOI) nanowires in the modeling. The waveguide propagation loss is assumed to be $\alpha = 5 \; dB/cm$. $t = 0.96$ is used for the transmission coefficient of the waveguide crossing [4]. The crosstalk of the waveguide crossing is better than -40 dB [4] and is ignored in the simulation. The power transfer coefficient of the couplers is assumed to be $T_c = 0.9$. A 2D finite-difference mode solver [5] is used for calculating the waveguide effective indices and the coupling coefficients of the directional couplers by following Ref. [6].

Optimal Coupling Coefficients

Critical to the design are the coupling coefficients of the directional couplers. Based on the dual-criteria of high reflectivity and high extinction ratio, it is found that the optimal coupling coefficients are $\kappa_{12} \simeq 0.84$ and $\kappa_{34} \simeq 0.77$.

MEASUREMENT AND PERFORMANCE

The device was fabricated at IMEC ePIXfab using a CMOS-compatible SOI technology [7]. The measurement schematic is shown in Fig. 2. Grating fiber couplers [7] are used for the input and output ports. A Y-branch power splitter, with an angle of 6 degrees between its two branches, is used to split the reflected light. The measured reflection spectrum, shown in Fig. 3, demonstrates that a Q factor of more than 10,000 and an extinction ratio of more than 25 dB can be obtained.

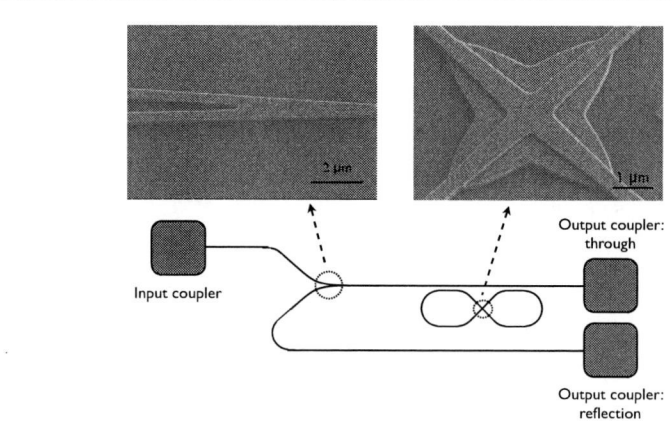

Fig. 2.   Layout design schematic. The insets are the SEM pictures of the Y-branch power splitter and the waveguide crossing.

CONCLUSION

A novel ring-resonator reflector with a waveguide

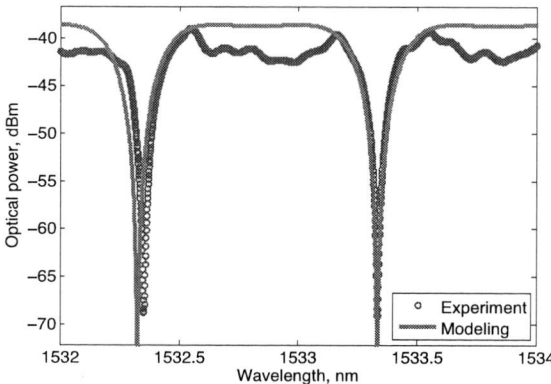

Fig. 3.   Measured and simulated reflection spectra at 25 ^{o}C. An estimated insertion loss of 36 dBm is included in the simulation data. The designed device has a coupler gap of 200 nm and a waveguide bend radius of 30 μm.

crossing has been designed and fabricated. It functions as a reflective-type notch filter. The high extinction ratio demonstrated experimentally shows that it has the potential to serve as a highly sensitive device for sensor applications and for optical communications.

ACKNOWLEDGMENT

The authors would like to thank CMC Microsystems and Lumerical Solutions Inc. for supporting this project and Dr. Nicolas Rouger at Grenoble University for fruitful discussions.

REFERENCES

[1] G. T. Paloczi, J. Scheuer, and A. Yariv, "Compact microring-based wavelength-selective inline optical reflector," *IEEE Photon. Technol. Lett.*, vol. 17, no. 2, pp. 390–392, 2005.

[2] Y. Chung, D.-G. Kim, and N. Dagli, "Reflection properties of coupled-ring reflectors," *Journal of Lightwave Technology*, vol. 24, no. 4, pp. 1865–1874, 2006.

[3] Haishan Sun, Antao Chen, Larry R. Dalton, "A reflective microring notch filter and sensor", *Optics Express*, vol. 17, pp. 10731–10737, 2009.

[4] Wim Bogaerts, Pieter Dumon, Dries Van Thourhout, and Roel Baets, "Compact microring-based wavelength-selective inline optical reflector", *Optics Letters*, vol. 32, Issue 19, pp. 2801–2803, 2007.

[5] A. Fallahkhair, K. Li, and T. Murphy, "Vector finite difference mode solver for anisotropic dielectric waveguides," *Journal of Lightwave Technology*, vol. 26, no. 11, pp. 1423–1431, 2008.

[6] Nicolas Rouger, Lukas Chrostowski, Raha Vafaei, "Temperature Effects On Silicon-On-Insulator (SOI) Racetrack Resonators: a Coupled Analytic and 2D Finite-Difference Approach", *Journal of Lightwave Technology*, vol. 28, no 9, pp. 1380–1391, 2010.

[7] W. Bogaerts, R. Baets, P. Dumon, V. Wiaux, S. Beckx, D. Taillaert, B. Luyssaert, J. V. Campenhout, P. Bienstman, D. V. Thourhout, "Nanophotonic waveguides in silicon-on-insulator fabricated with CMOS technology," *Journal of Lightwave Technology*, vol. 23, no. 1, pp. 401–412, 2005.

A MEMS DIGITAL MICROSHUTTER (DMS™) FOR LOW-POWER HIGH BRIGHTNESS DISPLAYS

J. Lodewyk Steyn, Timothy Brosnihan, John Fijol, Jignesh Gandhi, Nesbitt Hagood IV, Mark Halfman, Steve Lewis, Richard Payne, Joyce Wu

Pixtronix Inc., Andover, MA United States

ABSTRACT

A new display technology has been developed that overcomes the losses in conventional liquid crystal displays (LCDs). At the heart of this new class of displays is a digital microshutter (DMS™) that eliminates the need for a polarizer that is required in LCDs. The DMS™ also has a response time of approximately 20 times faster than that of an LCD pixel, making it possible to produce a bright field sequential direct view display with increased color gamut (up to 145% NTSC, CIE 1976).

INTRODUCTION

We live in a world where we rely on the daily use of mobile electronic devices such as cellular phones, portable navigation devices, portable game units, digital cameras and compact portable computers. A bright color display enhances the user experience for these devices. Conventional liquid crystal displays have a low optical throughput, typically less than 8%. The majority of these losses are incurred through the use of a polarizer (50% loss) and a color filter (67% loss). A recent alternative, Organic LED (OLED) displays, suffer from inefficient organic LEDs with poor high brightness performance. Here we present a display that requires neither a color filter, nor a polarizer. A digital microshutter (DMS™) modulates the light. Combined with a thin film transistor (TFT) active matrix backplane and a highly efficient backlight, this technology promises displays with a power consumption of 25% of that of conventional LCDs while maintaining the ability to produce high quality 24 bit images at video frames rates of 60 Hz.

THE DMS™ DISPLAY CONCEPT

The DMS™ display consists of three main components: a TFT backplane with microshutters, an aperture plate and a backlight unit, as shown in Figure 1. The aperture plate and backlight combine to form a waveguide that sends the light through the apertures located in each pixel where a microshutter opens or closes the aperture. The backlight is illuminated with one or more tri-color RGB LEDs.

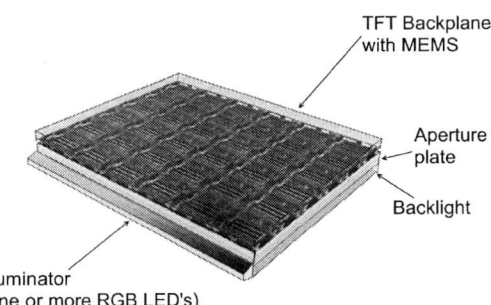

Figure 1: Structure of a DMS™ display.

The groups of red, green and blue LED's are driven sequentially and color mixing occurs in the time domain. For each image frame, each color is divided into multiple bits and the MEMS shutters are used to pass or block the light for each bitplane in this field sequential color driving scheme. Because each image frame requires multiple shutter movements, a fast response time is desired for the DMS™.

THE DIGITAL MICROSHUTTER

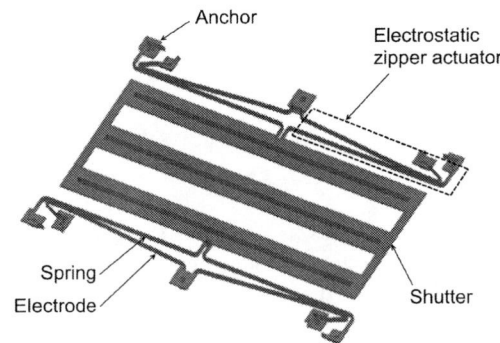

Figure 2: The Digital MicroShutter (DMS™).

Design considerations

Figure 2 is an isometric view of a typical DMS™. It consists of a shutter suspended by four springs and actuated by four electroquasistatic zipper actuators, similar as in [2]. We chose zipper actuators for their ability to deliver a large stroke whilst requiring relatively low voltage compared to other actuator options utilizing the same footprint (e.g. parallel plate or comb drive). As its name implies, the DMS™ is operated in a digital "on-off" mode using the pull-pull nature of the two opposing actuator pairs. A low actuation voltage is desired to minimize the electric power required to operate the display. Therefore, beam lengths, film thicknesses and actuator stroke are all chosen for low actuation voltage. Multiple slots are used to reduce the required stroke and thus the voltage required. A typical DMS^{TM} has a total travel of approximately 10μm, and a design actuation voltage of approximately 20V.

Fabrication

The fabrication process of the DMS^{TM} was designed to be compatible with conventional LCD manufacturing equipment [3]. A structural film is deposited over a sacrificial layer that is defined in two steps. The first step defines the anchors of the DMS^{TM} to the TFT backplane, and also the spacing between the moving portion of the shutter and the backplane. The second step defines the rest of the shutter, the springs and the electrodes. Referring back to Figure 2, the dimensions of the shutter and the anchor structures are defined lithographically, whereas the thicknesses of the springs and the electrodes are defined by the thickness of the deposited film. An anisotropic etch is used to ensure that the sidewall films remain while the top films are etched away. The sacrificial layer is removed in plasma process. Finally, an insulator is deposited to prevent shorting of the zipper actuator.

Typical characteristics

Digital microshutters have been built corresponding to pixel densities ranging from 120 to 300 pixels per inch (ppi). Actuation voltages of less than 15V and switch times of less than 100μs have been attained. Shutters have also been operated over extended periods of time and lifetimes in excess of 70 billion cycles have been attained, corresponding to more than 10000 hours of typical display operation.

DISPLAY PROTOTYPES UTILIZING DMS^{TM} TECHNOLOGY

Figure 3 is a photograph of a 2.5" diagonal QVGA (320x240) display prototype with DMS^{TM} technology. This prototype display combines the basic DMS^{TM} structure as shown in Figure 1 with external control electronics used to convert an incoming DVI video stream into the information needed to display images on the DMS^{TM} display. We have demonstrated crisp image quality and color gamut as high as 145% NTSC, CIE 1976, compared to less than 100% for typical LCDs. Contrast ratios in excess of 1000:1 have been obtained, as well as view angles of greater than 170°. For an on-axis brightness of approximately 200 nits, less than 50mW of optical power was

consumed, comparing favorably with LCDs requiring in excess of 200mW of optical power to attain similar brightness levels [4]. Furthermore, thanks to the digital, programmable nature of these displays, we have also demonstrated reflective modes with near-zero backlight power consumption [5] and high-brightness reduced color content modes for improved outdoor viewing of e.g. maps for navigation purposes.

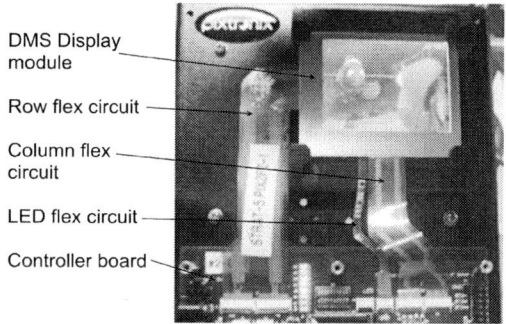

Figure 3: DMS^{TM} Display prototype attached to its controller board.

CONCLUSION

MEMS-based digital microshutters enable the fabrication of digital direct view displays with low power consumption, exceptional image quality and programmability – features not attainable with existing LCD or OLED technologies.

REFERENCES

[1] N. W. Hagood, R Barton, T. Brosnihan, J. Fijol, J. Gandhi, M. Halfman, R. Payne, J.L. Steyn, "A direct-view MEMS display for mobile applications", Vol 38, SID Symposium Digest, 2007, p 1278.

[2] R Legtenberg, J Gilbert, SD Senturia, M Elwenspoek, "Electrostatic Curved Electrode Actuators", Journal of Microelectromechanical Systems, Vol. 6, No. 3, September 1997, pp257-265.

[3] T. Brosnihan, R. Payne, J. Gandhi, S. Lewis, J.L. Steyn, M. Halfman, N.W. Hagood, "Pixtronix Digital Micro Shutter Display Technology – A MEMS Display for Low Power Mobile Multimedia Displays". Proceedings of the SPIE, 2010

[4] J. Gandhi et. al. High Image Quality of Ultra-Low Power Digital Micro Shutter Based Display Technology, Vol 40, SID Symposium Digest, 2009 p. 532.

[5] J. Gandhi et. al, "Sunlight Readability of Digital Micro Shutter based Display Technology", SID Symposium Digest, 2010

FLEXIBLE DISPLAY SYSTEM BASED ON MEMS FABRY-PEROT INTERFEROMETER

G.Tortissier[*1], C.-Y. Lo[2], H. Fujita[1], and H. Toshiyoshi[1,3]

[1]LIMMS/CNRS-IIS (UMI-2820), Institute of Industrial Science, The University of Tokyo, Japan
[2]Institute of NanoEngineering and MicroSystems, National Tsing Hua University, Taiwan
[3]Research Center for Advanced Science and Technology, The University of Tokyo, Japan

ABSTRACT

Previous works on MEMS actuated Fabry-Perot interferometer (FPI) highlighted promising results for flexible display applications. Three primary color pixels have indeed been obtained using both photolithography [1] and Roll-to-Roll printing process [2] with satisfying color purity and transmittance.

However both of these processes are expensive and time-consuming for preparing master micropatterns. For these reasons, a new process based on inkjet printing has been set up. It contributes in a more ecological-friendly, high reproducible and fast process development while targeting improved features.

INTRODUCTION

Recent market trend reports escalating interests of customers for portable devices and most particularly in display and entertainment field. In this context, new flexible devices intend to replace traditional publication media, such as papers and books, by using flexible polymer substrate gathering small size, light weight, low consumption and high integration capabilities.

FABRY-PEROT PRINCIPLE

Figure 1 shows the MEMS FPI layers structure. For constructive color interferences, reflection surfaces (Ag electrodes) and an intermediate dielectric material (SiO_2 with thickness between 160nm and 325nm) in between are needed. The transparent flexible substrates are PEN (polyethylene naphthalate) polymer films which present an optical transmittance between 80% and 85% in the visible range (400nm-800nm).

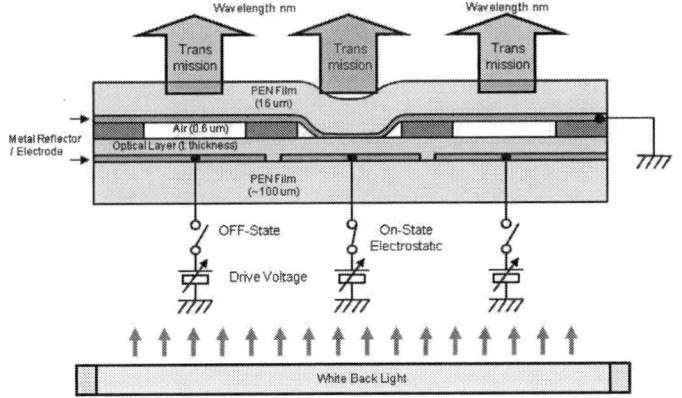

Figure 1. MEMS Fabry-Perot interferometer scheme

Under electrostatic Coulomb force, the upper layer will be attracted down to contact the isolation layer (SiO_2) and satisfy Fabry–Perot interference condition. The output light is then changed from input backlight (white light) to different colors according to different isolation thickness designs. In Fig. 1, the center pixel is in operation (ON state) which changed backlight from white to green.

ROLL-TO-ROLL PROCESS

From high speed production and large display area points of view, the "Roll-to-Roll" printing process seems to be suitable. The Figure 2 describes the complete process to obtain a "Rolled-to-Rolled" FPI. First the lower PEN was introduced into a flexography unit which printed commercial black ink with reverse patterns of lower electrode onto PEN. The black ink is then cured before coating electrode (Ag) and isolation layer (SiO_2) with sputter in vacuum chamber. The black ink is then removed in a chemical bath. Secondly, the obtained designed PEN film enters a gravure printing unit for spacer layer deposition. Finally, the structure on lower substrate was laminated with metal pre-sputtered upper layer in lamination unit under pressure.

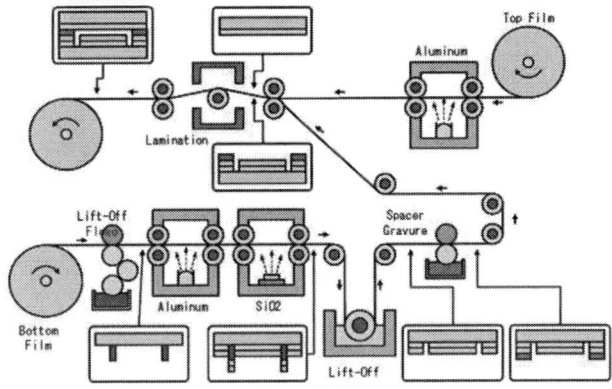

Figure 2. Roll-to-Roll system [3]

The FPI device has been tested under bending conditions and presents three primary colors pixels (Figure 3).

Figure 3. FPI red, green and blue pixel under bending conditions

978-1-4244-8926-8/10 $26.00 © 2010 IEEE

Thanks to high transmittance, bendability and vivid colors [3], "Roll-to-Roll" FPI device could be easily used in the display field and more particularly for decoration or flexible panel. With a view to improve the current FPI features, optimizations have been managed to integrate thinner PEN substrates and spacers to decrease the actuation power and improve the color purity. Figure 4 highlight the simulation results. Decreasing the PEN thickness from 16µm to 1µm enables to drastically reduce the applied voltage, from 42V to 1V, to obtain a 90% pixel contact area.

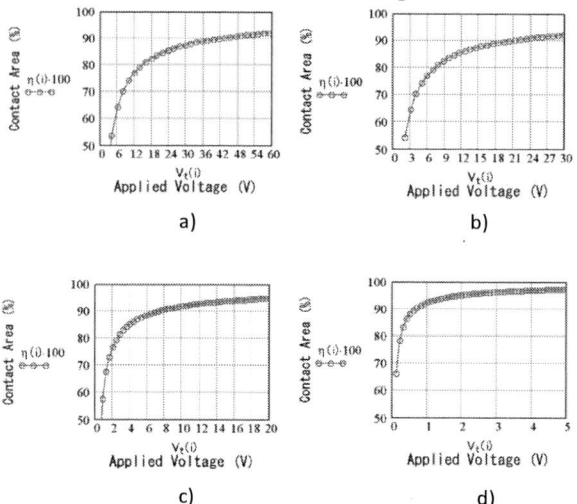

a) b)

c) d)

Figure 4. Contact area simulation results for 1mm^2 surface pixel with 16µm (a), 10µm (b), 5µm (c) and 1µm (d) thick PEN substrate

However, the "Rolling process" is expensive and does not enable fast new design development (which involves changing the master pattern rollers) in the research framework. From the other hand, inkjet printing becomes more and more suitable within display and MEMS technologies [4-6].

INKJET PRINTER UPCOMING CHALLENGE

To resolve the problem of low processing improvement of "Rolling process", we investigate the feasibility of replacing expensive and time consuming processes, as photolithography and "Roll-to-Roll", by inkjet printing (Figure 5).

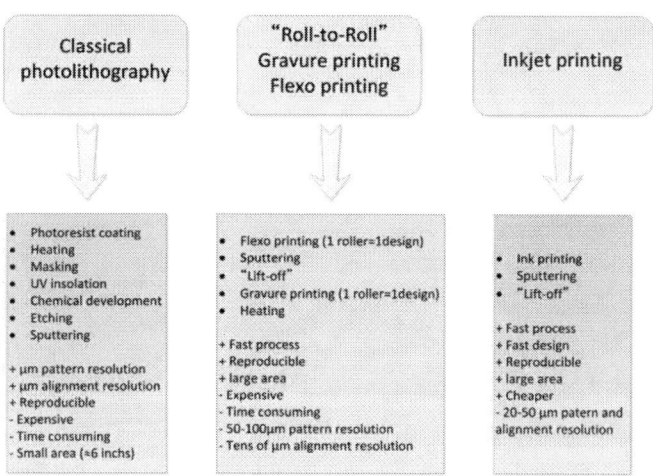

Figure 5. Comparison between Photolithography, "Roll-to-Roll" and inkjet printing processes for patterned multilayer

The main advantages of the inkjet printing lie in its fast design ability and user-friendly processing. For the FPI realization, the process is handling with Fujifilm Dimatix 2831 inkjet printer [7].

First the ink is deposited on the substrate (glass or PEN film) and then a thin metal layer (tens nm of Al or Ag) is coated by sputtering. Finally, the ink is removed using ultrasonic bath with appropriate solvent o dissolve the ink. Finally, thanks to the inkjet defined pattern, an easy "lift-off" process is realized using a very low amount of chemical products with micron resolution.

CONCLUSION

Thanks to new experimental set up, replacing photolithography or "Roll-to-Roll" processes by inkjet printing have been demonstrated. First tests are promising to realize inkjet printed Fabry-Perot intereferometer, faster and cheaper than conventional techniques. Moreover, using 1µm thick PEN film lowered the applied voltage to 1V, to obtain a 90% contact area for a 1000µm length pixel, without color purity degradation.

REFERENCES

[1] Y. Taii, A. Higo, H. Fujita and H. Toshiyoshi, "Transparent color pixels using plastic MEMS technology for electronic papers", IEICE Electronics Express, Vol.3, No.6, pp 97-101, 2006.
[2] C. Lo, J. Hiitola-Keinänen, O. Huttunen, J. Petäjä, J. Hast, A. Maaninen, H. Kopola, H. Fujita and H. Toshiyoshi, "Novel roll-to-roll lift-off patterned active-matrix display on flexible polymer substrate", Microelectronic Engineering 86, pp 979–983, 2009.
[3] C. Lo, H. Fujita, H. Toshiyoshi, "Toward realization of transmissive display by MEMS etalon", IEICE Electronics Express, Vol.5, No.9, pp 326-331, 2008.
[4] H. Kobayashi, S. Kanbe, S. Seki, H. Kigchi, M. Kimura, I. Yudasaka, S. Miyashita, T.Shimoda, C. R. Towns, J. H. Burroughes and R. H. Friend, "A novel RGB multicolor light-emitting polymer display", Synthetic Metals 111–112, pp 125–128, 2000.
[5] H. Sirringhaus, T. Kawase, R. H. Friend, T. Shimoda, M. Inbasekaran, W. Wu and E. P. Woo, "High-Resolution Inkjet Printing of All-Polymer Transistor Circuits", Science 290, 2000.
[6] O.Azucena, J. Kubby, D. Scarbrough, and Chuck Goldsmith, "Inkjet Printing of Passive Microwave Circuitry", IEEE MTT-S, Vol.2, 2008.
[7] http://www.dimatix.com/divisions/materials-deposition-division/printer_cartridge.asp

LOW VOLTAGE ELECTROSTATIC 90° TURNING FLAP FOR REFLECTIVE MEMS DISPLAY

Fabio Jutzi[1], François Gueissaz[2], Wilfried Noell[1], Nico F. de Rooij[1]
[1]Ecole Polytechnique Fédérale de Lausanne (EPFL), Switzerland
[2]Swatch Group Research & Development Ltd., Division Asulab, Switzerland

ABSTRACT

A low voltage electrostatically actuated 90° turning flap is modeled and fabricated. The application is a new kind reflective, low-power, MEMS-based display with high contrast and reflectance. The system consists of a poly-silicon flap fabricated onto a silicon substrate with a transparent counter electrode on glass and a device layer of a silicon-on-insulator (SOI) wafer as spacer in between them. Actuation voltages down to 20V are achieved.

INTRODUCTION

Commonly used reflective display systems, which are based on liquid crystals, have low contrast ratio between black and white pixels. Recently different new technologies have been developed, but still cannot satisfy either in angular response or contrast. A new MEMS-based reflective display concept is being developed using a 90° turning flap as shown in figure 1. When the flap is at flat state it reflects light and constitutes a white pixel. At the vertical state light is absorbed by an underlying absorptive substrate, therefore we have a black pixel. For this display system a low voltage electrostatically actuated 90° turning flap is modeled and fabricated. The device consists of a flap array with torsion bars, a transparent counter electrode and a spacer layer in-between them. By connecting the flaps in the columns and the electrodes in the rows line-column addressing can be used.

MODELLING

To analyze the dependence of the electrode distance to the actuation voltage finite element simulation are made. In Ansys 11.0 a 2D electrostatic model is used to calculate the capacitance C between the electrode and the flap for each turning angle θ and different electrode distances d. The electrostatic torque can be obtained by:

$$M_{el} = \frac{1}{2}\frac{dC}{d\theta}V^2$$

The torque M_{el} is independent of the flap length l_{flap}. The mechanical torque of a torsion bar can be expressed as: $M_{mec} = \theta \times K_\theta$, where K_θ is the spring constant of the torsion bar given by [1]:

$$K_\theta = 2 \cdot \frac{Gw_{bar}t^3}{3l_{bar}}\left\{1 - \frac{192}{\pi^5}\cdot\frac{t}{w_{bar}}\tanh(\frac{\pi w_{bar}}{2t})\right\}$$

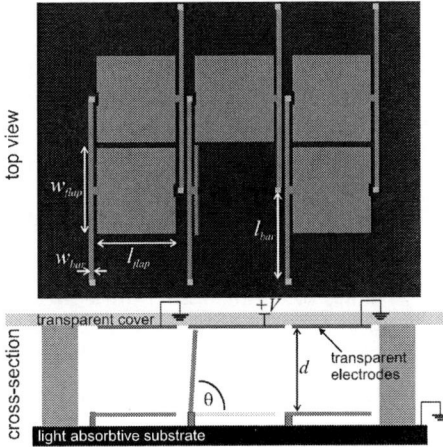

Figure 1 Schematic of a 3 x 2 pixel array consisting of flaps with torsion bars as flexure. One flap is bent to vertical position by applying a voltage V to the transparent counter electrode (black pixel). The other electrodes are grounded and the flaps at flat state (white pixel).

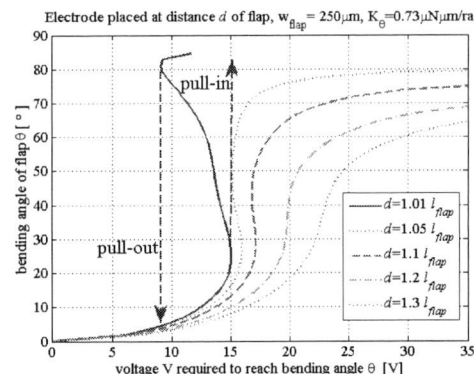

Figure 2 Finite element simulations of poly-silicon flap (dimensions given in table 1) attracted by electrode. The pull-in voltage can be obtained from this graph. It can be seen that for $d>1.2l_{flap}$ there is no pull-in anymore.

When equating both torques the required voltage V for arriving at any torsion angle θ is obtained:

$$V = \sqrt{2\frac{K_\theta \cdot \theta}{dC/d\theta}}$$

In figure 2 V for different electrode-flap distances d is shown for the dimensions given in table 1.
For thin t an actuation voltage between 15 and 25V is obtained. It can be seen that for small electrode-flap

distances there is a pull-in voltage, whereas at increasing d the actuation voltage raises and there is no pull-in anymore. To obtain an accurate actuation voltage the electrode-flap distance and thus the spacer requires an accurate thickness.

Table 1 Geometrical dimensions of flap

thickness of torsion bar and flap	t	100nm
width of torsion bar	w_{bar}	4µm
length of torsion bar	l_{bar}	235µm
width of flap	w_{flap}	250µm
length of flap	l_{flap}	120µm
shear modulus of poly-silicon	G	65GPa

FABRICATION PROCESS

The fabrication process is depicted in figure 3. It starts with anchor holes etched into 2µm thick SiO_2 on a silicon substrate. 100nm poly-silicon is deposited, doped, annealed, and patterned as flaps. Subsequently a 0.5µm thick gold layer is placed by lift-off around the flaps array. A SOI wafer patterned by deep reactive ion etching (DRIE) with a 120µm thick device layer and release holes in the handle layer is bonded to the wafer with the flaps. Au-Si eutectic bonding at 390°C is used. The bonded wafers are diced to chips size in the next step. In HF-vapor the flaps and the handle layer of the SOI wafer are released. The obtained result is an array of flaps suspended by torsion hinges, surrounded by a 120µm thick spacer layer consisting of the device layer of a SOI wafer. As a last step a glass chip with patterned Indium-tin-oxide (ITO) electrodes and gold connection lines and pads is micropositioned and glued with a conductive adhesive onto the flap-chip. The connection pads for the flaps and electrodes are on the glass chip. In this version all flaps are short circuited to the substrate. Future devices will be fabricated onto an insulating substrate, so line-column addressing can be used.

RESULTS

In figure 4 a flap array is shown, where one flap is actuated. Required actuation voltages range between 19 and 40V volts, which are partially higher than the calculated values for this geometry. The reason for the higher voltage is partial sticking of the flaps to the substrate. This is due to buckling of the flap poly-silicon layer because of intrinsic stress gradients. Thicker substrate-flap gaps and lower stress gradients in the poly-silicon layer can improve the reliability of the device.

ACKNOWLEGEMENT

This work was financially supported by Swatch Group Research & Development, Division Asulab.

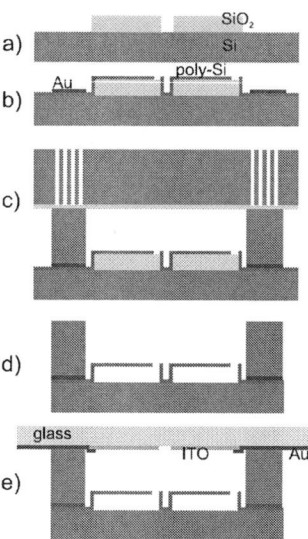

Figure 3 Cross section of fabrication procedure of poly-silicon flaps on 2µm thick sacrificial SiO_2 layer (a-b). A SOI-wafer is bonded by eutectic Au-Si bonding to the wafer with the flaps (c). The structures are diced and released in HF-vapor (d). A transparent electrode chip is glued to the flap chip (e).

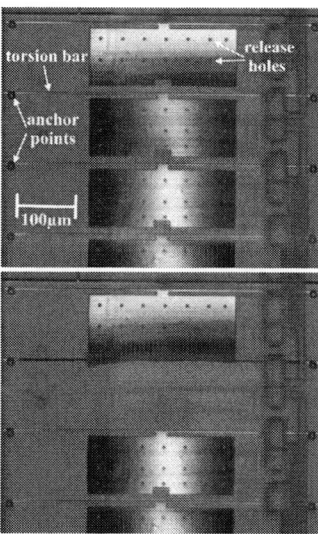

Figure 4 Partial view of a poly-silicon flap array assembled to a transparent electrode chip with dimension given in table 1. Top: No voltage is applied. Bottom: 35V applied between the substrate with the flaps and the electrode on second line. The flap turns to vertical position and is almost not visible anymore from the top.

REFERENCES

[1] H. Toshiyoshi, "Electrostatic Micro Torsion Mirrors for an Optical Switch Matrix", J. of Micromechanical Sys., Vol. 5 No. 4, pp. 231-237, 1996

GaN pitch-variable grating fabricated on Si substrate

H. Sameshima, T. Tanae, F. Hu, K. Hane

Tohoku University, Sendai, Japan

ABSTRACT

A tunable grating is fabricated by micromachining a GaN crystal layer grown on Si substrate. The tunable grating consists of grating lines, electro-static comb-drive actuator and connection springs. The grating consists of 85nm wide, 12μm long 24 grating lines with 674nm period. The crystallization stress of an HfO_2 layer deposited on the GaN crystal is used to compensate the residual stress of the GaN crystal grown on Si substrate. The freestanding GaN structure consisting of the grating and the actuator is fabricated by etching the Si substrate with XeF_2 gas. Applying the voltage of 140V, the grating is expanded by 600nm corresponding to the period change of 3.7%.

INTRODUCTION

GaN semiconductor is a powerful material for light emitting diode. Moreover, it is also promising for high electron mobility transistor. In addition to those GaN researches, the mechanical properties of GaN series materials are also valuable. GaN semiconductor can be applicable to the fabrication of MEMS. Although the research on GaN MEMS is still at the beginning stage, several demonstrations are reported for GaN sensors and microsytems [1-3]. However, there is no report on tunable grating. It is promising to integrate tunable grating with the light source for wavelength selective light source and tunable laser.

In this work, a GaN tunable grating with a comb-drive actuator is designed and fabricated form a GaN crystal grown on Si substrate. In order to compensate the residual stress of the grown GaN layer on Si wafer, an HfO_2 layer is deposited on the GaN crystal. The proposed device is operated well by the stress compensation.

DESIGN AND FABRICATION

Figure 1 shows the design of the GaN tunable grating with a comb-drive actuator. The grating consists of 24 grating lines with the period of 674nm. Each grating lines are 0.33μm wide, 12μm long, and 1.0μm thick. The comb fingers are 1.13μm long and 0.23μm wide, and the period of comb finger is 1.26μm. The number of the comb fingers is 19. The initial gap of the combs is 1.13μm.

The fabrication starts with a GaN/Si wafer. The GaN crystal layer is grown by metal organic chemical vapor deposition with a buffer layer on Si(111) wafer. The GaN crystal layer consists of 200nm thick AlN layer on Si substrate, 450nm thick $Al_xGa_{1-x}N$ (0.2<x<0.7) layer and Si doped 200-300nm thick GaN layer as top layer. The total thickness of the GaN crystal layer is approximately 1μm.

Due to the layered growth of the GaN crystal to decrease crystal defects in the hetero-epitaxy as well as the high temperature growth (1000deg.C) with the larger difference of the thermal expansion coefficients between GaN and Si crystals, a large stress was generated in the grown GaN crystal layer. Therefore, after etching the Si substrate, the freestanding slab of GaN crystal was bent by the stress. Therefore, it was difficult to fabricate a large flat freestanding structure with small supports. In our previous work, a tensile diaphragm was obtained by crystallizing an HfO_2 film deposited on GaN layer after removing Si substrate by inductively coupled plasma reactive ion etching (ICP-RIE)[4].

A 150nm thick HfO_2 film was deposited and it was crystallized at 700deg.C at the pressure lower than 10^{-4}Torr. In order to pattern GaN freestanding slab structure, a Cl_2 ICP-RIE was used at power of 100W, the frequency of 13.56MHz, Cl_2 pressure of 7.0-7.5 x 10^{-3}Torr, Cl_2 flow rate of 8.45sccm, and the bias voltage of 30V. The HfO_2 film was also used for a hard mask for etching GaN layer. The etching rate of GaN layer was 410nm/min while that for HfO_2 layer was 25nm/min. The 150nm thick HfO_2 film became 50nm thick after etching the GaN layer. The aspect ratio of the etching for the grating groove was 1000/85 by the HfO_2 layer mask in this experiment.

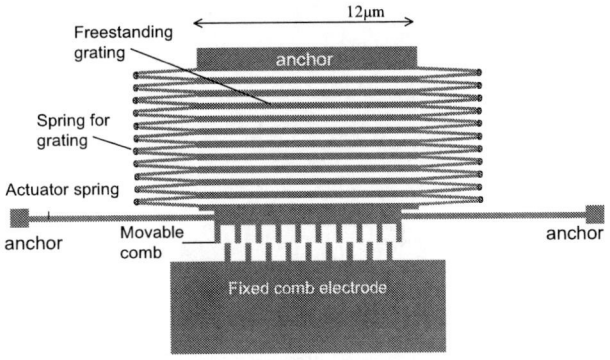

Fig. 1 Design of GaN pitch-tunable grating

FABRICATION RESULTS

Figure 2 shows the cross-sectional electron-micrograph of an example of grating groove etching by ICP-RIE. The GaN layer is completely etched into the Si substrate. Figure 3 shows the electron-micrograph of the fabricated device. After etching Si substrate with XeF_2 gas, the freestanding GaN grating with the electrostatic comb-drive actuator is fabricated well. The grating lines are about 85nm wide on the top surface and the comb-fingers are 230nm wide. Due to the small period of grating and a little over-etching, each grating line becomes narrower than the designed value. The deflection of the device is smaller than 0.1μm, which is evaluated by an optical interferometer.

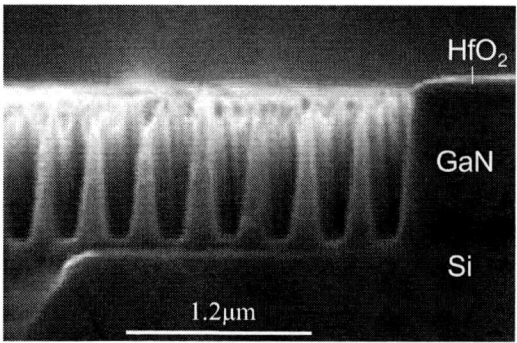

Fig.2. Cross sectional image after etching the GaN layer

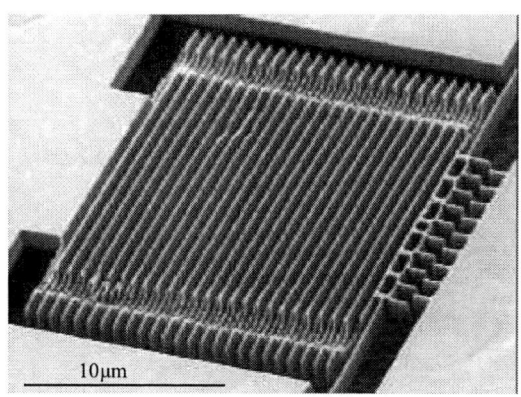

Fig.3. Fabricated freestanding GaN grating on Si substrate

MEASUREMET RESULTS AND DISCUSSION

The displacement of the actuator was measured by an optical microscope. Figure 4(a) shows the displacement of the actuator measured as a function of the applied voltage. The measured displacement increases monotonously in the measured region. The displacement of 0.6μm is obtained at the voltage of 140V, which corresponds to the period expansion of 3.7%. The expansion of the grating was also investigated by laser beam diffraction. The spot displacement of the first order diffraction beam of a He-Ne laser at the wavelength of 633nm is measured as a function of applied voltage as shown in Fig.4(b). At the voltage of 100V the diffraction angle changes by about 1.5 degrees.

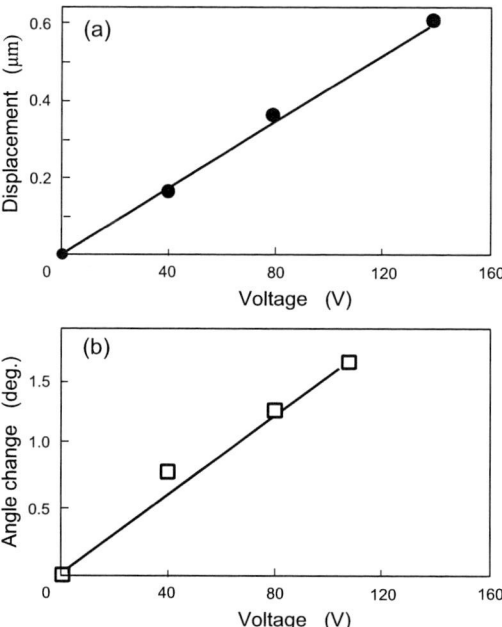

Fig.4. (a) Displacement and (b) diffraction angle change measured as a function of voltage.

ACKNOWLEGEMENT

The authors thank to Prof. T. Kuriyagawa for the interferometric measurement. The work is supported by JSPS.

REFERENCES

[1] V.Cimalla, J.Pezoldt, O.Ambacher, J.Phys. D:Appl. Phys. 40(2007) 6386-6434.

[2] R.Ito, M.Wakui, H.Sameshima, F.Hu, K.Hane, Microsys. Technol. 16 (2010) in press.

[3] Y.Wang, F.Hu, M.Wakui, K.Hane, IEEE Photon. Technol. Lett. 21(2009)1184-1186.

[4] H.Sameshima, M.Wakui, F.Hu, K.Hane, IEEE J. Sel.Top. Quant. Electron. 15(2009)1332-1337.

SYNCHRONIZED LASER SCANNING OF MULTIPLE BEAMS BY MEMS GRATINGS INTEGRATED WITH RESONANT FREQUENCY FINE TUNING MECHANISMS

[1,2]Yu Du, [1]Guangya Zhou*, [1]Kelvin Koon Lin Cheo, [2]Qingxin Zhang,
[2]Hanhua Feng and [1]Fook Siong Chau

[1]National University of Singapore; [2] Institute of Microelectronics, Singapore
*Corresponding author: Guangya Zhou; E-mail: mpezgy@nus.edu.sg

ABSTRACT

This paper presents an effective method to achieve synchronized laser scanning of multiple beams by using MEMS diffraction gratings with their resonant frequency fine tuning mechanisms. Multiple gratings are actuated in-plane by a common electrostatic comb-driven resonator and their resonant frequencies can be fine-tuned to compensate the micromachining process errors. Continuous and reversible resonant frequency tuning was achieved. The resonant frequency of one diffraction grating gradually dropped from 19870 Hz to 19588 Hz with its tuning voltages increased from 0V to 5V. Finally, synchronized laser scanning of multiple beams was demonstrated using stroboscopic method.

INTRODUCTION

Laser scanning using a diffraction grating [1-2] has the potential to achieve high scanning rate without optical performance degradation due to dynamic aberration. This is attributed to the adoption of the in-plane rotation of a diffraction grating instead of out-of-plane deflection of a reflective surface to scan the laser beam. The dispersive diffraction grating scanner with a single grating is highly suitable for narrow-band laser scanning applications, such as monochromatic laser scanning displays. For multi-wavelength collinear scanning applications, such as color displays, synchronized motion of multiple gratings has to be realized. This can be achieved by different methods. For example, multiple grating elements can be configured on a common platform [1]. We can also configure different grating elements separately on multiple grating platforms and synchronize their scanning.

In this paper, we present an effective method to achieve synchronized laser scanning of multiple beams by using diffraction gratings integrated with individual resonant frequency tuning mechanisms. Under this method, multiple gratings were actuated by a common driving resonator and vibrated at resonance. The vibrating amplitudes of each grating can be adjusted through tuning its resonant frequency by its frequency tuning mechanism. Different reversible frequency tuning methods have been investigated in literatures, such as

utilizing the electrostatic spring effect [3-4], localized thermal stressing effect [5] and mechanically stiffen effect of a torsional spring [6]. In this work, we utilize an electro-thermal bent-beam actuator to fine tune the stiffness of a T shape flexure through mechanical stretching.

DESIGN AND FABRICATION

A schematic illustration of two MEMS diffraction gratings (400nm grating pitch) with their integrated resonant frequencies fine tune mechanisms is shown in Fig. 1. Each diffraction grating was suspended and connected to the substrate through four circular folded-beam flexures. A common electrostatic comb-driven resonator was connected to each grating through a single beam flexure. Each grating was connected in turn individually to an electro-thermal bent-beam tuning actuator through a T shape spring.

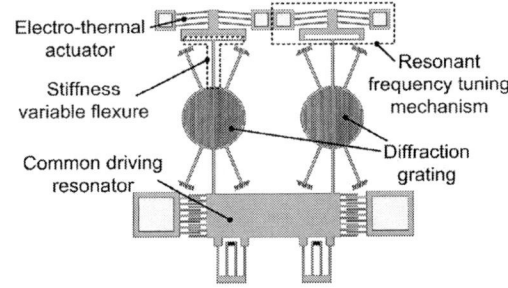

Fig. 1: Schematic view of the MEMS vibratory grating scanner with multiple diffraction gratings and individual resonant frequency fine tuning mechanisms.

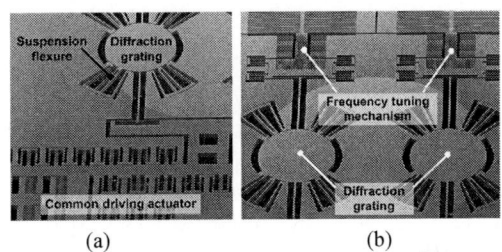

Fig.2: SEM images of the prototype device showing (a) common electrostatic comb-driven resonator with one diffraction grating and (b) diffraction gratings with their resonant frequencies fine tuning mechanisms.

978-1-4244-8926-8/10 $26.00 © 2010 IEEE

The prototype grating scanner was fabricated using Silicon-On-Insulator (SOI) micromachining process. The common electrostatic comb-driven resonator as well as the diffraction gratings and their resonant frequencies fine tune mechanisms are shown in Fig. 2 (a) & (b) respectively.

EXPERIMENTAL RESULTS

The variation of the resonant frequency of a grating with different tuning voltages was measured in atmosphere and shown in Fig.3. As tuning voltages increased from 0V to 5V, the resonant frequency gradually decreased from 19870 Hz to 19588 Hz and the tuning range is 282 Hz. The inversely proportional relationship between the resonant frequency and the tuning voltages is mainly due to the initial deformation of the T shape spring caused by residue stress during the micromachining process.

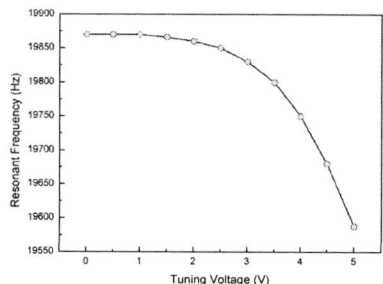

Fig.3: Measured variations of the resonant frequency with respect to different tuning voltages

The synchronized laser scanning of two beams were demonstrated by using the stroboscopic method and the experimental setup is shown in Fig. 4.

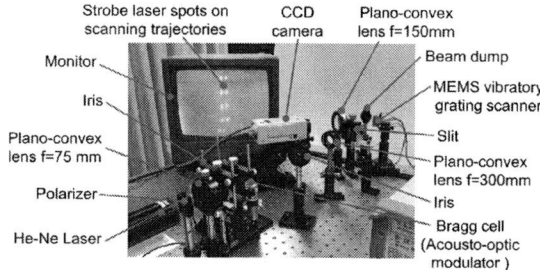

Fig. 4: Experimental setup to investigate the synchronized laser scanning of two beams using the stroboscopic method.

Fig.5 (a) & (b) show the Strobe spots from different positions of scanning trajectories before and after the resonant frequency fine tuning respectively. Before frequency tuning, the scanning amplitudes of the two gratings were not equal because their resonant frequencies were different due to fabrication errors. Upon applying tuning voltages of 0V and 3.87V for the two tuning actuators respectively, the scanning amplitudes were adjusted to be equal. Thus synchronized laser scanning of two laser beams was experimentally demonstrated.

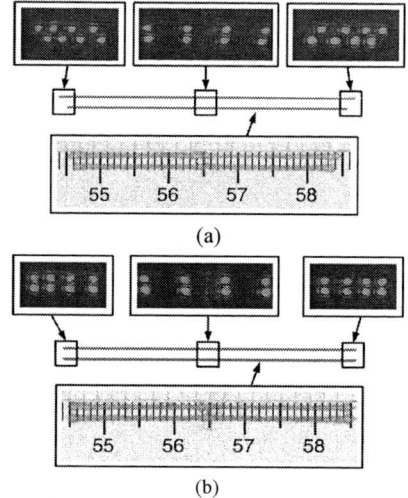

Fig. 5: Strobe spots from different positions of scanning trajectories (a) before and (b) after the resonant frequency tuning with tuning voltages of 0V and 3.87 V for two tuning actuators.

ACKNOWLEDGEMENT

Financial support by the MOE research grant R-265-000-306-112 is gratefully acknowledged.

REFERENCES

[1] G. Zhou and F. S. Chau, "Micromachined vibratory diffraction grating scanner for multi-wavelength collinear laser scanning", *Journal of Microelectro mechanical Systems*, 15 (2006), pp. 1777-1788.

[2] Y. Du, G. Zhou, K. L. Cheo, Q. Zhang, H. Feng and F. S. Chau, "A 2-DOF circular-resonator-driven in-plane vibratory grating laser scanner", *Journal of Microelectro mechanical Systems*, 18 (2009), pp. 892-904.

[3] K. Lee and Y. Cho, "A triangular electrostatic comb array for micromechanical resonant frequency tuning" *Sensors and Actuators: A. Physical*, 70 (1998), pp. 112-117.

[4] K. Lee, L. Lin and Y. Cho, "A closed-form approach for frequency tunable comb resonators with curved finger contour", *Sensors and Actuators: A. Physical*, 141(2008), pp. 523-529.

[5] T. Remtema and L. Lin, "Active frequency tuning for micro resonators by localized thermal stressing effects", *Sensors and Actuators: A. Physical*, 91(2001), pp. 326-332.

[6] J. Lee, S. Park, Y. Eun, B. Jeong and J. Kim, "Resonant frequency tuning of torsional microscanner by mechanical restriction using MEMS actuator", *2009 IEEE 22nd MEMS*, pp. 164-7, 2009.

Fluorometric Bio-Sniffer (Biochemical Gas Sensor) with UV-LED Light for Fomaldehyde Vapor as VOC (Volatile Organic Chemical)

Tomoko Gessei[1,2], Gen Itabashi[3], Yuki Suzuki[3], Daishi Takahashi[3], Takahiro Arakawa[3], Hiroyuki Kudo[3] and Kohji Mitsubayashi[1,3]*

[1]Graduate School of Medical and Dental Sciences, Tokyo Medical and Dental University, Tokyo, Japan

[2]Tokyo Metropolitan Industrial Technology Research Institute, Tokyo, Japan

[3]Inst. of Biomat. and Bioengineering, Tokyo Medical and Dental University, Tokyo, Japan

ABSTRACT

A fibre-optic biochemical gas sensor (bio-sniffer) for gaseous formaldehyde (FA) was developed and applied for *on-site* assessment of FA detoxification. The bio-sniffer measures FA as a fluorescence of reduced nicotinamide adenine dinucleotide (NADH) which is the product of formaldehyde dehydrogenase reaction. The detection limit of the bio-sniffer for gaseous FA was 2.5 ppb, which is enough lower for the purpose of evaluating FA detoxification. The bio-sniffer was then applied to measure phytoremediation with *N. E. "Bostoniensis"*. The FA level in the measurement chamber decreased down to the detection limit (2.5 ppb), which is lower level than FA level of room air (12 ppb), within 95 min and the value increased to FA level when the chamber was vented.

INTRODUCTION

Formaldehyde (FA) is one of the hazardous volatile organic compounds. Exposure to FA is associated with significant damage to human health such as sick building syndrome (SBS) [1-2] or cancers of blood and lymphatic system [3-4]. For this reason, the criterion of FA level is defined as 80 ppb by World Health Organization. Phytoremediation of environmental FA using foliage plants is one of the promising approaches to produce healthy atmosphere for its advantages in cost and beauty. However, *on-site* assessment of the detoxification process is still difficult, particularly in the relatively clean atmosphere (typically, several tens of ppb) for its low concentrations. In this study, an high-sensitive and fibre-optic biochemical gas sensor (bio-sniffer) with FA dehydrogenase (FALDH) was developed and applied for *on-site* monitoring of FA detoxification.

EXPERIMENTAL

2.1 Construction and Characterization of the FA bio-sniffer

The bio-sniffer measures FA as a fluorescence of reduced nicotinamide adenine dinucleotide (NADH) produced by following reaction.

$$FA + NAD^+ \xrightarrow{\text{FALDH}} \text{Formic acid} + NADH + H^+$$

An ultraviolet light emitting diode (UV-LED: $\lambda=340$nm) was employed as an excitation source. The excitation light was introduced to an optical fibre probe, which has a flow-cell with an enzyme immobilized membrane.

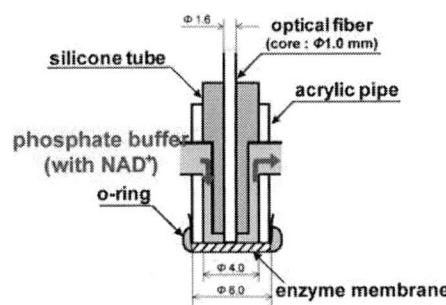

Fig. 1. Structure of the flow-cell at the probe.

Fluorescence of NADH was measured coaxially with a photomultiplier tube (PMT). Phosphate buffer was circulated into the flow-cell to maintain activity of enzyme, rinsing reaction products and supplying NAD$^+$ (Fig.1).

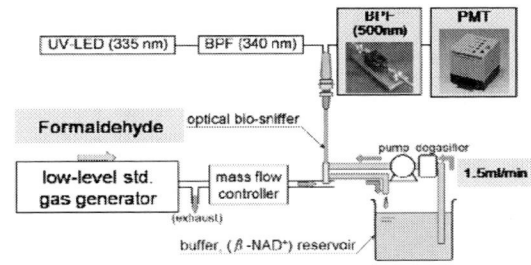

Fig 2. Gas measurement system with FALDH immobilized bio-sniffer

Assessment of the bio-sniffer was carried out using a standard gas generator (Fig. 2). Response, calibration range and selectivity to other chemical substances were investigated.

2.2 Application for FA Detoxification Assessment

After characterization, the bio-sniffer was then applied to measure phytoremediation of FA with *N. E. "Bostoniensis"*. First, a measurement chamber (Volume: 40L) was filled with 100 ppb FA. Then, *N. E. "Bostoniensis"* (wet weight: 24.0 g) was set in the chamber. The gas in the chamber was circulated using a diaphragm pump and FA level was monitored using the bio-sniffer continuously.

RESULTS AND DISCUSSION

3.1 Characteristic of the bio-sniffer

The calibration curve of the bio-sniffer for gaseous FA is shown in Fig. 3. The inset box shows the fluorescent response. As shown in the figure, fluorescent signal increased immediately when FA was exposed and the value decreased to the initial value by the effect of buffer circulation. This indicates that the bio-sniffer is useful for continuous monitoring. The steady state value was determined by the balance of NADH production and the rinsing effect of buffer circulation. The minimum detection limit was 2.5 ppb. The bio-sniffer also showed extremely high selectivity to formaldehyde, comparing with acetaldehyde, methanol, ethanol and benzene, due to specific activity of FALDH. Thus, both high sensitivity and selectivity of the bio-sniffer was confirmed.

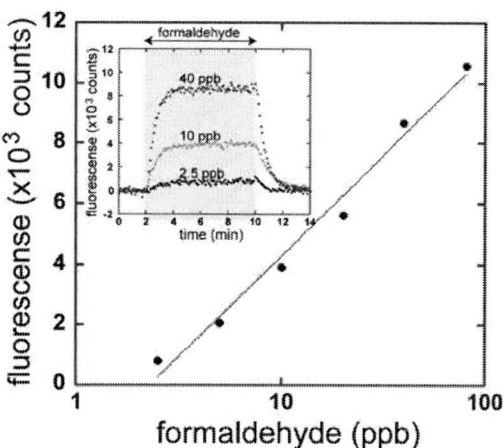

Fig. 3. Calibration curve for gaseous FA.

3.2 Assessment of FA Detoxification

Above all, the FA level in the vacant measurement chamber was confirmed to be stable at 100 ppb. When the *N. E. "Bostoniensis"* was set in the chamber, the FA level immediately decreased down to the detection limit (2.5 ppb). The time to reach the limit was 95 min. After the

FA level became stable, the chamber was opened and room air was taken into the chamber. As a result, the FA level in the measurement chamber increased to 12 ppb and it was again reached to detection limit when the chamber was closed. Thus, the bio-sniffer was confirmed to be useful for FA detoxification, even in the lower level than criterion value.

CONCLUSIONS

A high-sensitive bio-sniffer for FA monitoring was developed and confirmed to be useful for on-site monitoring of FA detoxification by a *N. E. "Bostoniensis"*. The bio-sniffer showed high sensitivity (2.5 ppb) by use of PMT as a photodetector. Continuous monitoring was realized by the flow-cell with FALDH membrane. Also, high selectivity due to specific activity of FALDH was confirmed. Owing to such characteristics, monitoring of phytoremediation process was successfully performed using the bio-sniffer.

ACKNOWLEDGEMENTS

This work was partly supported by Japan Society for the Promotion of Science (JSPS) Grants-in-Aid for Scientific Research System, by Japan Science and Technology Agency (JST) and by MEXT (Ministry of Education, Culture, Sports, Science and Technology) Special Funds for Education and Research "Advanced Research Program in Sensing Biology".

REFERENCES

[1] J.R. Beall and A. G. Ulsamer, "Formaldehyde and hepatotoxicity: a review," J. Toxicol. Environ. Health 14, 1984, pp. 1-21.

[2] L. E. B. Freeman, A. Blair, J. H. Lubin, P. A. Stewart, R. B. Hayes, R. N. Hoover, and M. Hauptmann, "Mortality From Lympho-hematopoietic Malignancies Among Workers in Formaldehyde Industries: The National Cancer Institute Cohort," J. Natl. Cancer Inst. 101, 2009, pp.751-761.

[3] A. Blair, P. Stewart, M. O'Berg, W. Gaffey, J. Walrath, J. Ward, "Mortality among industrial workers exposed to formaldehyde," J. Natl. Cancer Inst. 76, 1986, pp.1071-1084.

[4] I. M.Ritchie and R. G. Lehnen, "Formaldehyde-related health complaints of residents living in mobile and conventional homes," Am. J. Public Health 77, 1987, pp.323-328

HIGH-PRECISION OPTICAL & FLUIDIC MICRO-BENCH FOR ENDOSCOPIC IMAGING

Niklas Weber, Hans Zappe, and Andreas Seifert
University of Freiburg, Germany

ABSTRACT

An optical and fluidic micro-bench with integrated liquid-filled tunable micro-lenses is developed for an endoscopic probe. This novel approach offers not only dynamic focussing, but also a greatly increased working range for biomedical optical imaging. The fabrication and assembly method is highly precise and flexible and can be used for designing optical probes with high functionality.

INTRODUCTION

Recent years have shown a decisive shift towards minimally invasive methods in medical surgery. Medical diagnosis calls for miniaturized technical solutions for endoscopic imaging probes. Most approaches are based on micromachined steel [1] or polymer housings [2] in which the optical components are aligned by glueing or clamping. Due to tolerances in fabrication and assembly, alignment errors can hardly be avoided. Moreover, these probes suffer from a lack of functionality caused by fixed and rigid optics.

An alternative approach is the use of silicon optical bench (SiOB) technology. Using this method, micro-optical components can be aligned precisely. This technology has been enhanced, here for the first time, by adding fluid channels and vias to the bench. This enables the integration of fluidic components such as fluid-filled membrane lenses. The membrane curvature of these lenses can be tuned by applying fluidic pressure to the lens cavities.

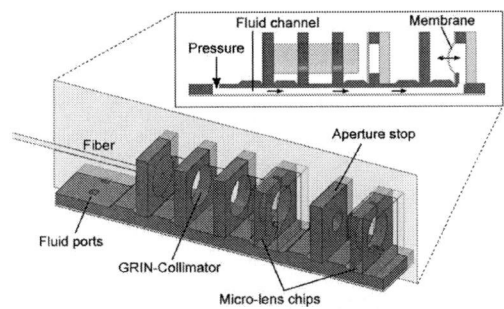

Figure 1: Schematic of the optical and fluidic micro-bench. Two adaptive membrane lenses are mounted on the micro-bench and can be individually tuned by fluidic pressure.

We present here a highly-functional micro-bench with integrated collimator and tunable micro-lenses for use in an endoscopic probe. Outer dimensions of the bare bench are $0.4 \times 1.5 \times 9\,mm^3$, whereas the mounted optical chips are $1.5 \times 1.5\,mm^2$. The layout of this micro-bench is illustrated in Figure 1. All components are precisely aligned by etch grooves of same depth.

The focal length of the micro-lenses can be tuned pneumatically and thus offers great flexibility in determining working distance and in beam shaping.

FABRICATION & PROCESS TECHNOLOGY

Micro-bench

As illustrated in Figure 2a, the micro-bench is fabricated by silicon bulk micromachining. The vias for the lenses and two 200 µm wide channels are realized by a two-step deep reactive-ion etching (DRIE) process on the backside of the substrate. The alignment grooves for the optical chips are defined by wet etching the frontside with KOH. Wet etching yields a very uniform etch depth and results in highly planar surfaces. After completion of the bulk micromaching, a 20 µm thick dry film resist is laminated onto the backside for forming buried channels.

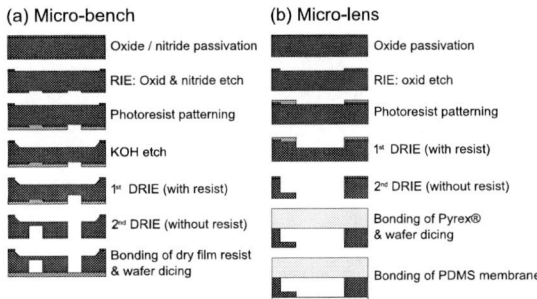

Figure 2: Schematic of the process technology for (a) the micro-bench and (b) fluid-filled micro-lenses.

Micro-lens

The micro-lens consists of a fluid-filled cavity with an aperture diameter of 1.2 mm and two fluid channels in a silicon chip. The silicon structures are defined by a two-step DRIE process and sealed with Pyrex® as shown in Figure 2b. After chip dicing, the cavity is sealed with an elastic and highly transparent, 50 µm thick polydimethylsiloxane (PDMS) membrane. These last two steps are critical, because the optical chips have to be mounted upright. For precise alignment, perfectly vertical walls as well as non-protruding membranes are of crucial importance .

Assembly

To assemble the system, a special stamp-and-stick bonding technique was developed. By using this method, the optical chips can be mounted upright in the alignment grooves and glued without occlusion of the fluidic interconnects. The applied adhesion film is about 5 µm thick. By aligning the chips at the groove edges, twist-free positioning can be guaranteed. Figure 3 shows an assembled micro-bench.

978-1-4244-8926-8/10 $26.00 © 2010 IEEE

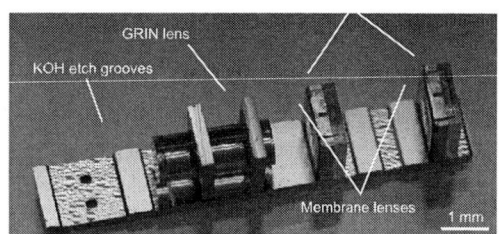

Figure 3: Assembled micro-bench with two membrane lenses. A GRIN lens is used for collimation.

CHARACTERIZATION & SIMULATION
Geometrical characterization

The characterization of the bench and micro-lenses was carried out by microscope image processing, as demonstrated in Figure 4. The images were analysed with respect to tilt, mis-positioning and offset caused by inaccurate dicing. Angular errors between the optical chips and the micro-bench were less than 0.5° and deviations from the optical axis below 5 μm.

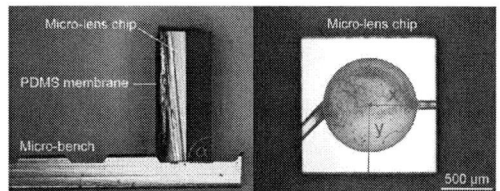

Figure 4: Optical determination of (a) angular error between membrane lens and micro-bench and (b) deviations from optical axis.

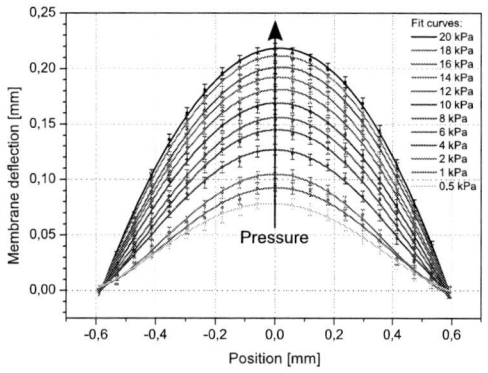

Figure 5: Measured surface profiles of the membrane lens as a function of the applied pneumatic pressure.

The surface profile of the micro-lens as a function of the applied pressure was obtained in the same manner. Figure 5 shows the measured data for pressures up to 2×10^4 Pa. The lens profiles were fit by 4th order polynomials, followed by optical simulations with ZEMAX, based on the polynomial coefficients.

Simulation

Modulation transfer function (MTF) analysis was performed to estimate the influence of the previously mea-

sured alignment errors. The obtained MTF cutoff frequencies (10% contrast) are illustrated in Figure 6 as a function of pneumatic pressure. As can be seen, there is almost no difference between setups with and without maximum misalignment. This clearly indicates that our approach is suitable for building adaptive optical probes with excellent accuracy and optical quality.

The MTF cutoff frequency shows an inverse correlation with focal length. With the measured surface profiles, focal length as a function of pressure was simulated. The resulting range, 2.0 - 6.1 mm, is suitable for endoscopic imaging.

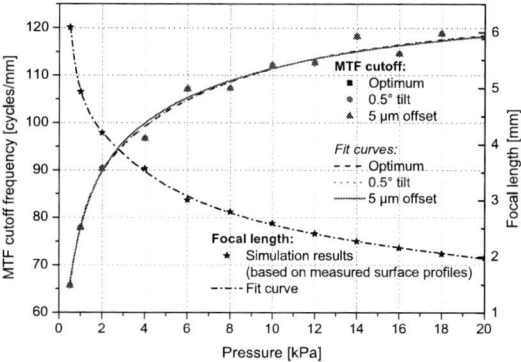

Figure 6: MTF cutoff frequency (10% contrast) as a function of applied pressure for best and worst case scenarios. Almost no difference exists between optimum and worst alignment. Focal lengths can be tuned between 2 and 6.1 mm with pressures from 0.5 - 20 kPa.

CONCLUSION AND OUTLOOK

A novel approach for fabricating optical silicon micro-benches with integrated fluidics has been demonstrated. A first prototype was a $0.4 \times 1.5 \times 9 \, \text{mm}^3$ bench with fiber, collimator and integrated $1.5 \times 1.5 \, \text{mm}^2$ fluid-filled micro-lenses which can be actuated by pneumatic pressure. A special bonding and mounting technique allows leak-proof assembly with such high precision that no optical impairment occurs. The device withstands pressures above 2×10^4 Pa before any leakage occurs.

The fabricated device is currently being tested extensively and further devices with additional components and functionalities are under construction.

REFERENCES

[1] W. Jung, D. T. McCormick, Y. C. Ahn, A. Sepehr, M. Brenner, B. Wong, N. C. Tien, and Z. Chen: *In vivo three-dimensional spectral domain endoscopic optical coherence tomography using a microelectromechanical system mirror.* Optics Letters, 32:3239–3241, 2007.

[2] J. Su, J. Zhanga, L. Yu, and Z. Chen: *In vivo three-dimensional microelectromechanical endoscopic swept source optical coherence tomography.* Optics Express, 15(16):10390–10396, 2007.

A FULLY INTEGRATED THERMO-PNEUMATIC TUNABLE MICROLENS

Wei Zhang, Khaled Aljasem, Hans Zappe, and Andreas Seifert
University of Freiburg, Germany

ABSTRACT

A thermo-pneumatically actuated tunable microlens has been developed as a completely integrated system with on-chip actuation and temperature sensing. The focal length of the microlens can be adjusted over a wide range without any external pressure controller. The module consists of a structured silicon chip, a spin-coated PDMS membrane, an optical fluidic chamber, a thermal cavity, and a heating and temperature sensing structure. The focal length is controlled by the applied voltage to the heater and could be tuned in the range between 3.3 and 18.2 mm, corresponding to a temperature variation of 37°C to 24°C, and a change in the numerical aperture from 0.303 to 0.055. At the same time, the cutoff frequency of the optical transfer function, referring to a contrast of 0.2, varies from 30 lines/mm at 27°C to 65 lines/mm at 34°C.

INTRODUCTION

Conventional optical lenses are made of solid material and have fixed focal lengths and lens shape; complex mechanical mechanisms are required to vary the optical focus. Adaptive fluidic lenses have been shown to be a suitable alternative for imaging objects which do not have a fixed position. Such structures make sensitive mechanical moving parts in optical imaging systems unnecessary. Thermal actuation has been shown to be useful for pneumatic MEMS structures, such as mirrors [1] and pumps. This paper introduces a thermo-pneumatically actuated tunable microlens with an aperture of 2 mm with fully integrated thermal actuation.

DESIGN AND FABRICATION

The lens consists of three main parts: a silicon lens chip, a PDMS layer and a heater chip including an integrated temperature sensor (Figure 1). The PDMS layer, which is formed by a casting process, is composed of a support ring and four thermal cavities. The support ring forms an inner volume which is filled with a dielectric optical liquid FC40 (3M™) with a high thermal expansion coefficient when compared to water or air. Thus, the liquid can be used as a suitable actuation medium and moreover, offers good optical properties [2]. In the current design, four identical thermal cavities with 2 mm diameter provide uniform heating of the optical liquid outside the lens cavity itself. This arrangement ensures a stable focal point during the heating process, since no undesired inhomogeneities of the refractive index occur due to thermal gradients. Due to the low expansion of the air in the heater cavities, the system is robust with respect to fluctuations of the applied voltage. Figure 2 shows a photo of such an integrated lens.

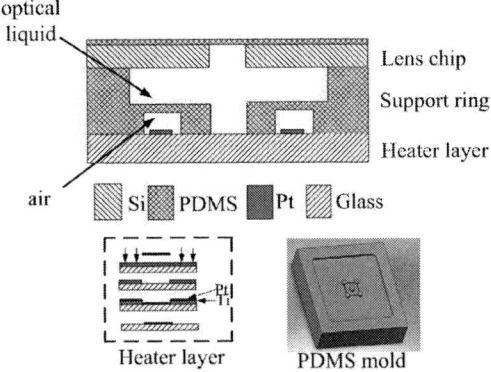

Figure 1: Process of the heater and schematic of the integrated system

The bottom glass chip seals one side of the PDMS layer to prevent leakage. An effective O2 plasma bonding is thus a critical step in the process chain. In addition, the glass supports the micro-hotplate structure as well as the micro temperature sensor as shown in Figure 3. Four heaters are symmetrically arranged on top of the glass chip and covered by the PDMS thermal cavities. A temperature sensor is integrated in one of the heating cavities, which are connected via thermo channels.

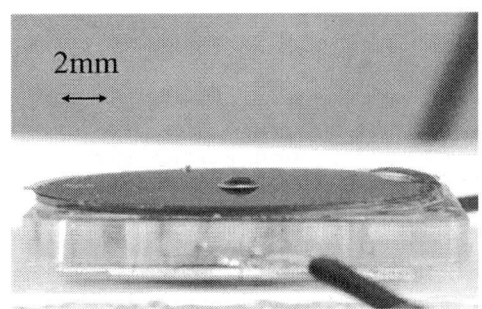

Figure 2: Photo of the integrated tunable thermo-microlens

The heater and sensor structure is made by an evaporation process with a 150 nm thick platinum coating. To improve the adhesion of the platinum on glass, a 20 nm thick titanium layer is pre-deposited. The resistance of the heater is 101 Ω, and the applied voltage ranges from 2.4 to 5 V. Hence the power consumption is only between about 60 and 250 mW for each of the four heaters.

MEASUREMENTS

Two hours of annealing at 250°C in an oven is required to obtain a stable behavior of the sensor and heater. As shown in Figure 4(a), there are three different kinds

978-1-4244-8926-8/10 $26.00 © 2010 IEEE

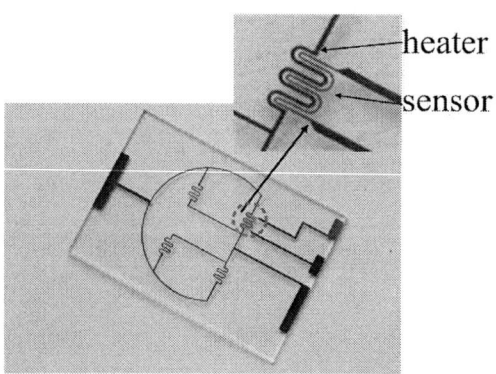

Figure 3: Structure of the glass layer with thermoelements

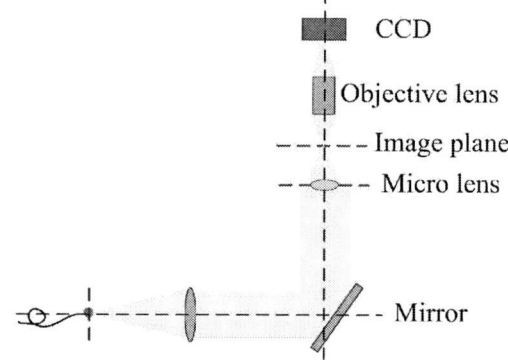

Figure 5: Optical measurement setup

of sensors to be tested. The resistance of each sensor drops rapidly during the first heating hour and saturates to a stable value. For the calibration of the temperature sensors, an oil bath method is used and a high sensitive commercial temperature sensor works as reference. As shown in Figure 4(b), the measurement result does not show any hysteresis during the heating and cooling process, and displays a linear increase of resistance between room temperature and 200°C.

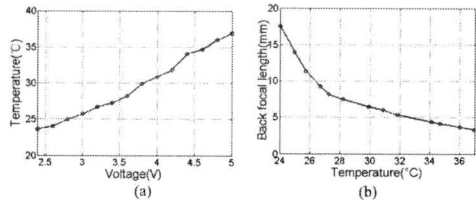

Figure 6: (a) Temperature dependence on voltage; (b) change of back focal length with temperature

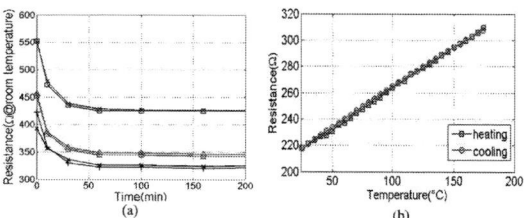

Figure 4: (a) Resistance of 3 different sensors at room temperature after different annealing time; (b) relationship between resistance and temperature

A modified microscope setup, which is shown in Figure 5, is used to measure the back focal length and the Modulation Transfer Function (MTF). Figure 6(a) shows the relationship between temperature and applied voltage, and Figure 6(b) demonstrates the influence of the temperature on the back focal length. The measurements were taken after a short delay for different constant voltages, in order to assure a stable equilibrium between the microlens and surrounding atmosphere. Figure 7 shows the behavior of the MTF at different temperature values. Due to different aspheric deformations with temperature, the optimal MTF curve appears at 34°C.

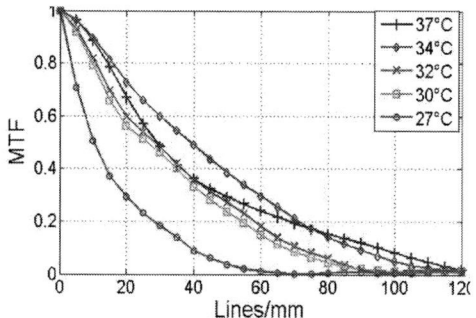

Figure 7: MTF curves for different temperatures

same time, the focal length can be controlled dynamically in a stable way using feedback from the integrated thermal sensor.

REFERENCES

[1] Werber, Armin and Hans Zappe: *Thermo-pneumatically actuated, membrane-based micro-mirror devices.* Journal of Micromechanics and Microengineering, 16(12):2524, 2006.

[2] Wang, Weisong and Ji Fang: *Design, fabrication and testing of a micromachined integrated tunable microlens.* Journal of Micromechanics and Microengineering, 16(7):1221, 2006.

SUMMARY AND OUTLOOK

We have fabricated and characterized a tunable microlens with integrated thermal structures for actuation and sensing. This lens achieves a wide range of variable focal lengths for low power consumption. Future work will focus on using easily expanding liquids instead of air in the thermal pump, and different optical liquids for optimizing the optical properties. At the

A TUNABLE LIQUID LENS WITH EXTENDED DEPTH OF FOCUS

[1,2]Jingran Kang, [1]Guangya Zhou*, [1]Hongbin Yu, [1]Fook Siong Chau, [2]Haiqing Chen

[1] Dept. of Mechanical Engineering, National University of Singapore, Singapore; [2]Huazhong Univerisity of Science and Techonolgy, P.R.China

Corresponding author: Guangya Zhou; E-mail address: mpezgy@nus.edu.sg

ABSTRACT

A liquid tunable lens with extended depth of focus (DOF) is presented. By integrating a rotational symmetric quartic function (QF) phase plate into the liquid lens' cavity, the lens can achieve higher tolerance to the defocus aberration. The liquid lens was fabricated through a convenient and low-cost process that combined single-point diamond turning (SPDT) with soft lithography using polydimethylsiloxane (PDMS). Experimental and theoretical results are in good agreement. Both focal length tunability and extended DOF have been demonstrated with the proposed liquid lens.

1. INTRODUCTION

Liquid lens is expected to play a significant role in minimizing the overall size of an optical zoom lens. In recent years, many novel liquid tunable lens prototypes are presented by researchers [1-3]. In this paper, we present a liquid tunable lens with extended depth of focus (DOF). With this novel design, a liquid lens can achieve both focal length tunability and an enhance DOF at the same time. This in combination with a simple fabrication process and potential mass production capability confer on the proposed lens wide applicability.

2. LIQUID LENS DESIGN AND FABRICATION

The proposed liquid lens is illustrated in Fig. 1. A liquid inlet and outlet are used to control the deformation of a PDMS membrane by changing the liquid injection volume. A quartic function (QF) phase profile [4] is integrated at the bottom of the cylindrical lens cavity. The phase profile is located within the effective aperture area, where the deformed membrane can be assumed to spherical.

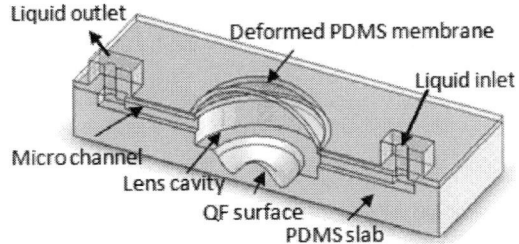

Fig. 1. Schematic of the hybrid tunable lens with a QF phase profile to extend the DOF

The depth of the liquid lens cavity with such a phase modulation structure can be expressed as:

$$d(r) = \begin{cases} d_0 + \dfrac{\alpha\lambda}{n_l - n_p}[(\dfrac{r}{R_1})^4 - (\dfrac{r}{R_1})^2], 0 \le r \le R_1 \\ d_0, R_1 < r \le R_2 \end{cases} \quad (1)$$

where λ is the incident wavelength, α represents the phase profile's modulate weight, d_0 is the depth of the lens cavity without the phase profile, n_l and n_p are the refractive indices of the liquid and PDMS, R_1 and R_2 denote the radiuses of the QF phase profile and the lens chamber, respectively. Parameters of our design are shown in Table 1.

Table 1 Parameters of the design

λ	α	R_1	R_2	d_0
632 nm	0.75 π	3.5 mm	5 mm	120 μm

Simulation results show that the proposed lens can achieve a much longer DOF than a conventional lens of the same aperture and focal length without significantly degrading the image resolution. Assuming the incident beam is collimated, the simulated intensity distributions of light spots along a transverse direction (TD) as a function of defocus along the axial direction (AD) are provided in Fig. 2. Compared with the conventional lens, it is clear that the DOF of the proposed lens is significant enhanced.

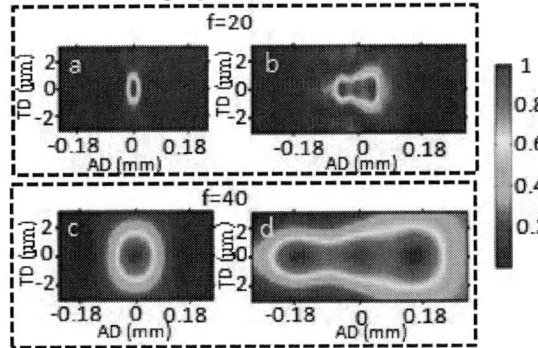

Fig. 2. Simulated light spot intensity distributions of a conventional liquid lens with focal length at (a) 20mm and (c) 40mm, and the proposed liquid lens with focal length at (b) 20mm and (d) 40mm

We developed a successful fabrication method for the proposed liquid lens, which combines single-point diamond turning (SPDT) with soft lithography processes [2,3]. The fabrication flow is summarized in Fig 3. A polymethylacrylate (PMMA) master mold for the lens device is fabricated by utilizing a four-axis SPDT machine. Then two cycles of soft lithography processes using PDMS are performed to get the desired lens structure. A PDMS membrane (about 70μm thick) made by spin coating PDMS on a clean silicon substrate is bonded to the PDMS slab by using the oxygen plasma machine. When the silicon wafer is detached, PDMS cubes for holding the inlet tube and outlet tube are bonded. Finally, the PDMS slab is bonded with the glass slide for easy handling.

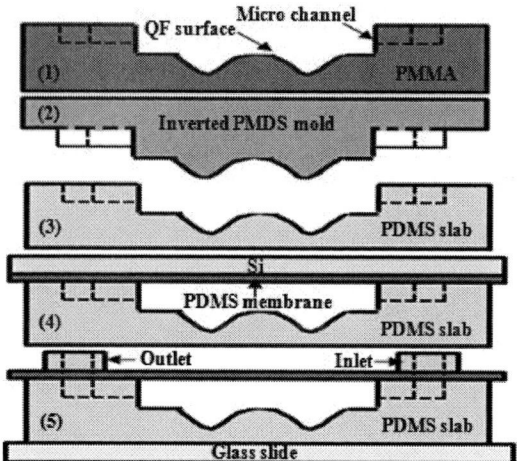

Fig. 3. Summary of the fabrication flow processes

3. RESULTS

The back focal length of the fabricated liquid lens is tested using collimated light from a red LED and results are depicted in Fig. 4. The focal length is tuned from 71.50 mm to 21.20 mm when injection volume is increased from 0.03 ml to 0.10 ml.

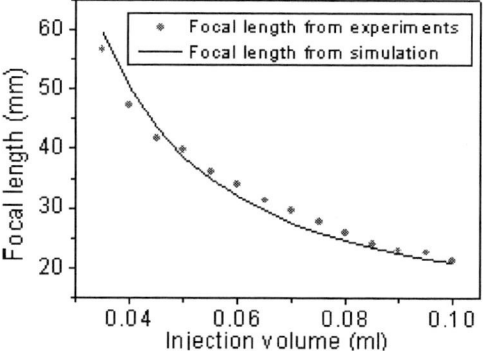

Fig. 4. Focal length versus injection volume

A setup schematically shown in Fig. 5 is used to demonstrate the extended DOF performance. A conventional liquid tunable lens having the same structural design and dimensions but with a flat bottom is also fabricated and tested. Object distance is fixed at 115 mm during the test. A CCD is fixed on a 2D translation stage. No other lens is placed in front of the CCD except the liquid lens under test. The focal lengths of both liquid lenses are tuned to 25.9mm. Images captured at 1 mm away from the best in-focus position are shown in Fig. 6. It can be

seen that the proposed liquid lens has a higher tolerance to defocus aberrations than the conventional one.

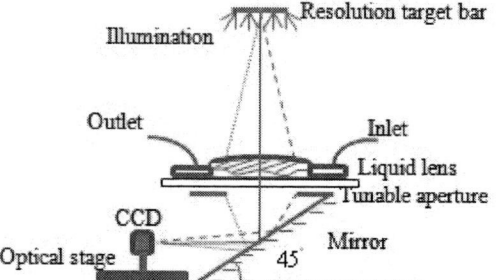

Fig. 5. Schematic of experimental setup to test imaging quality

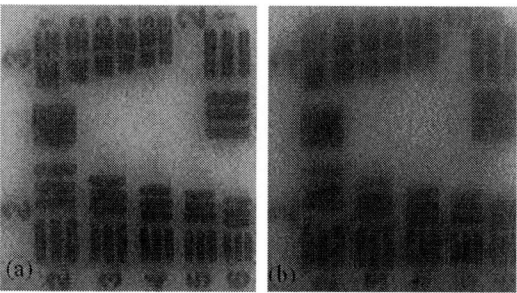

Fig. 6. Images captured at 1mm away from the best in-focus position for (a) the proposed liquid lens, and (b) the conventional liquid lens

4. CONCLUSION

A liquid tunable lens with extended DOF has been successfully fabricated with the process combined SPDT with soft lithography. It can be seen that both focal length tunability and an enhance DOF can be achieved from the experimental results.

ACKNOWLEDGEMENT

Financial support by the MOE research grant R-265-000-306-112 is gratefully acknowledged. The author also appreciates for the support of China Scholarship Council.

References:

[1] S. Reichelt, et al., "Design of spherically corrected, achromatic variable-focus liquid lenses," Opt. Exp., vol. 15, pp. 14146-14154, 2007.
[2] G. Zhou, et al., "Liquid tunable diffractive/refractive hybrid lens," Opt. Lett., vol. 34, pp. 2793-2795, Sep 2009.
[3] H. Yu, et al., "Lens with transformable-type and tunable-focal-length characteristics," IEEE Journal of Selected Topics in Quantum Electronics, vol. 15, pp. 1317-1322, 2009.
[4] S. Mezouari, et al., "Phase pupil functions for reduction of defocus and spherical aberrations," Opt. Lett., vol. 28, no. 10, pp. 771-773, 2003.

MICROMACHINED TWO DIMENSIONAL LENS SCANNER WITH LARGE APERTURE BEAM

Hyeon-Cheol Park, Cheol Song and Ki-Hun Jeong

Korea Advanced Institute of Science and Technology (KAIST), Korea

ABSTRACT

This work present a novel approach for miniaturized optical scanning using the two-axis MEMS lens scanning system with large beam diameter of 0.6mm. A millimeter aspheric glass lenses are integrated on electrostatic MEMS actuators to fold optical path. By integrating optical components perpendicularly on MEMS actuators, more compact device size along the optical axis is accomplished within 2mm. The scanning angle of 4.6° and 5.3° at the scanning speed of 276.5Hz and 294.4Hz for x-scanning and y-scanning respectively are achieved, when actuated by only DC 5V and AC peak to peak 10V biased resonance excitation.

INTRODUCTION

Compactness of the optical scanning system is very essential in developing *in-vivo* endoscopes or handheld projection devices. Current state-of-the-art MEMS techniques have been intensively studied for miniaturizing optical scanning system. However, the previous work requires relatively larger space than the effective clear aperture of the device and precise optical alignment during integration. Besides, conventional scanners are limited to have high NA, since they are designed to have all the optical and mechanical components with in-plane configurations. This work presents a new approach for miniaturized optical scanning using the two-axis MEMS lens scanning module. The system includes a fiber collimator (Light path, diameter: 1.25mm, beam diameter: 0.6mm) and aspheric glass lenses (Alps Electronics, diameter: 1mm, f = 0.5mm, clear aperture: 0.6mm) integrated on scanners using silicon lens holder. The two-axis lens scanning is achieved by implementing orthogonally moving two glass lenses on electrostatic MEMS actuators comprising a lateral comb drive for x-scanning and a vertical comb drive for y-scanning. The device size along the optical axis is only determined by the lens diameter and thickness of the substrate, since the glass lenses are just mounted on MEMS scanners perpendicularly. The optimized combinations of MEMS scanners and the lens can maximize the clear aperture of the device without any increment of unnecessary space so that can provide a capability for higher NA.

OPTICAL LAYOUT OF THE LENS SCANNER

Fig. 1 illustrates the optical layout of the lens scanning system. A light delivered from an optical fiber is coupled with a collimating lens and focused by an aspheric glass lens for x-scanning and re-collimated by an another lens for y-scanning. The light tilted by orthogonal lens scanning with electrostatic comb drives, eventually refocused on the image plane by an objective lens. The distance between scanning lenses s is set to be $2f$ to obtain the re-collimation where f is the focal length of the scanning lens. The objective lens is positioned at the sum of focal lengths f and f_0 behind the second lens. Then, the scanning angle θ_s and the scanning length at the image plane y_0 of the system can be determined by $-u/f$ and $-(f_0/f)u$, where u is the small displacement of the scanning lens.

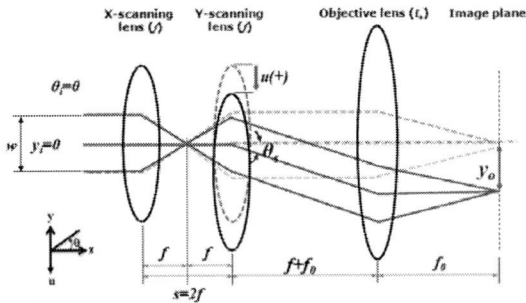

Fig. 1. Optical layout of lens based two-dimensional scanning module consisting of a collimating lens, x and y scanning lens and an objective lens.

MICROFABRICATION PROCESS

On-chip integration of vertical comb drive as well as in-plane comb drive has been achieved by employing a standard SOI (Silicon-On-Insulator) process. Firstly, microstructures for the scanners as well as lens holders and electrodes are defined at the top silicon layer by DRIE process using an etch mask of a negative photoresist. Next, the bottom silicon is also etched with masking layers of thick photoresist to provide spaces for lens moving. Then, the device is released by removing the buried oxide layer with vapor phased hydrofluoric acid (HF) to prevent the sticking problem of microstructures due to the surface tension during the evaporation of the rinsing liquid [1]. Fig. 2(a) shows the SEM picture of the microfabricated device.

After the microfabrication process, each chip is separated from the wafer by disconnecting silicon tether structures by thermal stresses of Joule heating [2]. Finally, integrating optical components using UV-curable epoxy resins, the two-dimensional lens scanner is completed. Fig. 2(b) shows the perspective view of the final device. The whole system is built in a dimension of 2 x 2.7 x 5.5mm^3.

978-1-4244-8926-8/10 $26.00 © 2010 IEEE

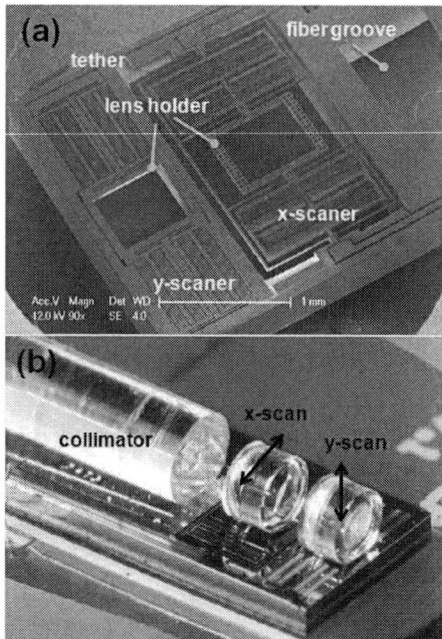

Fig. 2. (a) SEM image of the fabricated device. MEMS actuators, lens holders and fiber grooves and tether structures to separate device from the wafer are integrated on a chip (b) Perspective view of the 2D lens scanner. Fiber collimator, scanning lenses and are integrated on a chip within a dimension of 2 x 2.7 x 5.5 mm³.

DEVICE CHARACTERIZATION

The resonance frequency is 276.5Hz and 204.4Hz for lateral and vertical motion respectively. These are the result of the resonance frequency reduction by the added mass of the lens. Due to the heavy mass of the glass lens, 2mg which is about 70 times heavier than device mass, the resonance frequency after the lens integration is found to be 87% lower than before the lens integration [3]. Noticeably, while resonance frequencies are decreased, quality factor (Q-factor) is increased proportional to the reduction factor. Measured Q-factor is increased from 35 to 276.5 after lens integration. The high Q-factor induces that high confinement of energy at the resonance frequency thus reduces the actuation voltage. Under the resonance excitation, scanning amplitude of 40µm and 30µm for lateral and vertical motion respectively, corresponding to the scanning angle of 4.6° and 3.4°, is achieved with only DC 5V and AC 10V actuation voltage.

The optical scanning of the 2D MEMS lens scanner has been demonstrated. In the experiments, additional objective lens, which has NA of 0.25 and focal length of 20mm (Edmund Optics, NT46-403), and HeNe laser of 633nm is used for visualizing the beam scanning. The light source is directly coupled with scanning lenses instead of using fiber collimator, since the commercialized collimator is tuned at 1310nm. The scanning path is captured by CCD camera positioned at image plane. Fig. 3 (a) - (d) shows some scanning patterns produced from the

MEMS lens scanner. The initial beam spot size is measured to 17µm (calculated diffraction limit of 11µm). The lateral and vertical scan length is 800µm and 600µm, respectively.

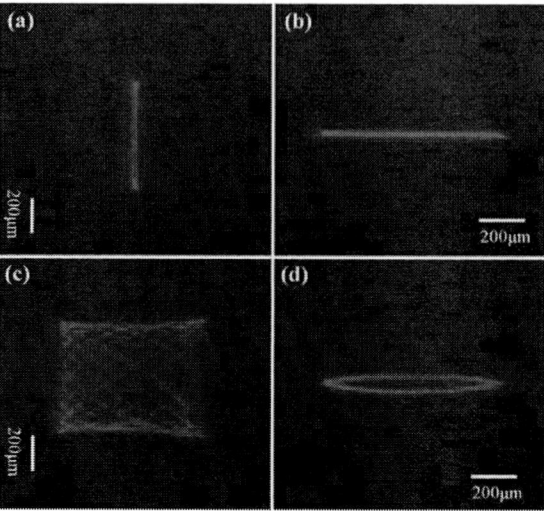

Fig. 3. Two-Dimensional optical scanning demonstration; (a) y-scanning, (b) x-scanning, (c) 2D Lissajous pattern. 800µm x 600µm scanning area is achieved under the resonance excitation. (d) Elliptical scanning pattern excited at same frequency. Since vertical motion is not excited at resonance, y-scanning is much smaller than x-scanning.

SUMMARY AND CONCLUSION

In summary, two dimensional optical scanning has been successfully demonstrated by implementing orthogonally moving two commercialized glass lenses on electrostatically actuated MEMS scanners at low operating voltages. The maximum scan angle is 4.6° in x-scanning and 3.4° in y-scanning. The device is integrated on a single chip with a dimension of 2 x 2.7 x 5.5 mm³ with the beam diameter of 0.6mm. The device can provide a new direction for miniaturizing laser scanning based endoscopes or handheld projectors.

ACKOWLEDGEMENT

This work was supported by the national research foundation (MEST) (No. 2010-0000210) and the Industrial Strategic Technology Program of the Ministry of Knowledge Economy (KI001889), grant funded by the Korea government.

REFERENCE

[1] Y. Fukuta, H. Fujita and H. Toshiyoshi, "Vapor Hydrofluoric Acid Sacrificial Release Technique for Micro Electro Mechanical Systems Using Labware", Jpn. J. Appl. Phys. **42**, 3690-3694 (2003)

[2] Y-S. Chiu, K-S. Chang, R. W. Johnstone and M parameswaran, "Fuse-tethers in MEMS", J. Micromech. Microeng. **16**, 480-486 (2006)

[3] S. S. Rao, *Mechanical Vibrations*, Reading: Addison-Wesley, 1990

Large-size Infrared Reflow Microlens Based on Stacked layers

Takuro Aonuma, Shinya Kumagai, and Minoru Sasaki,

Dept. of Advanced Science and Technology, Toyota Technological Institute,
Hisakata 2-12-1, Tenpaku-ku, Nagoya 468-8511, Japan
E-mail: mnr-sasaki@toyota-ti.ac.jp

Abstract

Large-size (~mm) reflow microlens is fabricated combining with UV curing process. Underlying stacked resist layers can assist the material movement of the photoresist in the reflow process generating the convex parabola shape. Stacking thickness shows the clear linearity. Infrared imaging performance is confirmed using 1.5mm-diameter microlens.

Keywords: Infrared microlens, Stacked layer, UV-curing, Large-size

1. INTRODUCTION

Si is frequently used as package or window material for infrared sensors. Microlens is sometime combined for gathering infrared. When MEMS thermopile sensor is used, lens size of ~1 mm will be useful for increasing the effective fill factor, since the sensing area is at the center of the diaphragm and the sensor chip size is ~1mm. Preparing such large microlens is sometimes technical. Resist reflow is well-known technique for realizing the rounded refractive microlens. The lens shape is basically determined by the surface tension of the melted resist. The obtained resist structure can be used as the dry etching mask. However, the material movement inside the film does not always show the convex shape due to the limited effect of surface tension. Figure 1 shows our example. The initial photoresist is 10 μm-thick disk (TMMR PW-1000PM, Tokyo Ohka Kogyo). The reflow condition is at 180°C for 3h showing the saturation of the material movement. The concave shape is obtained instead of the convex one. Some patterning methods for realizing 3D resist structure are reported including gray scale mask [1] and multiple-patterning [2].

In this study, UV curing is newly introduced for realizing large-size reflow microlens. The appropriate underlying layered structure is considered to assist the generation of convex shape. UV cured resist can be used for the base structure which is stable against the additional resist process.

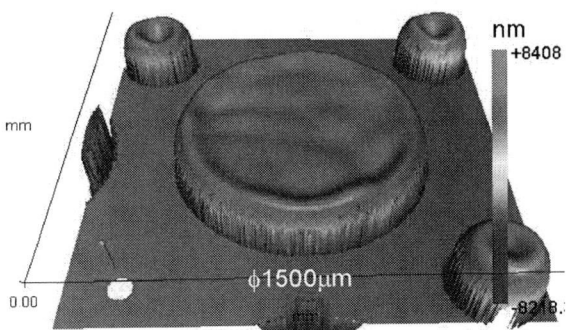

Fig. 1: Reflowed 10 μm-thick resist film showing the termination of the material movement.

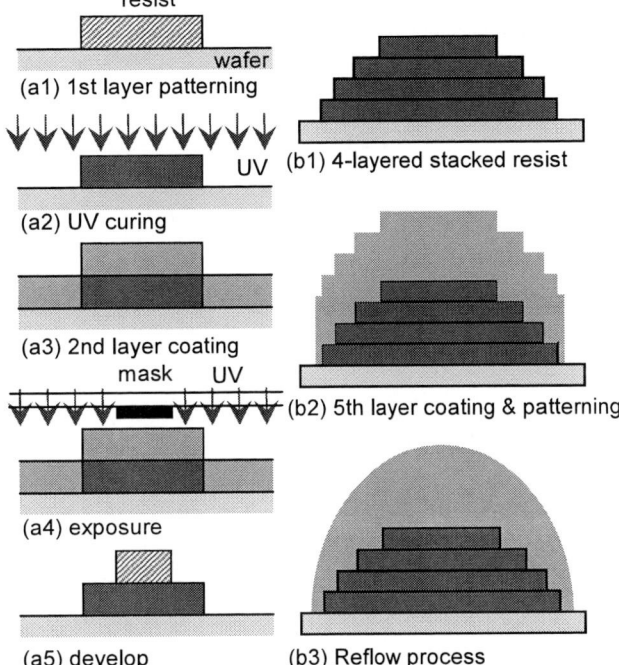

Fig. 2: (a) Process sequence for adding one-layer stack. (b) Process sequence for generating the rounded lens profile reflowing 10 μm-thick layer.

2. PRINCIPLE

Figure 2(a) shows the schematic drawing of the processing for adding one-layer stack. UV curing is the treatment realized by heating the photoresist under the vacuum and UV irradiation. Bridging molecules, the photoresist becomes the larger molecule. UV cured photoresist changes to be stable against the thinner, and additional spin-coating of photoresist. Patterning is possible keeping the underlying structure. Figure 2(b) shows the stacked layers. Disks can be obtained repeating the process shown in Fig. 2(a). The profile can be designed to be similar to the parabola shape. The final step is covering the layered

978-1-4244-8926-8/10 $26.00 © 2010 IEEE

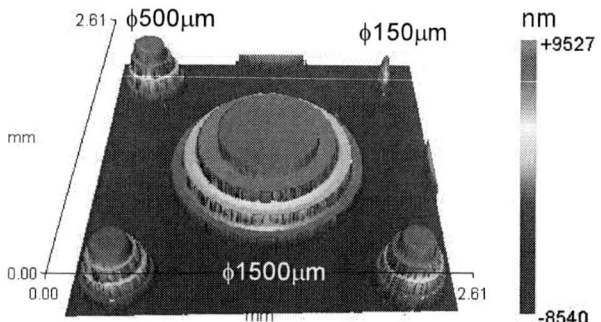

Fig. 3: Stacked resist film using UV curing. Structures with 3 different sizes are fabricated at the same time.

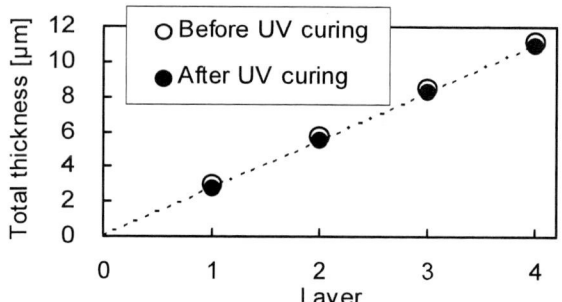

Fig. 4: Total thickness of the stacked resist structure as a function of number of layers.

structure with the thicker resist and reflowing resist making the profile rounded.

3. EXPERIMENTS

Underlying stacked resist is AZ1500, 38cp. UV cure is at 100°C, ~3kPa, 40min. The light (main wavelength 240-380 nm, ~1000mW/cm^2) of mercury lamp is irradiated through the quartz window. Figure 3 shows the structure after 4 layers are stacked. Figure 4 shows the total thickness of the layered structure as a function of the number of layers. Each layer is 2.75+/−0.05μm thick after UV curing, which generates the thickness shrinkage by ~0.2μm. Final 5th layer shown in Fig. 2(b2) is TMMR PW-1000PM. This layer is first heated at 100-140°C for 30 min for removing the gas inside, and then heated up to 160-180°C for 3h, higher than the glass transition temperature.

4. RESULTS

Figure 5 shows structures after the reflow process. The lens diameters are (a)500 and (b)1500 μm. The cross-section profile is similar to the parabola curve having R^2 value of 0.97 and 0.93 against the approximated curve obtained using the least square root method.

Imaging function against infrared is examined. Light source is the thermal lamp illuminating the objective of metal mask having holes as shown in Fig. 6(a). The transmitted infrared light passes through the resist microlens

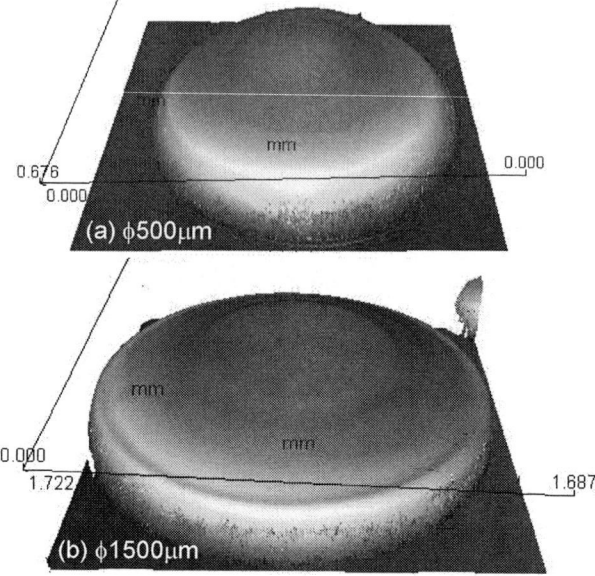

Fig. 5: Reflowed 10 μm-thick resist film covered on the underlying stacked film structure.

Table 1: Data of rounded resist profile after the reflow.

Diameter [um]	150	500	1500
Radius of curvature [mm]	0.28	3.08	26.43
R^2 value against approximated parabola	0.997	0.972	0.934

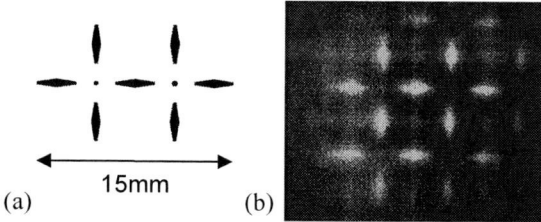

Fig. 6: Thermograph IR image generated through 1.5mm-diameter microlens.

making the image as shown in Fig. 6(b). The focus length is measured to be 20.5mm, which reasonably agrees with the calculation of 24.3mm based on the radius of curvature.

This research was supported by a MEXT program for forming strategic research infrastructure for private Universities from 2008.

REFERENCES

[1] K. Totsu, K. Fujishiro, S. Tanaka, M. Esashi, Sensors and Actuators A, Vol. 130, p.387 (2006).

[2] Nikon Corporation, Japanese patent 2005-72152.

OPTICAL WAVEGUIDE DEVICES FOR BIOANALYSIS

James S Wilkinson
University of Southampton, UK

ABSTRACT

Integrated optical waveguides offer great potential as versatile platforms for constructing advanced biosensors, optical cell-sorters and integrated optofluidic systems, exploiting the technological approaches of microelectronics and guided-wave optics to realise low-cost on-chip systems. Progress towards optical integration in microsystems for bioanalysis will be discussed, with examples in key applications, and challenges and opportunities will be described.

INTRODUCTION

Optical waveguides are ideal structures for the integration of advanced optical functions in microsystems. Waveguides are normally formed by deposition and structuring of a thin dielectric film on the surface of a substrate and the waveguide power is confined near the surface yielding a strong controllable interaction. Microfabrication of optical waveguides enables low-cost mass-production of compact, robust, multianalyte chemical sensor chips. The fabrication techniques which revolutionised electronics, making possible hugely complex microelectronic systems at very low cost, are enabling a similar transformation in optical devices. This renders optical circuits particularly well suited to mass-produced bio/chemical sensor arrays exploiting surface chemistry, optical cell-sorters which discriminate on the basis of optical properties and for integration in microfluidic systems for advanced micro-cytometry. These planar optical chips are compatible with microfluidic systems for sample delivery and with optical fibre for solid-state connection to instrumentation.

OPTICAL WAVEGUIDE BIOSENSORS

In addition to low cost and the potential for a high level of integration, optimised monomode waveguides offer ultimate sensitivity, ultimate stability and the smallest sample volume, due to strong, localised evanescent interaction of light with liquid samples. They are ideal for excitation and interrogation of molecules at surfaces, being inherently surface sensitive devices [1].

Several highly sensitive and specific waveguide biosensors have been successfully demonstrated, and examples of biosensors based on SPR and on fluorescence will be described. The waveguide SPR immunosensors, which demonstrated detection limits for pesticides below 0.05 ppb [2], employed phase-matched excitation of propagating plasmons at a gold film surface from an underlying dielectric waveguide by tunnelling through the gold film. Biochemical specificity was achieved by modifying the gold surface with a molecule complementary to the target. Combined electrochemical and optical sensing, exploiting the gold SPR film as an electrode, has also been demonstrated [3].

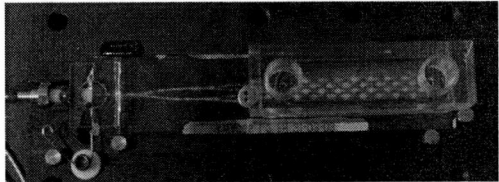

Fig. 1 Fluorescence biosensor array chip

For applications requiring lower detection limits, fluorescent labelling has been employed with fluorescence multisensor chips to yield a detection limit of 1ppt for oestrone, for example [4-6].

OPTICAL TRAPPING AND PROPULSION

Optical tweezers are well-established as a tool for non-contact, non-destructive handling of biological materials. Recently, interest has grown in optical manipulation at surfaces [7] as part of the toolbox of the "lab-on-a-chip". In particular, advances have been made in trapping and propulsion of dielectric microparticles and biological cells in the evanescent fields of optical waveguides [8], which may form part of a planar microsystem into which optical detection and spectroscopy of separated species could also be integrated.

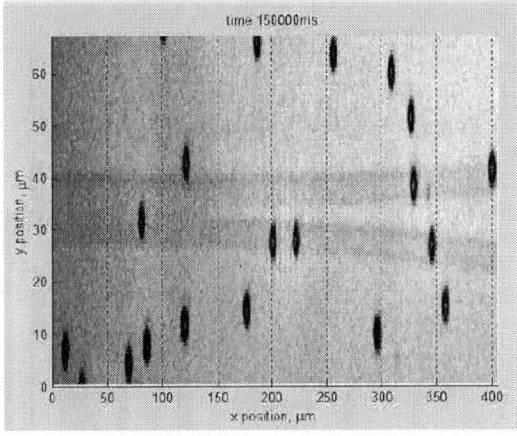

Fig. 2 Particle sorting on Y-branched waveguide

Optical waveguides embedded in surfaces represent a powerful means of controlling the distribution of optical intensity and intensity gradient at such surfaces, for particle control, and recent results on the manipulation of biological cells will be described [9].

INTEGRATED OPTOFLUIDICS

Evanescent fields supported by optical waveguides are ideal probes for surface interactions. However, in applications where the volume of a liquid or of a micron-scale object such as a biological cell is to be interrogated, in-plane optofluidic approaches must be adopted [10,11]. Optical waveguide devices must be integrated with microfluidic channels to excite a defined volume in the fluid and to collect the scattered spectrum, in terms of angle or wavelength.

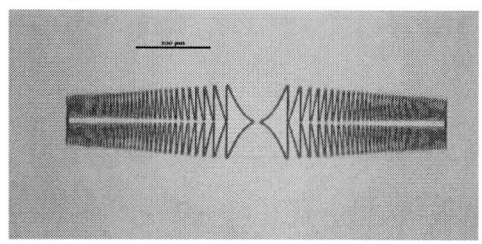

Fig. 3 Integrated kinoform lens

A key example function is that of the in-plane lens, which can be used for trapping, scattering, fluorescence or Raman measurements. Kinoform lenses are promising candidates for in-plane lenses, due to their compactness and flexibility in design [12]. Advances in technologies incorporating integrated lenses for microcytometry will be described.

CONCLUSIONS

The lab-on-a-chip presents great benefits in terms of reagent and sample consumption, speed, precision, and automation of analysis, and thus cost and ease of use, resulting in growing adoption of microfluidic approaches for practical measurements, and the potential for their widespread use in society. Optical techniques are ubiquitous in bio/chemical analysis and in the manipulation of biological cells, and optical waveguide devices show promise for on-chip integration with microfluidic systems for ultra-high performance, if low-cost approaches to fabrication and deployment can be achieved.

REFERENCES

[1] J.E. Midwinter "On the use of optical waveguide techniques for internal reflection spectroscopy" *IEEE Journal of Quantum Electronics,* 7, 339 ,1971.

[2] R.D.Harris, B.J.Luff, J.S.Wilkinson, J.Piehler, A.Brecht, G.Gauglitz & R.A.Abuknesha, "Integrated Optical Surface Plasmon Resonance Immunoprobe for Simazine Detection", *Biosensors & Bioelectronics*, 14, 377 ,1999.

[3] A.K. Sheridan, P. Ngamukot, P.N. Bartlett & J.S. Wilkinson, "Waveguide surface plasmon resonance sensing: electrochemical desorption of alkane thiol monolayers", *Sensors & Actuators B*, 117, 253-260, 2006.

[4] P. Hua, J.P. Hole, J.S. Wilkinson, G. Proll, J. Tschmelak, G. Gauglitz, M.A. Jackson, R. Nudd, H.M.T. Griffith, R.A. Abuknesha, J. Kaiser & P. Kraemmer, "Integrated optical fluorescence multisensor for water pollution", *Optics Express*, 13, 1124 ,2005.

[5] J. Tschmelak et al., "Automated water analyser computer supported system (AWACSS) Part I: Project objectives, basic technology, immunoassay development, software design and networking", *Biosensors & Bioelectronics*, 20, 1499-1508, 2005.

[6] J. Tschmelak et al., "Automated water analyser computer supported system (AWACSS) Part II: Intelligent, remote-controlled, cost-effective, on-line, water-monitoring measurement system", *Biosensors & Bioelectronics*, 20, 1509-1519, 2005.

[7] K. Dholakia and P. Reece, "Optical micromanipulation takes hold", *Nanotoday* 1, 18-27, 2006.

[8] S. Kawata and T. Tani, "Optically-driven Mie particles in an evanescent field along a channeled waveguide", *Optics Letters*, 21, 1768-1770, 1996.

[9] B.S. Ahluwalia, O.G. Hellesø, A.Z. Subramanian et al., "Integrated platform based on high refractive index contrast waveguide for optical guiding and sorting", *Proc. Photonics West*, San Francisco, CA USA, Jan. 23-28, 2010, 76130R.

[10] T.N. Buican, M.J. Smyth, H.A. Crissman, G.C. Salzman, C.C. Stewart, and J.C. Martin JC "Automated single-cell manipulation and sorting by light trapping", *Appl. Opt.*, 26, 5311-5316, 1987.

[11] K.B. Mogensen, N.J. Petersen, J. Hübner, and J.P. Kutter "Monolithic integration of optical waveguides for absorbance detection in microfabricated electrophoresis devices", *Electrophoresis* 22, 3930-3938, 2001.

[12] H.C.Hunt and J.S.Wilkinson, "Integrated lenses for microfluidic systems", *Proc. 15th Micro-optics Conference MOC'09*, Odaiba, Japan, Oct. 25-28, 2009, J14.

Submicron silicon waveguide Mach-Zehnder interferometer using micro electro-mechanical phase-shifter

T. Ikeda, Y. Kanamori, and K. Hane

Tohoku University, Sendai, Japan

ABSTRACT

Mach-Zehnder interferometer (MZI) using submicron silicon waveguide with micro electro-mechanical phase-shifter is studied. The phase-shifter consists of freestanding submicron-wide silicon waveguide with two waveguide couplers and an ultra-small silicon comb-drive actuator. The freestanding waveguide is moved by the actuator to change the optical path length. The optical phase-shift was calculated theoretically by changing the waveguide cross-section. The phase-shift caused by actuation was measured to be 3.0π-4.5π at the actuator displacement of 1.0μm at the voltage of 31-32V. The maximum contrast of the MZI signal was -10.8dB.

INTRODUCTION

Submicron silicon waveguides are promising for dense integration of optical circuits with silicon electronics in the fields of optical telecommunications and interconnections [1]. The optical-phase controlling devices such as interferometric switches and filters were studied. Although time response is slower than electronic phase-shifter [2], phase-shifter based on micro electro-mechanical systems [3] is also promising for a large phase-shift with low power consumption.

In this paper, the MZI using submicron silicon waveguide with micro electro-mechanical phase-shifter is studied to investigate the interferometric performance. Applying the voltage of 32V, the phase-shift of 4.5π with the contrasts of -10.8dB is obtained in the device size of 50μm x 85μm.

PRINCIPLE AND DESIGN

Figure 1 shows the schematic diagram of the proposed MZI using micro electro-mechanical phase-shifter, which consists of a fixed waveguide, a movable waveguide and a micro electro-mechanical actuator. The movable waveguide includes waveguide couplers at two ends. The movable waveguide is connected to the actuator with low optical loss suspension bridges and suspended in air by the springs of actuator. The fixed waveguide is also suspended in air by low optical loss suspension bridges. The waveguide is designed to be single mode. Two types of the waveguides with different cross-sections are designed, which are the 300nm wide and 260nm thick and the 320nm wide and 340nm thick, respectively. The air surrounding waveguides works as clad of waveguide. The fixed waveguide and the movable waveguide are parallel to each other with 250nm wide air-gap. The coupling regions are 7μm and 10μm long for 300nm wide and 260nm thick waveguide and 320nm wide and 340nm thick waveguide, respectively. The movable waveguide is designed to move parallel to the fixed waveguide by the displacement of 1μm at the voltage of 30V while keeping the air gap and the coupling length. The optical phase-shift is caused by the change in optical path length. The optical path length is determined by the effective refractive index of waveguide depending on the waveguide cross-section.

Figure 2 shows the calculated optical phase-shift ϕ as a function of waveguide cross-section in TE-00 mode at the wavelength of 1.55μm when the actuator moves 1μm. Beam propagation method was used for the calculations. When the waveguide cross-section becomes smaller, the phase-shift decreases.

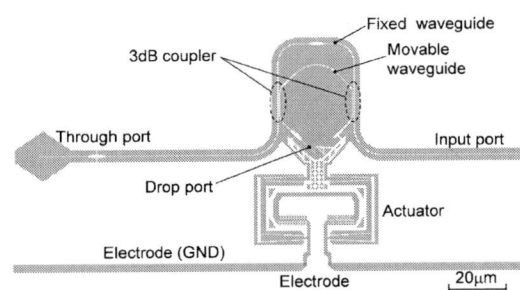

Fig.1 Design of the proposed MZI using micro electro-mechanical phase-shifter

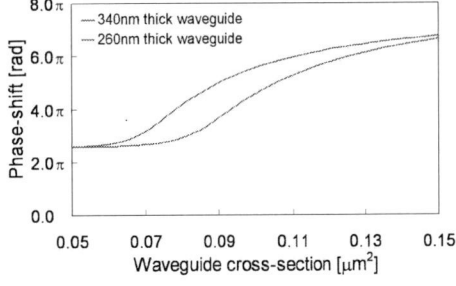

Fig.2 Optical phase-shift by 1μm displacement as a function of waveguide cross section

978-1-4244-8926-8/10 $26.00 © 2010 IEEE

FABRICATION AND MEASUREMENT SETUP

Two SOI wafers were used for the fabrication. Top silicon layers were 260nm and 340nm thick, which equaled to the waveguide thicknesses, the buried oxide layer was 2µm thick on 625µm thick silicon substrate. First, an electron-beam patterning machine (JEOL JBX-5000LS) was used for device patterning on the positive resist polymer (ZEON ZEP-520A). Using the resist polymer as mask, the top silicon layer was etched by fast atom beam (Ebara FAB-60ML) of an SF_6 gas. Next, the resist polymer was removed by O_2 plasma ashing. After the lithographic processes, the SOI wafer was cleaved to obtain a facet of input waveguide for coupling light. Finally, the SiO_2 layer was etched by hydrofluoric acid vapor to obtain the freestanding structure of MZI.

In the measurement of the fabricated MZI, a tunable infrared laser (Agilent 81682A) was used as light source at the wavelength around 1.5µm. The light intensities at the through and drop ports were measured from the spot images of scattered light at the ends of waveguides using an IR camera (Goodrich SU320KTS-1.7RT). The ends of the waveguides were rounded to measure the scattered light efficiently.

RESULTS AND DISCUSSION

Figure 3 (a) shows a SEM image of the whole view of the fabricated MZI. The waveguides and the actuator are fabricated with high accuracy. From the SEM observation, the designed cross-sections of the 300nm wide and 260nm thick waveguide and 320nm wide and 340nm thick waveguide were measured to be 245nm wide and 260nm thick and 290nm wide and 340nm thick, respectively. Figure 3 (b) shows the light emission from the through and drop ports of the MZI for the 245nm wide and 260nm thick waveguide. The light wavelength is 1.459µm and the voltage applied to the actuator is 15V. As seen in Fig.3 (b), light is mostly emitted from the through port under the condition.

Figures 4 and 5 show the light intensities measured as a function of the displacement of actuator for the MZIs with the 245nm wide and 260nm thick waveguide and the 290nm wide and 340nm thick waveguide, respectively. From the light intensity modulation, the phase-shifts are 3.0π and 4.5π for the MZIs, respectively, at the actuator displacement of 1µm. Measured values almost agree with the calculated results. The maximum contrast of the output signals are -3.9dB and -10.8dB for MZIs with the 245nm wide and 260nm thick waveguide and the 290nm wide and 340nm thick waveguide, respectively.

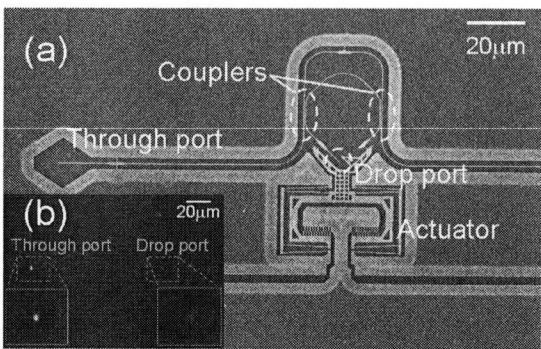

Fig. 3 (a) SEM image of the fabricated device and (b) IR image of scattered light.

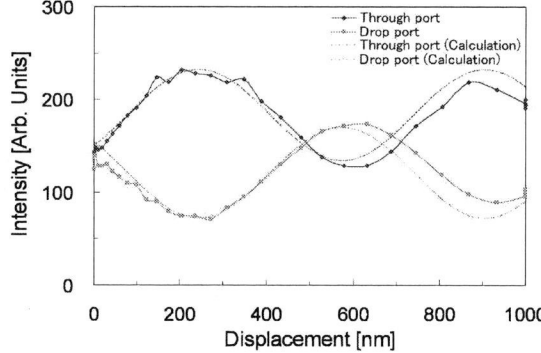

Fig. 4 Light intensities measured as a function of the actuator displacement for MZI with 245nm wide and 260nm thick waveguide

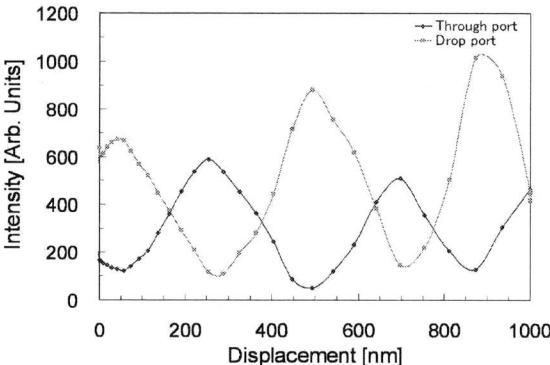

Fig. 5 Light intensities measured as a function of the actuator displacement for MZI with 290nm wide and 340nm thick waveguide

REFERENCES

[1] B. Jalali, et. al., J. Lightwave. Thchnol. 24 (2006), 4600.

[2] W. M. J. Green, et. al., Opt. Exp. 15 (2007), 17106.

[3] T. Ikeda, et. al. Opt. Exp. 18 (2010), 7035.

MEMS-ACTUATED WAVEGUIDE PHASE MODULATORS

Chun-Che Chang[1], Wei-Chao Chiu[1], Jiun-Ming Wu[1], and Ming-Chang M. Lee[1]
Jia-Min Shieh[2]

[1]Institute of Photonics Technologies, National Tsing Hua University, Taiwan R.O.C.
[2]National Nano Device Laboratories, Taiwan R.O.C.

ABSTRACT

An optical phase modulator for integrated optics was proposed by monolithic integration of deformable silicon photonic wires and MEMS actuators. Via applying voltage, the silicon wire is deformed and the wire length is extended, resulting in change of optical phase. A phase shift of 0.18 π is achieved at a bias of 35 V for a 150 μm wire according to numerical simulation. By cascading many of the MEMS-actuated wires, a π-phase shift can be realized at very low operation voltage. Based on this idea, a MEMS-actuated Mach-Zehnder interferometer was demonstrated.

INTRODUCTION

Silicon planar lightwave circuits (PLC) fabricated on silicon-on-insulator (SOI) permits high-dense integration of various optical devices on a single chip. Moreover, integration of microelectromechanical systems (MEMS) and PLC is an intriguing idea to realize many tunable optical functions. Prominent examples include reconfigurable optical add/drop multiplexers [1] via tunable optical coupling by MEMS actuators, optical phase modulators with controllable optical path difference by micromirror actuation [2], and optical switches driven by a comb actuator [3]. Among these tunable devices, phase modulators are the key element and are widely applied. Although other mechanisms such as thermo-optics [4] and free-carrier injection [5] are commonly utilized in varying optical phase, power consumption is usually an issue. In this paper, we propose a tunable optical phase shifter though deforming a silicon wire waveguide by MEMS actuators. By incorporating the Mach-Zehnder interferometer (MZI) configuration, the phase modulation can be converted to the amplitude modulation. Unlike other designs, our proposed device is very compact and could be integrated in a large-scale PLC.

DEVICE DESIGN AND FABRICATION

Fig. 1 (a) shows the schematics of a Mach-Zehnder interferometer monolithically integrated with MEMS-actuated waveguide phase modulators. Two 3-dB multimode interference (MMI) couplers are utilized to split and combine input wave. The relative phase difference between two arms can be controlled by deforming the suspended wires via the MEMS actuator as shown in fig. 1 (b), yielding variable optical intensity at the output end. While the voltage is applied to the electrode and the wire is

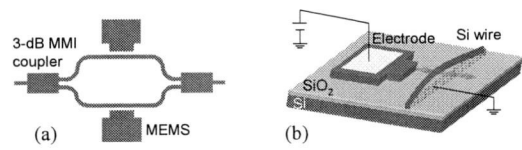

Fig. 1. Schematics of MEMS waveguide phase modulators: (a) Mach-Zehnder interferometer configuration, and (b) a MEMS actuator.

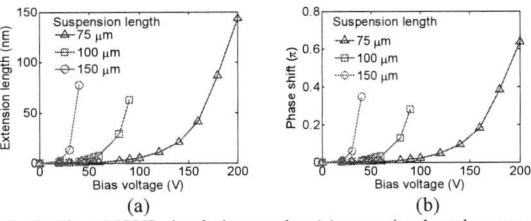

Fig. 2. The ANSYS simulation results: (a) extension lengths versus bias voltages and (b) optical phase shift versus bias voltages.

electrically ground, the wire is pulled toward the electrode through electrostatic force. The optical wave propagating along this deformable wire could experience a phase change due to physically wire length extension. For the optical wavelength of 1550 nm, a π-phase shift requires only 224.6 nm wire length elongation. To avoid the pull-in of wires and electrodes, the wire-electrode gap spacing was design to be 6 μm, and the width of electrodes is half of the suspension length of wires. To quantitatively analyze the wire deformation, a commercial finite element simulator (ANSYS) was used to calculate the corresponding phase shift. A phase shift of 0.18 π is attained as the suspension lengths are set to be 75 μm, 100 μm and 150 μm as the bias voltages are 150 V, 75 V and 35 V, respectively. Detailed simulation results are shown in Fig. 2. If multiple MEMS-actuated suspended wires are cascaded in series, a π-phase shift can be achieved with the bias voltage lower than 35 V and a proper choice of the suspension length.

To fabricate the device, the electrodes and the waveguides of 5 μm high were first patterned on SOI by a conventional dry etching process. Next, wet thermal oxidation was applied to reduce surface roughness and further shrink the waveguide width to 0.2 μm. Then a metal Au was deposited on the electrodes defined via photolithography. Finally the

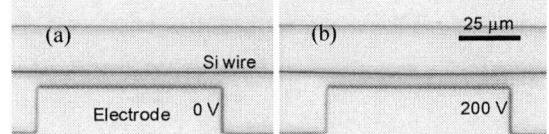

Fig. 3. Optical micrographs of MEMS-actuated deformable Si wire: (a) before deformation, and (b) after deformation.

silicon wires were released through buffered oxide etching (BOE). Fig. 3 displays the optical micrograph of a 150 μm long suspending wire subjected to MEMS actuation.

OPTICAL MEASUREMENT

The experimental setup for measuring the tunable phase shift is described as follow. First, a tunable laser source at the wavelength of 1548.6 nm was coupled to the silicon wire via a polarization-maintaining (PM) lensed fiber. To increase the coupling efficiency, two tapered waveguide structures are connected at the two ends of the silicon wire prior fiber coupling. The polarization state of the incident light was adjusted to be linear and vertical (TM-polarized) in order to minimize the insertion loss. The output optical signal was then coupled out using another PM lensed fiber. First, we examined variable optical transmittance of the MZI by simultaneously actuating six wires with a suspension length of 150 μm in one arm. The optical transmission was linearly attenuated with the bias voltage, as shown in Fig. 4 (a). The maximum attenuation reaches to 1.4 dB at the bias voltage of 250 V. To verify the dynamic operation speed of this device, an alternating current (AC) square wave signal of 10 Hz with a 250 V peak-to-peak voltage was applied to the same device. Fig. 4 (b) displays the measured dynamics of output signal. The signal rise time was estimated to be 1.5 ms. The resonate frequency of a 150 μm long silicon wire was calculated to be 130 kHz according to the ANSYS simulation. The inset shows the modulated optical attenuation. The dynamic range of modulated signal is near 1 dB which is close to 1.4 dB measured in direct current (DC) operation. The relatively small

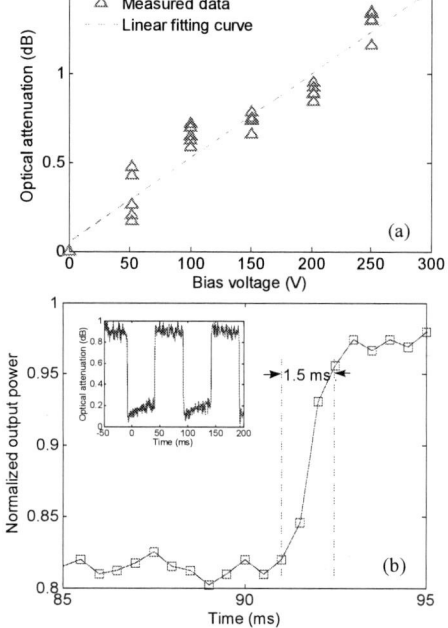

Fig. 4. (a) Measured optical attenuation versus the applied voltages. (b) Dynamic response of modulated output signal (measured from photodiodes).

dynamic range could be due to imbalance of optical transmittance of the two arms or multimode phase shift.

CONCLUSION

In conclusion, we have designed and demonstrated a MEMS-actuated waveguide phase modulator on SOI. In this study, the phase modulation is examined by deforming Si wires in a Mach-Zehnder interferometer configuration. According to the simulation result, a π-phase shift can be accomplished through cascading six MEMS-actuated suspending wires with a length of 150 μm, and the operation voltage could be as low as 35 V. However, our preliminary experiment results show that the maximum attenuation only reaches to 1.4 dB at the bias voltage of 250 V. The low optical attenuation could be attributed to the optical power imbalance between two arms which is strongly correlated with fabrication condition. Other possibilities may due to multimode phase shift. An optimal design of device dimension and a better control of fabrication process are required. Despite that, this MEMS phase modulators show potential of large-scale integration of devices in photonic integrated circuits

ACKNOWLEDGEMENT

The work is supported by National Science Council (NSC98-2622-E-007-002-CC1) and National Nano Device Laboraories (NDL98-C02M3C-040) in Taiwan.

REFERENCES

[1] M. C. M. Lee and M. C. Wu, "Variable bandwidth of dynamic add-drop filters based on coupling-controlled microdisk resonators," *Optics Letters,* vol. 31, pp. 2444-2446, Aug 2006.
[2] D. T. Fuchs, H. B. Chan, H. R. Stuart, F. Baumann, D. Greywall, M. E. Simon, and A. Wong-Foy, "Monolithic integration of MEMS-based phase shifters and optical waveguides in silicon-on-insulator," *Electronics Letters,* vol. 40, pp. 142-143, Jan 2004.
[3] E. Bulgan, Y. Kanamori, and K. Hane, "Submicron silicon waveguide optical switch driven by microelectromechanical actuator," *Applied Physics Letters,* vol. 92, Mar 2008.
[4] J. S. Xia, J. Z. Yu, Z. C. Fan, Z. T. Wang, and S. W. Chen, "Thermo-optic variable optical attenuator with low power consumption fabricated on silicon-on-insulator by anisotropic chemical etching," *Optical Engineering,* vol. 43, pp. 789-790, Apr 2004.
[5] A. Cutolo, M. Iodice, P. Spirito, and L. Zeni, "Silicon electro-optic modulator based on a three terminal device integrated in a low-loss single-mode SOI waveguide," *Journal of Lightwave Technology,* vol. 15, pp. 505-518, Mar 1997.

INERTIAL FORCE SENSOR USING OPTICAL MACH-ZEHNDER INTERFEROMETER AND MULTI MODE INTERFEROMETER

Masato Suzuki[1], Gou Kawai[1], Kouji Nishioka[1], Tomokazu Takahashi[1], Seiji Aoyagi[1],
Yoshiteru Amemiya[2], Masataka Fukuyama[2], and Shin Yokoyama[2]

[1] Faculty of Engineering Science, Kansai University, Japan
[2] Research Institute for Nanodevice and Bio Systems, Hiroshima University, Japan

ABSTRACT

In this paper, a novel inertial force sensor which uses a Mach-Zehnder Interferometer (MZI) type optical waveguide made of crystal silicon is proposed. In this sensor, one branched waveguide of the MZI have floating beam structure, and it is intersected with a cantilever for supporting a proof mass in same plane. To prevent optical loss at the intersection, only the intersectional point of floating waveguide has multi-mode interference (MMI) structure. Efficacy of the MMI waveguide is confirmed by simulation and experimental results. Conclusively, it has succeeded in changing in output of the fabricated initial force sensor by applied force.

INTRODUCTION

Currently, almost of micromachined inertial force sensors which detects an inertial force electro-statically, electro-magnetically or piezo-resistively [1]-[3]. However, sensitivity of these sensors tends to degrease with those downsizing. Optical interference has a potential to solve the above mentioned problem. J. Zhou *et al.* reported an accelerometer using an optical Fabry-Perot interferometer [4]. However, fabrication cost of the senor becomes expensive because it needs accurate adjustment of optical axis at assembling process.

Therefore, we proposed a novel inertial force sensor which uses a MZI type optical waveguides as shown in Fig 1. In this sensor, a part of the waveguide is expanded by applied force, and then output of the MZI is changed by changing in its interference condition. This sensor has advantage that it does not needs optical axis adjustment because light is propagated only in the waveguide. In this paper, optical optimization by using simulation and evaluation results of an actually fabricated sensor are reported.

MEASURING PRINCIPLE

The proposed inertial force sensor consists of the MZI and cantilever; and there is a proof mass on the tip of cantilever (Fig 1). The MZI and the cantilever are fabricated by using device layer of the silicon on insulator (SOI) wafer which is crystal silicon (c-Si). Here, one blanched waveguide in the MZI have

floating structure (air-bridged type), and it is crosses with the cantilever in same plane. When inertial force is applied to the mass, the cantilever and floating waveguide is bended and expanded (Fig. 2). As a result, output of the MZI is changed by applied inertial force (Fig 3).

However, there is large optical loss at the cross point of a single-mode waveguide and cantilever. Moreover, this optical loss increases with width of the cantilever (Fig. 4). In order to decrease the optical loss, a part of the floating waveguide changes to MMI structure; and then the cantilever is crossed with the waveguide at MMI part (Fig. 5). When design of the MMI is optimized, there is rarefaction of a compression optical wave at the cross point with cantilever (Fig. 6). As a result, optical loss at the cross point drastically decrease (Fig. 7).

EXPERIMENTAL AND RESULTS

Proposed sensor is actually fabricated (Fig. 8). Here, sacrifice laver for the floating waveguide and cantilever made of SiO_2 is removed by a vapor of fluoric acid. Input and output light of the MZI are coupled with hemispherical lensed fiber (Fig. 9). Here, wavelength of input light is 1550 nm. As a measurement results, it is found that optical insertion loss of the MMI is approximately 0.2 dB (Fig. 10) and optical loss at the cross point of MMI and cantilever is approximately 0.5 dB. Since the sensor must be fixed on a sample stage of the measurement system, force is directly applied to the cantilever of the sensor by using tip of a needle in this study. As a result, it is recognized that the output of MZI is changed by application of force (Fig. 11).

ACKNOWLEDGEMENT

This study was supported by Grant-in-Aid for Young Scientists (B) (No. 20710102) from the Ministry of Education, Culture, Sports, Science and Technology of Japan, and the Kansai University Grant-in-Aid for progress of research in graduate course, 2009-2010.

REFERENCES

[1] O. N. Tufte, P. W. Phanpman, and D. Long "Silicon Diffused-Element Piezoresistive Dia-

phragms" J. Appl. Phys., Vol.33, pp.3322-3327, 1962

[2] N. Yandi, F. Ayazi, and K. Najafi "Micromachined Inertial Sensors" Proc. IEEE sensors, Vol. 86, pp. 1640-1659, 1998

[3] LC. Spangler and CJ. Kemp : "ISACC-Integrated Silicon Automotive Accelerometer" Proc.

Transducers '95, pp. 585-588, 1995

[4] J. Zhou, S. Dasgupta, H. Kobayashi, J. M. Wolff, H. E. Jackson, and J. T. Boyd "Optically Interrogated MEMS Pressure Sensors for Propulsion Applications" Optical Engineering, Vol. 40, pp.598-604, 2001

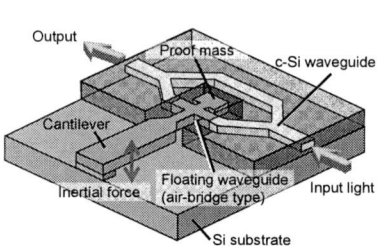

Fig. 1. Concept of inertial force sensor using Mach-Zehnder interferometer using optical waveguides made of crystal silicon.

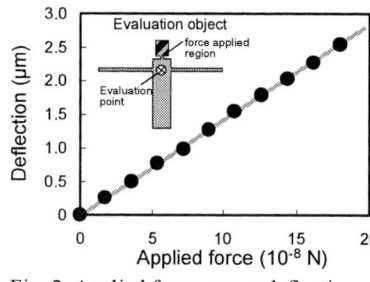

Fig. 2. Applied force versus deflection at cross point of floating waveguide and cantilever (simulated result using finite element method).

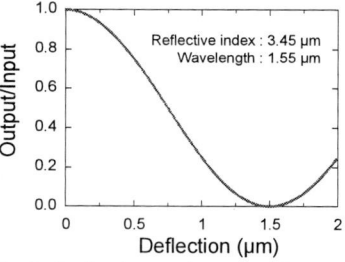

Fig. 3. Deflection versos output/input of MZI (calculated result).

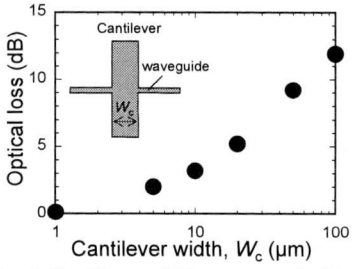

Fig. 4. Cantilever width versus optical loss at cross point when cantilever crosses with single-mode waveguide (simulated result).

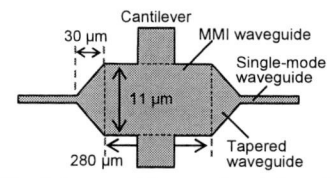

Fig. 5. Proposed intersectional form for decrease of optical loss. Here, single mode waveguide connects by tapered waveguide with MMI. Cantilever crosses with MMI at middle of MMI.

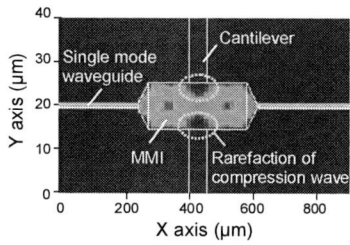

Fig. 6. Optical simulation around cross point of MMI waveguide and cantilever. There are rarefactions of compression wave at intersection region.

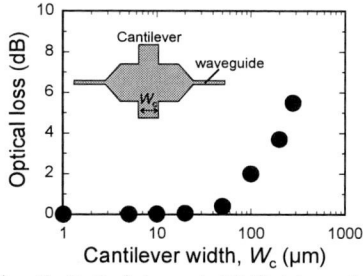

Fig. 7. Optical loss at MMI intersection versus width of cantilever (simulated result).

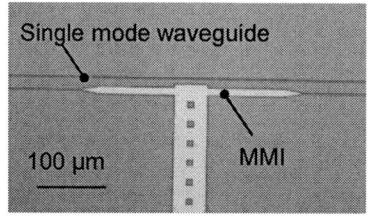

Fig. 8. Optical image of fabricated MZI type inertial force sensor (magnification around cross point of cantilever and MMI).

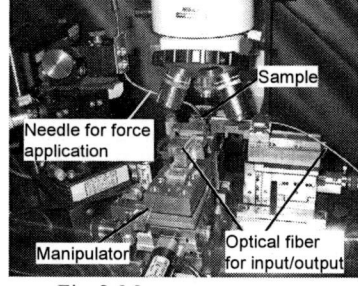

Fig. 9. Measurement system.

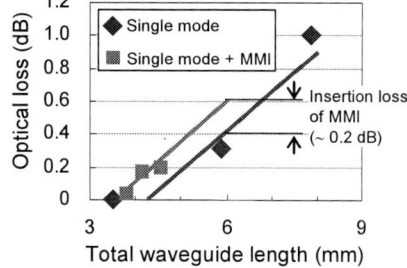

Fig. 10. Optical loss versus total length of waveguide including MMI (experimental result).

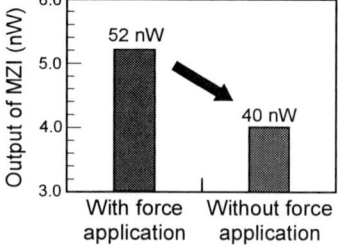

Fig. 11. Change in output of MZI by force application. In this measurement, the force is applied to middle of cantilever using needle.

978-1-4244-8926-8/10 $26.00 © 2010 IEEE

PURE PISTON MOTION OF OPTICALLY FLAT MICROMIRRORS IN A FULLY PROGRAMMABLE MICRO DIFFRACTION GRATING

R. Lockhart, R.P. Stanley, M. Tormen

CSEM SA (Swiss Center of Electronics and Microtechnology), Switzerland

ABSTRACT

We present a fully programmable micro diffraction (F-PMDG) grating in which each micromirror in a 1D array can be individually deflected out-of-plane by up to 1 µm. The maximum operating voltage is <30V and the lowest resonant frequency is >10kHz. The 120 µm wide and 700 µm long mirrors are designed to remain optically flat during actuation. The peak-to-valley curvature of a fully actuated micromirror is shown to be 10 nm equal to a radius of curvature of 7m. Fabrication of the F-PMDG requires only three photolithography masks.

INTRODUCTION

Programmable Micro Diffraction Gratings (PMDG) are one dimensional arrays of micromirrors whose height can be adjusted to produce a desired spectrum at a given output angle. They can be operated both as spatial light modulators and as reconfigurable generators of high-resolution spectra. PMDGs are useful for their size, speed and optical performance and have been implemented in projection displays, microspectrometers, optical communication systems and external cavity lasers [1-3]. In current PMDGs, the diffraction efficiency of the device is reduced due to bending of the micromirrors during actuation [1]. Burns *et al.* demonstrate the detrimental effects of mirror curvature in diffractive devices [4]. The diffraction efficiency drops by more than 50% for peak-to-valley curvatures of $\lambda/4$. Polychromix has solved the problem of mirror bending by separating the mechanical actuators from the optical micromirrors in a multi-level design [5]. This, however, requires a complex fabrication procedure involving the deposition and patterning of several thick layers. In addition, polysilicon, which suffers from surface roughness and residual stress, acts as both the optical and mechanical layers.

In this work, we present the design, fabrication and characterization of a fully programmable micro diffraction grating (F-PMDG) in which each of the micromirrors in an *n*-element array can be individually displaced out-of-plane. The device was designed to operate at wavelengths up to 2 µm, requiring a maximum out-of-plane stroke of 1 µm. The F-PMDG was designed to have mirror lengths of at least 700 µm with a maximum distortion of $<\lambda_{min}/10$ (40 nm), an optical fill factor >90%, mirror reflectivity >80% and a response time of 1 ms.

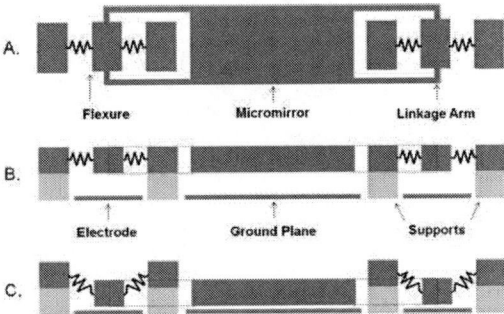

Fig. 1. 2D schematic of a single micromirror a) from above; b) from the side in its rest state and c) in its actuated state. An applied potential acting between the flexure and the underlying electrode pulls the flexures towards the substrate causing the central micromirror to be vertically displaced.

As illustrated in Fig. 1, rigid micromirrors are connected to the center of two compliant mechanical flexures via linkage arms. The applied potential between the flexures and their underlying electrodes pulls the flexures towards the substrate. The separation of the optical micromirror from the mechanical flexures vastly reduces bending of the micromirrors throughout actuation.

Finally, the use of serpentine flexures reduces the dependence of the spring stiffness on the thickness of the device layer. This allows rigid micromirrors and compliant flexures to exist side by side in the same optomechanical layer.

The F-PMDG fabrication requires only three photolithography masks. The device layer of a SOI wafer is transferred to a pre-patterned glass substrate by anodic wafer-bonding. After removal of the SOI handle and buried oxide layers, the device layer is patterned to define the micromirrors, flexures and linkage arms. The gratings were diced and wirebonded for electromechanical and optical characterization.

EXPERIMENT AND RESULTS

An optical image of a 4 mirror F-PMDG is presented in Fig. 2. Vertical displacement and flatness of the micromirrors versus applied voltage was measured using a white light interferometer. It was shown that by incorporating serpentine flexures, the mirrors can be vertically displaced up to 1 µm with <30V while maintaining resonant frequencies >10kHz.

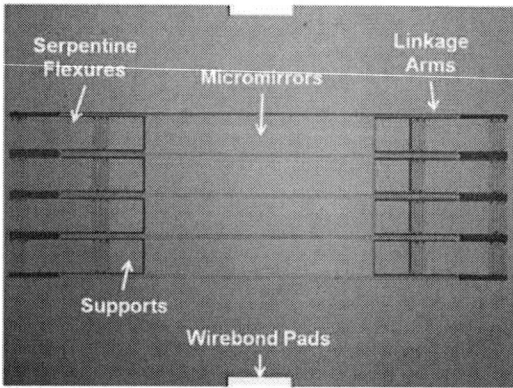

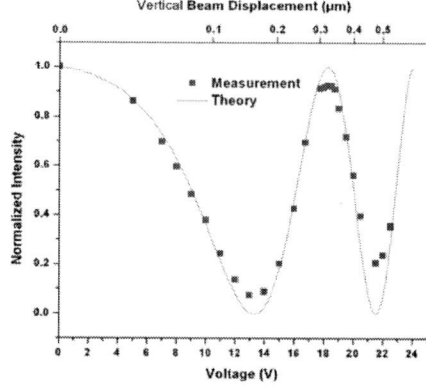

Fig. 2. Optical image of a FPMDG with 4 micromirrors. The mirrors are 120 μm wide and 700 μm long with 300 μm serpentine flexures situated on either side. The wirebond pads connect to the electrodes lying beneath the flexures.

Fig. 5. The intensity in the zeroth order for a monochromatic light source (λ=632.8 nm) reflecting off the FPMDG while varying the vertical displacement of every second micromirror. At λ/4 the intensity drops below 10% recovering to >90% at λ/2.

The cross section of a single mirror measured at several deflected positions between 0 and 1 μm is presented in Fig. 3. The measured height difference between the center and the edges of the 700 μm long mirror is less than 10 nm. This small amount of deflection is equivalent to λ/40 at 400 nm which corresponds to a radius of curvature of 7m (Fig. 4).

The device was tested in an SLM configuration and the optical performance is shown in Fig. 5. Here, the intensity of the zeroth order was recorded while displacing every second mirror by an amount d. The measured intensities closely follow that of the simulated theoretical response. Both a high efficiency and contrast ratio was achieved using the F-PMDG.

In conclusion, a high performance F-PMDG has been designed, fabricated and tested. Based on the current design, FPMDGs are being fabricated with up to 1024 individually controllable mirrors and narrower mirror widths (20μm). The presented device opens new possibilities for compact spectroscopic and imaging systems. Future tests will use this device in complete optical systems.

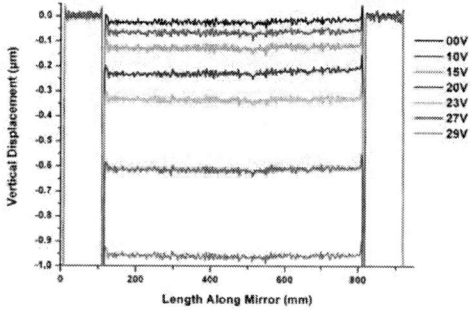

Fig. 3. Surface profile of a single micromirror for applied voltages between 0-29V. The mirror undergoes pure piston motion up to 1 μm while remaining optically flat (peak-to-valley deflection of 10 nm).

REFERENCES

[1] O. Solgaard et al., "Deformable Grating Optical Modulator," Opt. Lett., vol.17, pp. 668-690, 1992.

[2] S.D. Senturia et al., "Programmable diffraction gratings and their uses in displays, spectroscopy, and communications," J. Microlith., Microfab., Microsyst., vol. 4, 2005.

[3] R. Lockhart et al., "High efficiency MEMS tuneable gratings for external cavity lasers and microspectrometers," IEEE/LEOS Int. Conf. Optical MEMS and Nanophotonics, pp. 33–4, 2008.

[4] D.M. Burns, V.M. Bright, "Development of microelectromechanical variable blaze gratings," Sensors and Actuators A, vol. 64, 1998.

[5] E. R. Deutsch, Achieving Large Stable Vertical Displacement in Surfaced Micromachined MEMS, PhD thesis, MIT, Cambridge, MA, 2002.

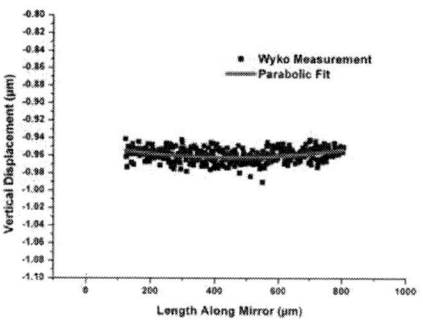

Fig. 4: Parabolic fit of the surface profile of a single mirror that has been vertically deflected by 1 μm. The radius of curvature (ROC) of the mirror is found to be 7m (10 nm of peak-to-valley deflection).

LAMELLAR GRATING BASED MEMS FOURIER TRANSFORM SPECTROMETER

Hüseyin R. Seren[1], N. Pelin Ayerden[1], Jaibir Sharma[1], Sven TS Holmström[1], Thilo Sandner[2], Thomas Grasshoff[2], Harald Schenk[2], Hakan Urey[1]

[1]Koç University, Department of Electrical Engineering, Istanbul-Turkey
[2]Fraunhofer Institute for Photonic Microsystems (IPMS), Germany

ABSTRACT

A Lamellar grating interferometer based Fourier Transform Infrared Spectrometer (FTIR) with out-of-plane resonant mode is implemented and characterized. Device has 10mm^2 clear aperture. Dynamic diffraction grating is comb-actuated and a maximum p-p deflection of 355 μm is obtained at 76 V. The excitation frequency is 971 Hz and deflection frequency of 485.5Hz.

INTRODUCTION

Fourier Transform Spectroscopy is a widely used method for color measurement, quality and process control, gas detection and chemical analysis. FTIR uses a single photodetector instead of an array of detectors and offers several advantages over other spectroscopy methods such as high signal to noise ratio, high throughput, compact form-factor and low cost. Lamellar grating interferometer (LGI) is based on MEMS dynamic diffraction gratings and well-suited for FTIR applications [1]. LGI offers significant advantages compared to conventional Michelson interferometer configuration that use a translating mirror as it eliminates the reference mirror, beam splitter, and the critical alignment requirements. Our group developed the comb-drive based out-of-plane moving lamellar grating device and previously achieved 53um deflection [2-3] and recently established the optical limitations of LGI based FTIR [4].

The target of the current project is to achieve compact handheld FTIR device with spectral sensitivity in the range 5-10cm^{-1}, operation frequency >500Hz, and spectral operation range of 2.5um-16um. The required spectral resolution translates to a deflection amplitude >500um (10 fold increase compared to previous performance), a novel compact IR source with high radiance (developed by Bruker Optics), and a novel IR source with several MHz bandwidth, broad spectral response range, and low-power cooling (developed by VIGO systems).

DEVICE STRUCTURE AND OPERATION

Figure 1 shows the fabricated MEMS device SEM picture after it was deflected by mechanical force. The device is designed to survive mechanical deflections up to 600-700um. The design is composed of stationary and movable parallel fingers, which serve as a variable depth diffraction grating. These fingers are carried by a backbone of mirror and the motion is transferred to the backbone through four pantograph structures. Each pantograph carry two comb structures either side which is used to excite the device. The width of fingers and gap between fingers for grating are kept constant. The device is actuated via both grating fingers and comb fingers that are located on the pantographs to maximize the electrostatic force.

Higher translational amplitude and better spectral resolution were achieved by using pantograph type suspensions. Maximum translational amplitude is expected to be ±500 μm at its first mode. Fig. 2 shows the mechanical simulation results. The device was designed for first out of plane mode to occur at 500 Hz and the other modes to remain sufficiently away from it. According to the mechanical simulations, the length of the grating fingers was limited to 1 mm in order to avoid dynamic deformations, which was calculated needs to be limited to about 250nm (Fig 2b). To maintain the flatness of grating fingers during deflection, deformation-absorbing suspensions were introduced.

CHARACTERIZATION

Characterization on wire-bonded devices with constant pitch grating fingers was performed in a vacuum chamber at 1.5×10^{-4} Torr. Out of four pantographs, only two opposite were used to excite the device. The excitation voltage was a square wave with 50% duty cycle. A Laser Doppler Velocimetry device was used to measure the deflection. A maximum deflection of 355 μm was obtained at excitation frequency of 971 Hz and excitation voltage of 76 V. Due to the nonlinear nature of out-of-plane comb devices, mechanical deflection is achieved at half of the excitation frequency [2]. It should be noted that the device was excited with a minimum excitation voltage of 15.6 V and a maximum deflection of 291 μm was observed. Also, the device was excited with a 10% pulse square wave and maximum deflection more than 200 μm was obtained.

In order to find the minimum excitation voltage for the devices, the measurements were performed with excitation voltages from 10 to 15 V with an interval of 1 V. It was observed that devices were excited at 10 V with maximum deflection of 23 μm and there was a small increment in deflection till 14 V. Fig. 3 shows the up- and down-sweep of frequency for the device at 62 V. The device shows the well known hysteresis and tongue type of shape for the frequency sweep. The device is also excited with very low excitation voltage of 10 V and a maximum deflection of 23 μm is achieved. On an earlier device, 400um deflection was obtained in ambient pressure at 150V before pull-in was observed.

Figure 1: SEM picture of the fabricated deflected device

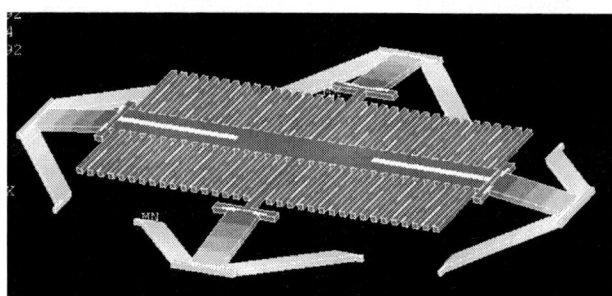

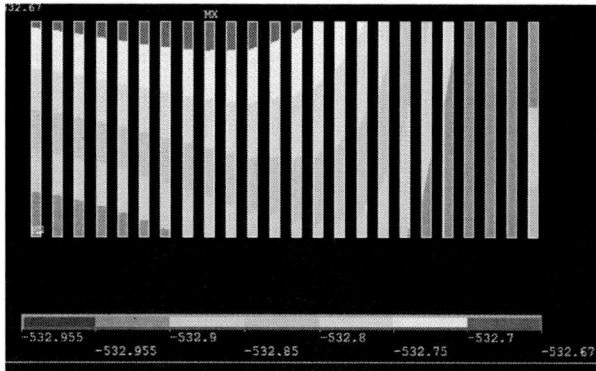

Figure 2: (a) first vibration mode at 494 Hz, total clear aperture is 2.5mm x 5mm with 80% usable area; (b) dynamic deformation at 532um deflection is about 285nm across the clear aperture.

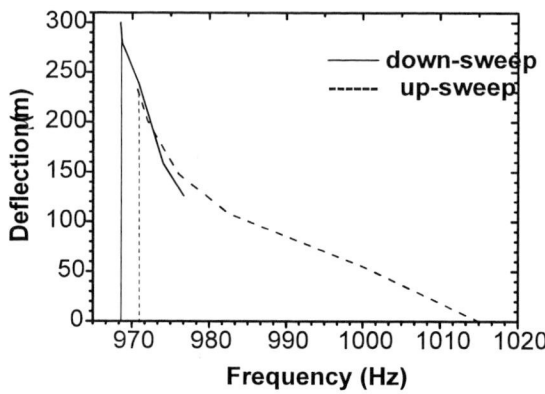

Figure 3: Frequency sweep at 62 V p-p excitation with a positive 50% duty-cycle square-wave. Vibration is sinusoidal and at half the frequency of oscillations.

CONCLUSION

FTIR Spectrometer with out-of-plane resonant mode is implemented and characterized. Maximum peak-to-peak deflection of 355 μm is obtained at 76 V in vacuum at an excitation frequency of 971 Hz. It is expected that pp deflection of up to 1000um is possible using all four pantographs and by applying voltage to the grating fingers as well. Spectrum measurements will be reported during the conference.

ACKNOWLEDGEMENT

This project is sponsored by MEMFIS project, which is supported by EC FP7 program grant no: 224151. The authors would like to thank all MEMFIS partners for their contributions.

REFERENCES

[1] O. Manzardo, R. Michaely, F. Schadelin, W. Noell, T. Overstoltz, N. F. de Rooij, H. P. Herzig, "Miniature lamellar grating interferometer based on silicon technology", *Opt. Lett.*, Vol. 29, p. 1437–1439, 2004.

[2] Caglar Ataman, Hakan Urey, Alexander Wolter, "MEMS-based Fourier Transform Spectrometer," J. Micromechanics and Microengineering, Vol.: 16, Pages: 2516-2523, 2006

[3] C. Ataman, H. Urey, "Compact Fourier Transform Spectrometers using FR4 Platform," accepted to Sensors and Actuators: A. Physical, Aug 2008.

[4] O. Ferhanoglu, H. R. Seren, S. Luttjohann, H. Urey, "Lamellar grating optimization for miniaturized fourier transform spectrometers", *Optics Express*, Vol. 17, Issue 23, p.21289-21301, 2009.

A Study on Color-tunable MEMS Device based on Plasmon Photonics

Taelim Lee[1], Akio Higo[1], Hiroyuki Fujita[2], Yoshiaki Nakano[1], and Hiroshi Toshiyoshi[1,2]

[1]Research Center for Advanced Science and Technology (RCAST), The University of Tokyo

[2]Institute of Industrial Science, The University of Tokyo

ABSTRACT

Oscillation of electrons on the metal surface makes surface plasmon polariton, which works as a guide for the waveguide when an array of holes is perforated on the metal surface. From its plasmonic properties, that kind of structure has a wavelength filtering property. Focusing on this filter characteristic, we have designed a novel MEMS tunable filter based on plasmon photonics. The idea is to make line and space structures and to put one on another vertically such that one would change the gap to control the wavelength filtering effect. By using this structure, we would electromechanically control the surface plasmon. As a preliminary research, we obtained optical properties of line-and-space structures that are crossed, by using the FDTD simulator.

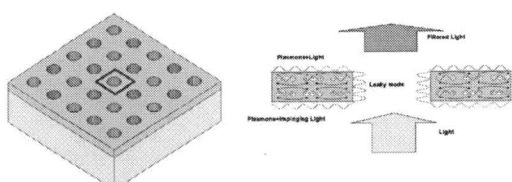

Fig1. Light traveling through the perforated structures.

I. Introduction

From Ebbesen's first paper on the extraordinary optical transmission through perforated metal, many theoretical and experimental studies have been performed [1]. The main reason of these kind of transmission are surface plasmon polaritons (SPPs), which are due to the EM wave on the metal surface caused by electron oscillation of the free electrons. If the light comes into the metal surface with the resonant frequency, SPPs are coupled to the light, and it allows the light to travel through the perforated

holes even if the size of spaces is smaller than the wavelength, as shown in Figure 1. Also, after the light transmitted through the holes, lights are filtered by resonance frequency of SPPs. Colors of transmitted light depends on the metal, the size of the perforated hole and its periodicity [2~4].

II. MEMS Design

As we were interested in this filter property, and designed the tunable SPPs filters as shown in Fig. 2. Two line-and-space structures of aluminum were made on a Quartz substrate. The periodicity was set to be in 150 nm to 350 nm. When these structures were overlapped perpendicularly to each other, we could expect this worked as a twisted-hole structure. On the contrary, if two structures are separated, it will work as just a line structures. Consequently, it works as a color filter. To ensure this really works, we simulated the structures with FDTD method.

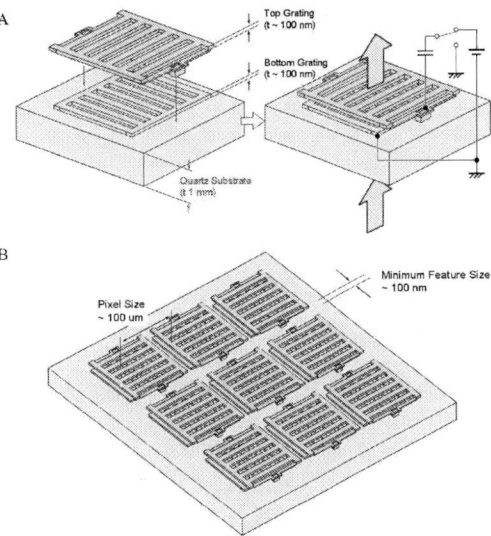

Fig2. Concept of tunable SPPs filter.

III. Simulations

We used the structure shown in Figure 3 for the simulation of the transmitted light, by FDTD (Fullwave, RSOFT Co.) simulator, and got encouraging results. Following results were obtained with the aluminum crossed structures as shown in Fig. 4. Aluminum thickness was 150 nm, line widths were 100 nm to 400 nm. Line width and space had the same length. Distance between two structures were zero (bold line) to 1 μm (dotted line). Other conditions were kept the same for control.

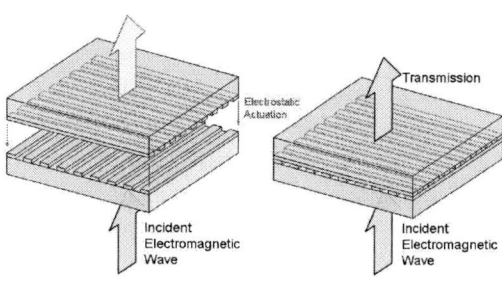

Fig3. Simulation Designs

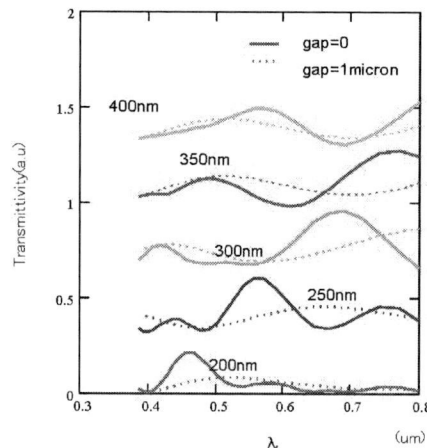

Fig.4 Transmission spectra. Line width 100 nm to 40 0nm. Al (150 nm) only, period / line width=2, gap=0 (line) to 1000 nm (dotted line)

IV. Simulation Results

We got significant peaks when line width was

set to be 200 nm, 250 nm, and 300 nm. They were recognized as Blue in color to human eye, Yellow, Red regions, respectably. Furthermore, the gap between the two aluminum structures was found to have significant effect in the transmission rates, implying the control of the color shift by means of the electromechanical motion. When the line width and space width widen, light can travel with guided mode.

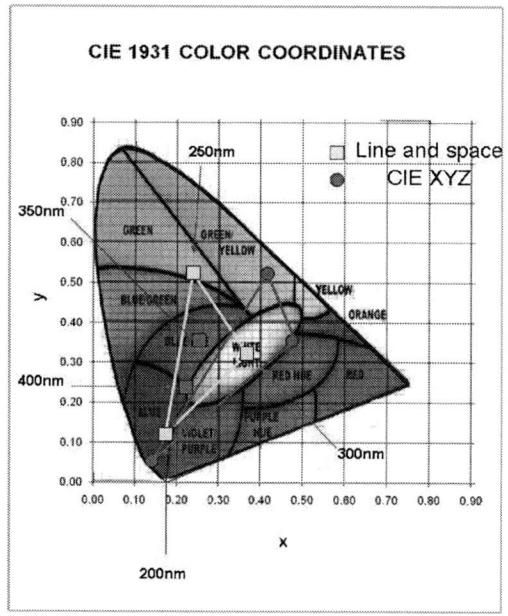

Fig5. CIE 1931 color coordinates: line and space, circle: CIE XYZ [5]

References

[1] T. W. Ebbesen "Extraordinary optical transmission through sub-wavelength hole arrays," Letters to Nature 391, 667-669, 1998

[2] Degiron, A. "Effects of hole depth on enhanced light transmission through subwavelength hole arrays," Applied Physics Letter **81**, 4327-4329 ,2002

[3] Rodrigo, S. G., F. J. Garcia-Vidal, et al. "Influence of material properties on extraordinary optical transmission through hole arrays," Physical Review B **77**(7): 075401, 2008

[4] Inoue et al., "RGB Color Filter comprising Aluminium Film with Surface Plasmon Enhanced Transmission through Sub-wavelength Hole-Arrays," IEEE/LEOS Optical MEMS and NanoPhotonics 2009, Florida, USA.

[5] JKL COMPONENTS CORPORATION http://www.jkllamps.com/

A Mixed-signal Analysis for Tilted MEMS Torsion Mirror Devices

Satoshi Maruyama[1], Akio Higo[1], M. Nakada[1], K. Takahashi[3], T. Takahashi[1], M. Mita[4],
Hiroyuki Fujita[2], Yoshiaki Nakano[1], and Hiroshi Toshiyoshi[1]

[1]Research Center for Advanced Science and Technology, The University of Tokyo

[2]Institute of Industrial Science, The University of Tokyo

[3]Department of Electrical and Electronic Engineering, Toyohashi University of Technology

[4]Institute of Space and Astronautical Science, Japan Aerospace Exploration Agency

E-mail: higo@hotaka.t.u-tokyo.ac.jp (Akio Higo)

ABSTRACT

We have already proposed a simulation model for a 90-deg-tilted MEMS mirror, but in some reasons, numerical simulation methods for several angle counter electrode devices are not established. In order to solve these problems, we propose a cascade drive voltage signal technique and demonstrate a 51.5 deg tilted angle counter electrode and assure the effectiveness of the equivalence circuit.

Keywords: simulation, mixed-signal analysis, equivalent circuit model

I. INTRODUCTION

Different types of electrostatic torsion mirrors/shutters have been developed by a silicon bulk/surface machining technology and mainly focused on optical micro electromechanical systems (OMEMS) applications, such as a projector[1], shutter arrays[2] and optical endoscope[3]. We have successfully demonstrated skewed deep reactive ion etching (Deep-RIE) process with angled shape counter electrodes. Commercially available MEMS simulators are nowadays more powerful and useful by accumulated co-solver modules. However, Long simulation time is required if you design 3D complex structure. Avoiding several bottomless results, estimating simple actuation results are required with a handy simulation tools on an open-source circuit simulator[5,6].

II. SIMULATION MODELS

Figure 1 shows the sub-circuit models for an 51.5 degree tilted torsion mirror. Each icon box has input and output on the both sides, where signals such as force, displacement and velocity are exchanged between neighborhood modules for co-solving displacement-voltage. The

equivalent circuit model for the equation-of-motion is based on a two-step electrical integration using an ideal capacitor and modified to fit the Qucs representation. Figure 2 (a) shows the tilted angled counter electrodes for torsion mirror. To achieve the 51.5 degree tilted pull-in phenomena, we employ a cascaded drive voltage signals as an input. We have already reported 90 degree motion with a single drive voltage signal by a previous report [2], however, the model doesn't work less than 90 degree electrode.

III. RESULTS

Figure 2 (b), (c) show the simulation result and measurement result of pull-in phenomena. Figure 2 (b) shows pull-in voltage as 51.5 degree at 91 V and measurement result is 51.5 degree movement at 52 V. These results presented this model has a beneficial effect on estimating the pull-in voltage. Measurement sample works low voltage compared with the simulation because initial mirror position is down after a release.

978-1-4244-8926-8/10 $26.00 © 2010 IEEE

References

[1]M. Tani, et al., "Two-Axis Piezoelectric Tilting Micromirror with a Newly Developed PZT-meandering Actuator," in Proc. 20th IEEE Int. Conf. on Micro Electro Mechanical Systems (MEMS 2007), Jan. 21-25, 2007 (Poster M35)

[2]T. Takahashi, et al., "A Mixed-signal Analysis Tool for MOEMS based on Circuit Simulator," in Proc. IEEE Optical MEMS and Nanophotonics 2009, Clearwater Beach, Florida, USA, 17-20 August 2009, ThB1.

[3]M. Nakada, et al., "Optical coherence tomography by all-optical MEMS fiber endoscope," IEICE Electronics Express, vol. 7, no. 6, 2010, pp. 428-433.

[4] M. Nakada, et al., "DEVELOPMENT OF SKEWED DRIE PROCESS AND ITS APPLICATION TO ELECTROSTATIC TILT MIRROR," in Proc. 22nd IEEE Int. Conf. on Micro Electro Mechanical Systems (MEMS 2009), pp.1087-1090, 2009

[5]Makoto Mita and Hiroshi Toshiyoshi, "An Equivalent-circuit Model for MEMS Electrostatic Actuator using Open-source Software Qucs," IEICE Electronics Express, vol 6, no. 5, 2009, pp. 256-263.

[6]S. Maruyama, et al., "A Mixed-Signal Equivalent Circuit Model for MEMS Digital Mirror," in Proc. 16th Int. Display Workshop (IDW'09), Dec. 9-11, 2009, Miyazaki, Japan, session MEMS5-4.

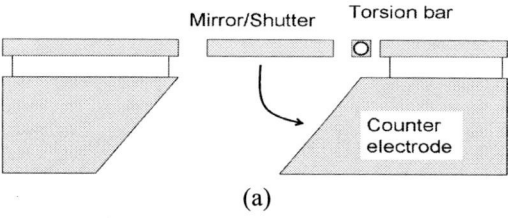

(a)

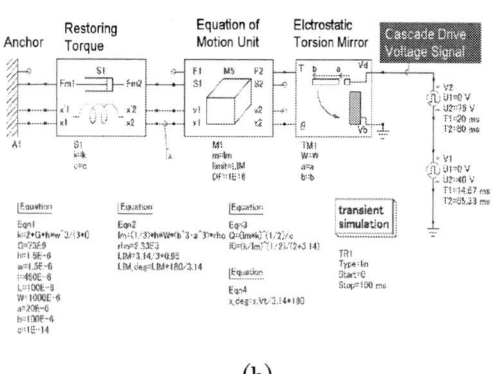

(b)

Figure 1. Torsion mirror model and sub-circuit models for electrostatic mechanism

(a) Torsion mirror model, (b) sub-circuit models

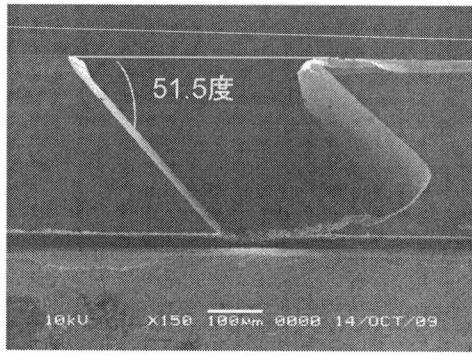

(a)

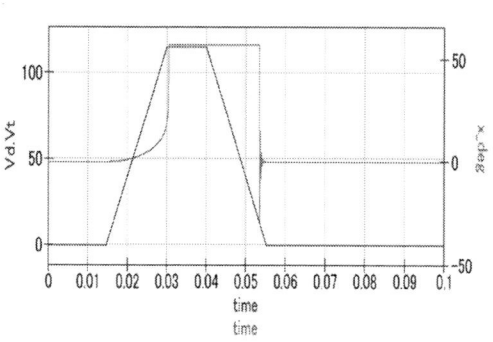

(b)

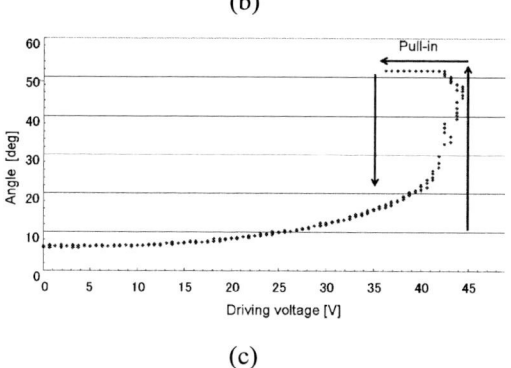

(c)

Figure 2, SEM cross-sectional picture and results

(a) Counter electrode SEM picture, (b) Simulated result, (c) Measurement result

A 2-axis MEMS Scanner for the Landing Laser Radar of the Space Explorer

M. Mita[1], T. Mizuno[1], M. Ataka[2], H. Toshiyoshi[2]
[1]ISAS/JAXA and [2]University of Tokyo, Japan

ABSTRACT

We have developed a novel 2-axis MEMS scanner of the small laser radar for landing to the planet. The scanner has to overcome launching vibration and impact and can be used in harsh environment of the space.

INTRODUCTION

For the deep-space missions that require cost-reduction, small satellites and explores of small payload are highly appreciated, where MEMS devices find important roles of attitude sensing of the satellite, altitude measurement and RF application. Conventional scanning laser radars were too large to mount on a small satellite or explorer [1-2]. Therefore we have developed small scanning laser radar system by using 2-axis MEMS scanner.

CONCEPT

Figure 1 shows concept of our landing laser radar system for the space explorer [3]. A Laser radar is used to measures the altitude of the space explorer when the explorer land to the planet.

A robust scanner is one of the most important elements of the landing laser radar. The scanner has to overcome launching vibration and impact and live in the harsh environment of the space. In the case of long distance measurement such as several tens kilometer, quite strong laser is used for altitude measurement. The power of the laser reaches MW order. Then the dielectric multilayer reflector is required for the mirror surface of the scanner due to very high reflectivity. However, the stress of the reflector is too strong to keep flatness of the mirror. We have developed robust MEMS scanner, which overcome the launching vibration and impact and the stress of the multilayer reflector.

STRUCTURE AND DRIVING PRINCIPLE

Figure 2 illustrates the structure and driving principle

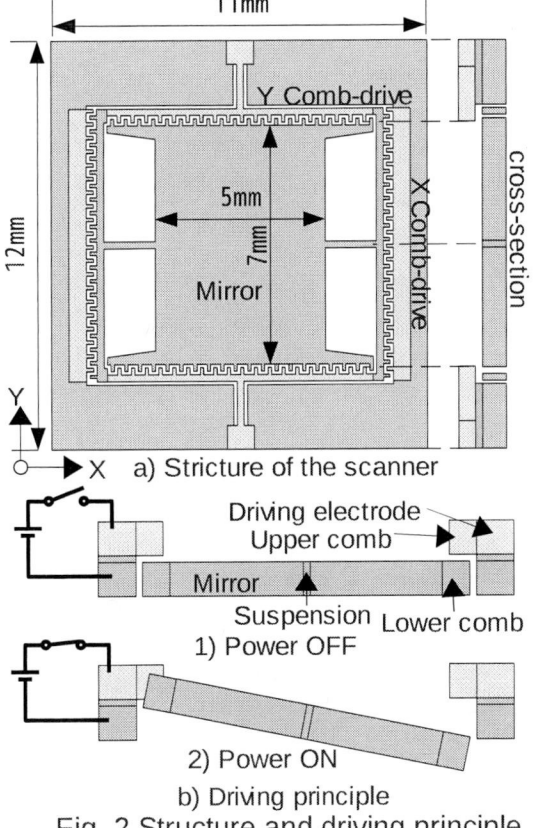

Fig. 1 Concept of the system

Fig. 2 Structure and driving principle

978-1-4244-8926-8/10 $26.00 © 2010 IEEE

of our 2-axis MEMS scanner.

We use vertical comb-drive to drive the mirror. The vertical comb-drive is suitable for our scanner thanks to the large generation torque and large receptivity for harsh environment. The spot size of the laser is large as 5mm to reduce the extent angle of the laser. Then we designed the mirror size to 5mm x 7mm because the mirror is tilted at 45 degree to optic axis of the laser. To overcome the launching vibration and impact and the stress of the multilayer reflector, we used 200-micron-thick SOI wafer for the scanner structure.

FABRICATION

Figure 3 shows the fabrication process using deep RIE process on a 200-micron-thick SOI wafer: (a) An aluminum layer (Al) is formed on the top and the bottom surfaces, respectively, to be used as etching masks. (b) The rear surface etched with deep RIE, which forms suspensions and a mirror. (c) The front SOI surface is deep RIE-processed to form the driving electrodes and the upper comb of the vertical comb-drive, and (d) finally the structure is released by removing the buried oxide in HF.

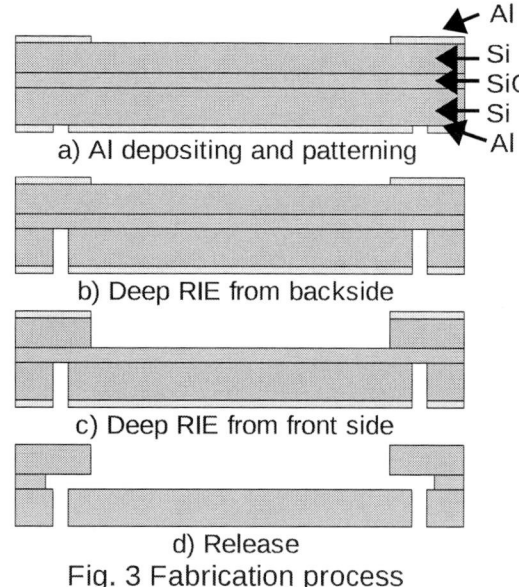

Fig. 3 Fabrication process

Fabricated device is shown in Figure 4. The chip dimensions are 12 mm by 11 mm in area; suspensions formed in the substrate (200 microns thick) are 40 microns wide and 1.5 millimeters long. The mirror size is 5mm x 7mm and thickness is 200um.

DRIVING RESULT

The resonant frequency of X-axis and Y-axis of the scanner were found to be 340 Hz and 360 Hz,

respectively. When we applied 40 V sine voltages at resonant frequency, the scanner scanned approximate 10 degrees (X-axis) and 12 degrees (Y-axis). The result is smaller than the designed angle (15 degrees). The reason of this result is from damping effect between the mirror and the package. We will improve the driving angle by changing setup of the device.

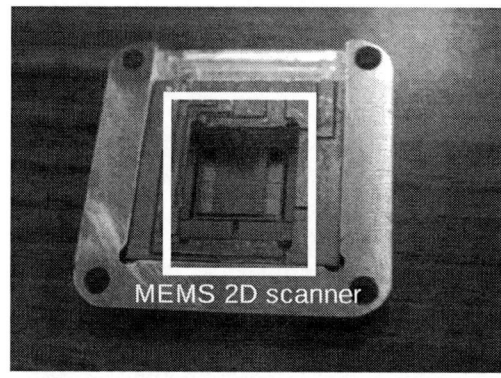

Fig. 4 Device photographs

References

[1] B. Moebius, et al., "Miniaturized 3D LIDAR for Lunar Landing", Proceedings of the 26th International Symposium on Space Technology and Science, 2008-d-02 (2008).

[2] M. Nimelman, et al., "Spaceborne scanning lidar system(SSLS)", Proceedings of the SPIE - Spaceborne Sensors, Vol.5798, pp73-82 (2005).

[3] T. Takahara, et al., "Study of 2D Scanning LIDAR Optics for Planetary Explorer", IEEJ Trans. SM, vol. 126, no.8, 2006, pp. 476-480, (in Japanese).

VACUUM OPERATION CHARACTERISTICS OF TWO-DIMENSIONAL MICRO-MIRROR

Hoang Manh Chu and Kazuhiro Hane
Tohoku University, Japan

ABSTRACT

We present design, fabrication and characteristics of two-dimensional micro-machined scanner. The resonant frequencies of horizontal and vertical axes are 40 kHz and 162 Hz for the inner mirror and gimbal frame, respectively. The optical scanning angles are obtained to be 11.5 and 14 degrees at the low driving voltages of 12 and 10 V in 1 Pa vacuum for the horizontal inner mirror and vertical gimbal frame, respectively. The scanner can be actuated simultaneously and independently in the two orthogonal axes using a slanted electrostatic comb-drive and silicon conductive V-shaped torsion hinges. The dependence of quality factor on pressure for the inner mirror and gimbal frame was also experimentally investigated and compared with the theoretical calculation based on air-friction models.

INTRODUCTION

Optical scanners have been interested for scientific and industrial applications such as laser displays, printers and barcode scanners. To drive the scanner, one can use several actuation mechanisms such as magnetic, piezoelectric and electrostatic. The electrostatic mechanism is interested due to mass production with high repeatability and reliability. However, the operation voltage for the electrostatic comb-drive is still high (~ 100 V).

It was recently reported that the operation voltage was decreased substantially by decreasing the air friction of comb-drive micro-mirror at a high resonant frequency when the mirror was operated at low pressure. In order to use this advantage for mobile micro-mirror, vacuum packaging is a key technology. The micro mirrors were recently packaged in vacuum by anodic bonding [1] and metal can packaging [2]. The air friction influence and the characteristics of vacuum operation of one-dimensional micro-mirrors were investigated quantitatively in the experiments and theories, specially for the high-frequency-resonant display mirror [3]. However, the influence of air friction on the gimbal frame of two-dimensional comb-drive micro-mirror and the characteristics optimization between the high-frequency resonant mirror and the gimbal frame in vacuum have not been investigated well experimentally and theoretically.

In this work, we describe the design, fabrication and characteristics of a two-dimensional scanner electrostatically driven in vacuum.

DESIGN AND THEORY

Design of two-dimensional micro-scanner

The designs of 2-D scanners are two types, Type I and Type II. The electronic connections are realized by separating the top silicon layer into five regions: one region is for the movable combs of scanner, the four others are for the four fixed comb electrodes of inner mirror and gimbal frame, respectively. For Type I, we use rectangular torsion hinges for the gimbal frame. For Type II, V-shaped torsion hinges are used instead of rectangular torsion hinges. The V-shaped torsion hinge is not only to increase stability of scanner but also to supply the electronic connections for the inner mirror and gimbal frame. In order to investigate the effect of torsion hinge of gimbal frame, we designed four types of the gimbal frame with different dimensions.

Motion equations for resonant micro-scanner

When the scanner is excited at its natural frequency, it oscillates resonantly. To characterize the dynamic behaviors of the resonant scanner with damping, a second-order mechanical system under the harmonic excitation is analogous to the translational spring-mass-damper system. At resonance, mechanical rotation angle $\theta(f,Q,V)$ is evaluated as,

$$\theta(f,Q,V) = \alpha Q V^2 / f^2 \qquad (1)$$

where, α is a dimension constant.

The energy performance of oscillating object is usually evaluated by quality factor Q, $Q = 2\pi E_{st}/L$. Here, E_{st} and L are the stored energy and loss in a single cycle, respectively.

From (1), since Q is inversely proportional to the energy loss, the driving voltage decreases if the energy loss of scanner is decreased. The rotation angle θ increases when f decreases.

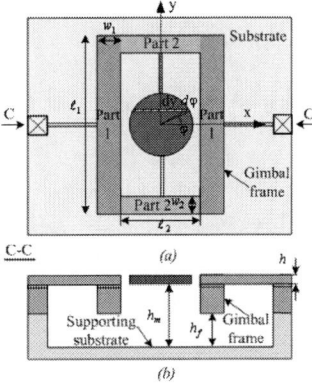

Fig. 1. Schematic design of 2-D scanner used for the theoretical calculation.

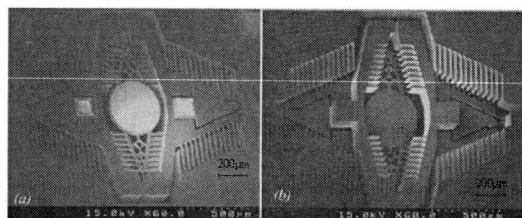

Fig. 2. Electronmicrographs of the fabricated micro-scanners: (a) and (b) are the top and back side views of type II.

Air friction model for two-dimensional scanner

In the energy performance approach, we consider the air friction loss to be a dominant mechanism which decides the quality factors Q of 2-D scanner. We replaced the complicated structure of 2-D scanner by a simpler rectangular structure as shown in Fig. 1 (a) and (b). To find out the quality factor of the gambal frame Q_f, we considered the individual contributions of parts, the inner mirror and gimbal frame. The gimbal frame was divided into two parts as shown in Fig. 1 (a).

The quality factor Q_f at the high vacuum of pressure less than 1000 Pa is defined by the formula, $Q_f = \omega I / G$, where I is the mass moment of rotation inertia of scanner, and G is the total damping coefficient of parts and equals $G_m + G_{p1} + G_{p2}$.

Next, we consider the case of low vacuum from 1000 Pa to 10^5 Pa. If we neglect minor influences such as acoustic radiation, the loss due to viscous friction is estimated approximately as

$$L_{vis} = 2A\left(\omega\rho\eta/2\right)^{1/2}\left\langle v^2 \right\rangle, \qquad (2)$$

where A is damped area, ρ is the air density, and η is the dynamic viscosity of air in Nsm^{-2}. $\left\langle v^2 \right\rangle$ is the mean square velocity of oscillating object.

Stored energy is calculated by the rotational motion energy of scanner as follows,

$$E_{st} = I_\theta \theta_0^2 \omega^2 / 2 \qquad (3)$$

The Q_f of the gimbal frame in the low vacuum is expressed as

$$Q_f = 2\pi E_{st} / \left(L_m + L_{p1} + L_{p2}\right) \qquad (4)$$

where L_m, L_{p1} and L_{p2} are the losses of inner mirror, part 1 and part 2, respectively.

FABRICATION AND EXPERIMENT

The fabrication was started from a SOI wafer. The devices are fabricated using inductive coupled reaction ion etching (ICP-RIE).

To investigate the characteristics of the fabricated micro-scanners in air as well as in vacuum, the micro-scanners were fixed in a vacuum chamber with a glass window. The chamber was also equipped with two voltage supplies, vacuum pumping system and a pressure controlling valve.

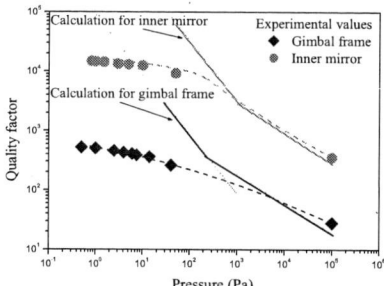

Fig. 3. Quality factors measured and calculated as a function of pressure for the inner mirror and gimbal frame, respectively.

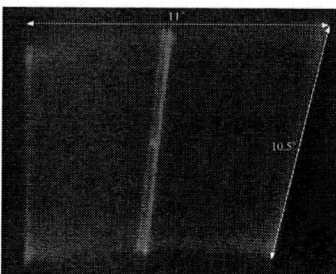

Fig. 4. Scanning image along two axes of 2-D scanner in vacuum.

RESULTS AND DISCUSSION

Fig. 2 shows the scanning electron micrographs of the fabricated 2-D scanners.

We measured the quality factor Q as a function of pressure in the range from 0.1 Pa to atmosphere. The measured values of Q are shown in Fig. 3 for the inner mirror and gimbal frame, respectively. The maximum value of Q is about 15000 for the inner mirror of 40kHz resonant frequency and 524 for the gimbal frame of 162 Hz resonant frequency below 1.5 Pa. The calculated values of Q roughly explain the experimental values when pressure is equal to 10^5 Pa. Due to the vacuum operation, the voltages for the inner mirror and gimbal frame decreased approximately by the factors of 21 and 3, respectively.

We carried out the 2-D scanning of laser beam in vacuum using the fabricated device as shown in Fig. 4.

REFERENCES

[1] H. Tachibana, K. Kawano, H. Ueda, and H. Noge, "Vacuum wafer level packaged two-dimensional optical scanner by anodic bonding" The 22nd IEEE MEMS, pp. 959-962, 2009

[2] H. M. Chu, T. Tokuda, M. Kimata, and K. Hane, "Compact low-voltage operation micro-mirror based on high vacuum seal technology using metal can" J. Microelectromech. Syst, 2010 (Submitted)

[3] C. H. Manh and K. Hane, "Vacuum operation of comb-drive micro display-mirrors" J. Micromech. Microeng. Vol. 19, 105018 (8pp), 2009

978-1-4244-8926-8/10 $26.00 © 2010 IEEE

A TWO-AXIS HYBRID MEMS SCANNER INCORPORATING ELECTROTHERMAL AND ELECTROSTATIC ACTUATORS

Gordon Brown, Li Li, Ralf Bauer, Jinsong Liu and Deepak Uttamchandani
University of Strathclyde, UK

ABSTRACT

A hybrid two-axis MEMS scanning mirror has been fabricated incorporating two different actuation mechanisms for its orthogonal rotation axes. The 2 mm diameter mirror is rotated about one axis using an electrostatic comb-drive for fast resonance movement and about the other axis using two out-of-plane electrothermal actuators. This hybrid actuation arrangement enables fast sweeps (at the mechanical resonant frequency) to be generated on one axis, with a slower sweep (from a few Hz to a few tens of Hz) to be generated on the orthogonal axis, raising the possibility of complex scan patterns being produced using relatively simple driving schemes. The characteristics of the first iteration of our device have been experimentally determined, and simultaneous driving of both actuation mechanisms has been demonstrated.

INTRODUCTION

MEMS scanners have increasingly found application in a range of fields such as portable and compact imaging or projection systems that take advantage of small packaging size and low power consumption. In the case of dual-axis scanning mirrors the same type of actuation mechanism (typically electrostatic, thermal or magnetic) is usually used to drive the two orthogonal axes of the device. In this paper we report a dual-axis device fabricated in SOI that simultaneously uses two different actuation mechanisms for its orthogonal scanning axes. The device has been fabricated using a commercial multi-user SOI process (SOIMUMPS).

It is well known that thermal actuators can allow arbitrary control of scan angle/position, but at a relatively slow rate of angular change owing to the limitations imposed by the thermal time constants. The overall framing rate for a MEMS optical scanner driven by electrothermal actuation alone is therefore limited to a few tens of Hz. Electrostatic actuators (comb-drive) can be used to excite large amplitude oscillations with low power requirements, but are restricted to operation at the mechanical resonant frequencies of the device. A 2-axis scanner using resonant electrostatic drives for both axes would generate a Lissajous scan pattern and require complex control systems for the light source (or detector) in order to construct an image.

In the MEMS scanner described here, the mirror is made to oscillate about one axis, driven at mechanical resonance by vertically offset comb-drive actuators to produce a line scan pattern. That line is then swept up and down the field of view at a lower frequency by means of the electrothermal actuator attached to the orthogonal axis. By combining two types of actuation into the same scanner, video framing rates (few $10s$ Hz) can be achieved with a less complex control system than that required for a dual electrostatic (resonant) scanner. While the use of both electrothermal and electrostatic actuation mechanisms within one actuator has been reported for switching purposes [1-3], to the authors' knowledge, the implementation of two different actuation mechanisms within a MEMS optical scanner has not been reported before.

DESIGN

The scanner described in this paper is fabricated using a commercial SOI multi-user process (SOIMUMPs, available from MEMSCAP, USA) with the moving parts of the device defined in the 10μm thick top (SOI) layer. Employing this standard commercial process allows rapid development time and provides cost effective fabrication. The layout of the device, with the 2 mm diameter mirror in the center, can be seen in the SEM image of Fig. 1. The two thermal actuators allow rotation about the y-axis through the out-of plane movement of the three beam actuator which has been thoroughly evaluated in [4]. They are connected via serpentine springs to the mirror and the beam containing the moving combs. Rotation about the x-axis is excited by the electrostatic comb-drive with 10

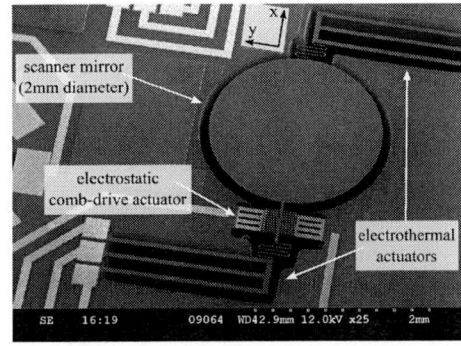

Fig. 1: SEM image of the fabricated hybrid scanner

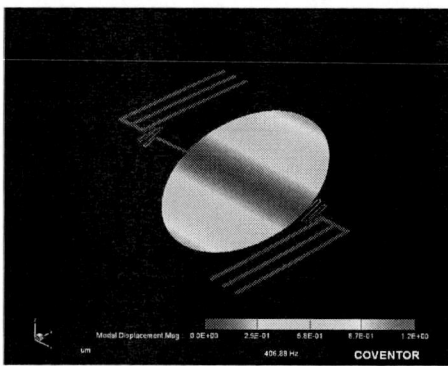

Fig. 2: FEM simulation showing the first order mechanical resonance mode excited by the electrostatic actuator at 407 Hz

pairs of interdigitated fingers. The comb-drive exerts a torque about the axis because the opposing sets of fingers exhibit a small vertical offset due to residual stresses arising from the gold coating applied over part of the length of the fingers.

The mechanical behaviour of the device has been modelled using finite element analysis (FEA) software (ConventorWare) in order to obtain the mode shapes and the corresponding resonance frequencies. The desired resonance mode shape excited by the comb-drive can be seen in Fig. 2, with a calculated resonance frequency of 407 Hz. The resonant frequency for the rotation around the orthogonal axis is calculated to be 1.6 kHz.

MEASUREMENTS AND RESULTS

The angular deflection and resonance frequencies of the fabricated mirrors have been experimentally determined from the scan pattern of a HeNe laser spot reflected from the mirror onto a screen. The resonance frequency of the electrostatic axis was found to be 380 Hz, in good agreement with that predicted by FEA. The optical peak-to-peak deflection in this axis was

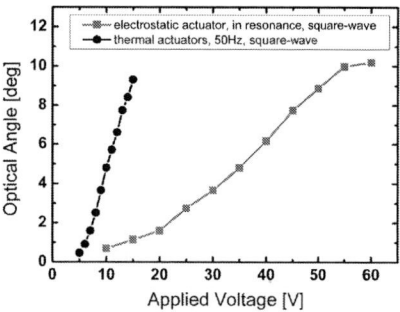

Fig. 3: Optical deflection angle of the mirror vs applied voltage for the electrostatic and thermal actuators

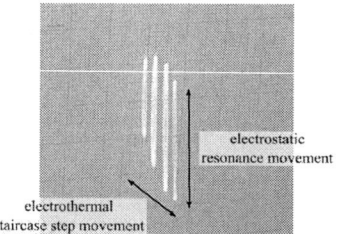

Fig. 4: Screen image of the raster-like scan pattern

measured as a function of a pulsed driving voltage with 50 % duty cycle (Fig. 3). The maximum deflection obtained was 10.2 ° for a drive voltage of peak amplitude 60 V. In the case of the orthogonal axis, the two thermal actuators were driven in anti-phase by synchronised pulsed driving voltages with 50 % duty cycle applied to the outer beams of the thermal actuators (Fig. 3). A maximum optical deflection of 9.3 ° was found when 15 V was applied.

A raster-type scan pattern was generated by driving the electrostatic axis at resonance to generate a fast line scan and then using a staircase-type drive signal for the thermal axis operating at several tens Hz in order to cause the line to sweep over the field of view. A photograph of such a scan pattern is shown in Fig. 4. For clarity in the image, the staircase drive signal had many fewer steps (4) than would be the case for an imaging application (15 − 25).

CONCLUSION

The hybrid scanner operated successfully, with a raster-type scan pattern observed when both axes were driven. It was observed that the two scanning axes were not orthogonal, resulting in a scan pattern with a trapezoidal envelope, rather than the desired rectangular shape. Initial investigations indicate that the action of the thermal actuators is not vertical but follows an arc, leading to a twisting of the structure. Revisions of the design will be carried out, verified by FEA, in order to minimise the twisting effect and render the scan pattern more regular.

REFERENCES

[1] A. Alwan and N. R. Aluru, "Analysis of Hybrid Electrothermo-mechanical Microactuators With Integrated Electrothermal and Electrostatic Actuation", *J. Microelectromech. S.*, vol. 18, no. 5, pp. 1126–1136, 2009.

[2] P. Robert, D. Saias, C. Billard, S. Boret, N. Sillon, C. Maeder-Pachurka, P. L. Charvet, G. Bouche, P. Ancey, and P. Berruyer, "Integrated RF- MEMS switch based on a combination of thermal and electrostatic actuation", in *Transducers '03*, (Boston), pp. 1714–1717, 2003.

[3] I.-J. Cho, T. Song, S.-h. Baek, and E. Yoon, "A low-voltage and low-power RF MEMS series and shunt switches actuated by combination of electromagnetic and electrostatic forces", *IEEE T. Microw. Theory*, vol. 53, pp. 2450–2457, July 2005.

[4] L. Li and D. Uttamchandani, "Dynamic response modelling and characterization of a vertical electrothermal actuator", *J. Micromech. Microeng.*, vol. 19, no. 7, p. 075014, 2009.

MEMS SCANNING MIRROR USED AS AN LASER EXTERNAL MODULATOR FOR PHOTOACOUSTIC SPECTROSCOPY

Li Li, Graham Thursby, George Stewart and Deepak Uttamchandani,
University of Strathclyde, Scotland, UK

ABSTRACT

We present a novel MEMS application aimed at laser based gas sensing. A low-cost MEMS based external intensity modulator for a laser diode source has been realised and applied to photoacoustic spectroscopy. By using a MEMS based modulator, pure intensity modulation of the laser emission is achieved without the accompanying wavelength modulation which occurs with diode current modulation. This reduces measurement error. We describe the use of the optical MEMS modulator/photoacoustic technique to recover the profile of the 1535.4nm absorption line of acetylene at 100pm concentration in the photoacoustic cell. Based on initial results, we predict a sensitivity of ~1ppm with this system.

INTRODUCTION

Many applications have been described in the literature that use optical MEMS devices [1],[2]. In this paper we report on the application of an optical MEMS based modulator in a photoacoustic spectroscopy based gas sensing system. In this gas sensing technique a wavelength scanned laser beam transmits through a sealed photoacoustic gas cell containing the target gas as shown in Figure 1.

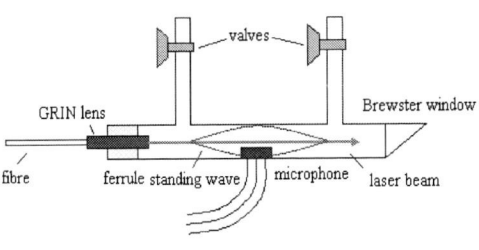

Figure 1. Photoacoustic gas cell

When the wavelength of the laser matches the absorption lines of the gas, some of the transmitted light energy is absorbed by the gas molecules and heat is generated. This, in turn, produces a change in pressure inside the cell which is detected by a microphone. The amount of light absorbed is determined (from Beer's law) by a combination of factors: the gas absorption coefficient at a given wavelength; the gas concentration; the cell length and the power of the propagating light beam.

In our photoacoustic system, a distributed feedback (DFB) laser is thermally tuned (using a thermoelectric unit) to a wavelength close to that of the gas absorption line, Figure 2. Low frequency (a few Hz) linear ramping of the drive current allows the laser wavelength to be scanned across the gas absorption line. The output of the laser is amplified using an erbium doped fibre amplifier (EDFA) to a level of typically 100-500mW. If, in addition to ramping the laser current, a sinusoidal current

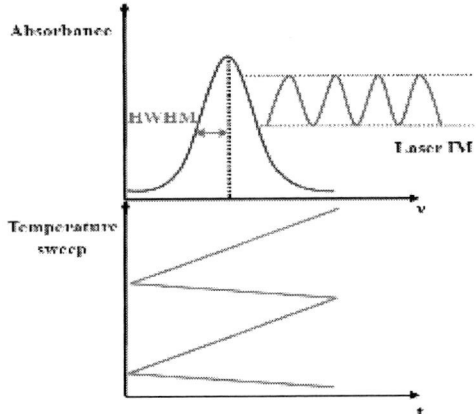

Figure 2. Laser scanning and current modulation

modulation (a few kHz) is simultaneously applied, then the sensitivity of detection can be enhanced by the use of a lock-in amplifier set to detect the acoustic signal at the modulation frequency. A major problem encountered in the above system is that the desired sinusoidal optical intensity modulation generated by the laser diode drive current is accompanied by an unwanted wavelength modulation, which will distort the results if recovery of the shape of the gas line is required.

An alternative approach to obtaining intensity modulation without the unwanted wavelength modulation is to use an external modulator. Conventional electro-optic lithium niobate modulators are a possibility, but they are both expensive and optically lossy. Since only low

978-1-4244-8926-8/10 $26.00 © 2010 IEEE

modulation frequencies are usually required in photoacoustic optical gas sensing systems a compact, MEMS based opto-electro-mechanical intensity modulator can be used instead (figure 3).

from the return path and is not recaptured by the lens. In this way, the retroreflected light can be intensity modulated.

A dc bias voltage with superimposed ac modulation

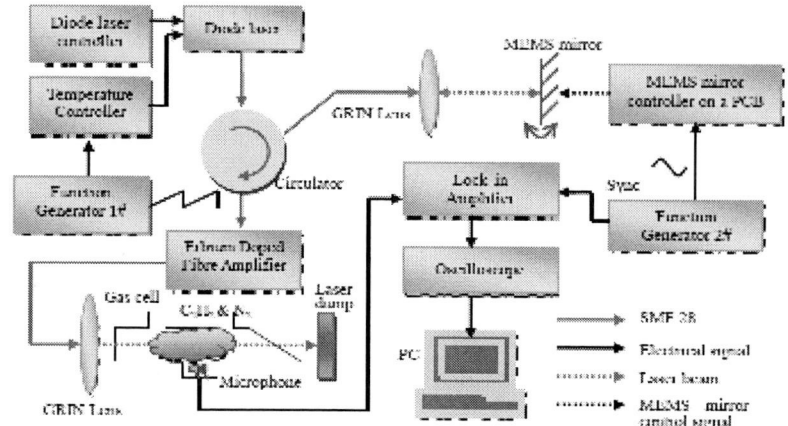

Figure 3. Schematic diagram of photoacoustic equipment

EXPERIMENTAL METHOD AND RESULTS

The external modulator used here is based on a MEMS scanning mirror aligned to a single mode optical fibre. A 2mm diameter gold coated MEMS mirror fabricated from 10μm thick SOI layer and with 100 nm thick gold coating is attached by double-loop serpentine springs to four electrothermal actuators that are fixed to the substrate, Figure 4. Unmodulated laser light

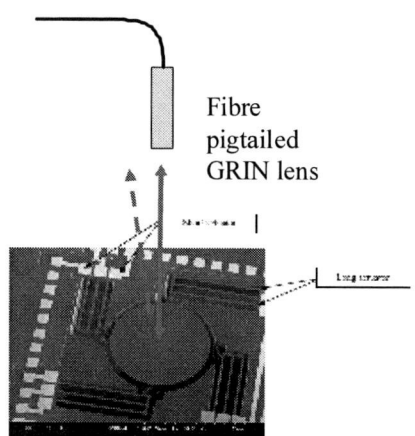

Figure 4. MEMS mirror used as an external modulator

emerging from an optical fibre is collimated using a gradient index (GRIN) lens and is directed onto the MEMS mirror. When the mirror surface is normal to the fibre, the reflected light is recaptured by the GRIN lens and coupled back into the fibre. When the mirror is fully tilted, the reflected light deviates

voltage (at a few kHz) was empirically determined to give the best optical modulation characteristics.

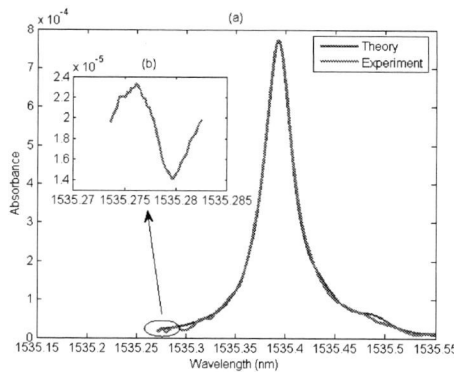

Figure 5. Recovered acetylene lineshape

Figure 5 above shows the absorption line of 100ppm acetylene obtained using 500mW power from the EDFA. The off-line distortion may be due to absorption of light at the cell end face. Based on initial results, we predict a sensitivity of ~1ppm acetylene using this system.

REFERENCES

[1] O Solgaard, Photonic Microsystems, Springer, 2008
[2] Ai-Qun Liu, Photonics MEMS Devices, Boca Raton: CRC Press, 2009

TORSIONAL MIRROR DRIVEN BY A CANTILEVER BEAM INTEGRATED WITH 1x10 INDIVIDUALLY BIASED PZT ARRAY ACTUATOR FOR VOA APPLICATION

Kah How Koh[1], Takeshi Kobayashi[2], and Chengkuo Lee[1]*, *Member, IEEE*

[1] Department of Electrical & Computer Engineering, National University of Singapore,
Singapore 117576
[2] National Institute of Advanced Industrial Science and Technology (AIST),
1-2-1 Namiki, Tsukuba, Ibaraki 305-8564, Japan

ABSTRACT

A gold coated silicon mirror (5mm x 5mm) actuated by 1x10 piezoelectric Pb(Zr,Ti)O$_3$ (PZT) array actuator integrated on a cantilever beam has been demonstrated for variable optical attenuator (VOA) application. Torsional attenuation based on the difference in the dc biasing voltage applied to the 1x10 PZT array actuator was investigated. The dynamic attenuation range of 40dB at 1.8V DC bias is observed in the attenuation curve.

INTRODUCTION

Micro-electro-mechanical systems (MEMS) VOA has been an enabling tool for modern optical network system based on dense-wavelength-division-multiplexed technology. Most of the designs of VOAs reported in literature adopted attenuation schemes which are generally classified into three groups: shutter-type [1], planar reflective-type [2] and 3-D reflective type [3-4]. Design of optics and their integration with large reflective mirrors are typically realized in three-dimensional (3-D) configurations. In conjunction with large micro-optics such as dual-core-fiber collimators, the enabled 3-D VOA devices can gain excellent data of return loss, PDL and WDL under reasonable driving voltage. To achieve large mechanical rotation angle of a large mirror, gimbaled mirrors using torsion bars or springs are the popular design for integration with various actuators such as, electrostatic [3] and piezoelectric [4].

Limited research effort has been reported in 3-D MEMS VOAs in contrast to the reported activities in planar MEMS VOAs [1-2]. Thus, in this work, we explore a new 3-D MEMS VOA driven by 1x10 piezoelectric Pb(Zr,Ti)O$_3$ (PZT) array actuator integrated on a cantilever beam. A novel torsional attenuation based on different dc biasing voltage applied to individual piezoelectric cantilever actuator on a cantilever beam is demonstrated. This actuation mechanism differs greatly from those actuation mechanisms of gimbaled mirror in torsional mode, i.e., the mirror rotation is generated against to torsion spring.

This work was supported by URC Tier 1 Fund R-263-000-475-112 at the National University of Singapore. *E-mail: elelc@nus.edu.sg.

DEVICE DESCRIPTION

A schematic diagram of this torsional mode based 3-D MEMS VOA is shown in Fig. 1(a). The device is micromachined from a SOI substrate of 5μm thick Si device layer, with multi-layers of Ti/Pt/LaNiO$_3$/PZT/LaNiO$_3$/Ti/Pt deposited as electrode materials and the piezoelectric actuator. A silicon mirror plate (5mm long x 5mm wide x 0.4mm thick) and a mechanical supporting silicon beam (3mm long x 5mm wide x 5μm thick) integrated with 1x10 arrayed PZT cantilever actuators (3mm long x 0.24mm wide x 5μm thick) arranged in parallel are formed after the release process. The ten cantilevers are designed to be electrically isolated from one another, with individual bonding pads connected to each of the top and bottom electrodes of the cantilevers. The inset in Fig. 1(a) illustrates the device in torsional mode, where for example, cantilevers 1-5 are biased in such a way that they bend down while cantilevers 6-10 are biased in an opposite way and bend in the opposite direction i.e. bending upwards. The difference in bending directions for the two sets of actuators causes the mirror to rotate along the x-z plane. Fig. 2 shows the device bonded onto a dual in-line package (DIP) and the bond pads on the device were connected by gold bond wires to the metal pins of the dual-in-line (DIP) package as shown in Fig.2.

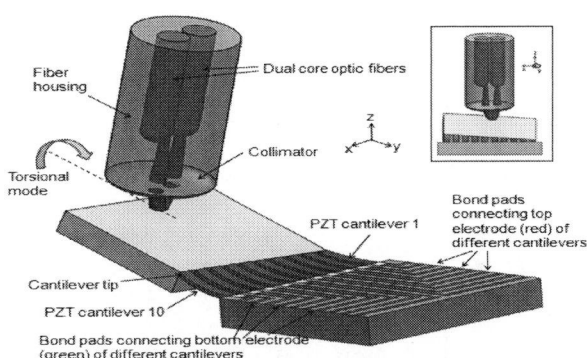

Fig. 1. Schematic drawing showing PZT MEMS VOA with dual core collimator arranged in 3-D configuration. Inset shows torsional mode where a set of five cantilevers bends in one direction while the other set of five cantilevers bends in the opposite direction.

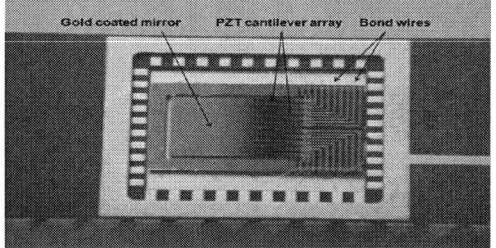

Fig. 2. A closed-up photo showing the packaged PZT MEMS VOA with a gold-coated silicon mirror.

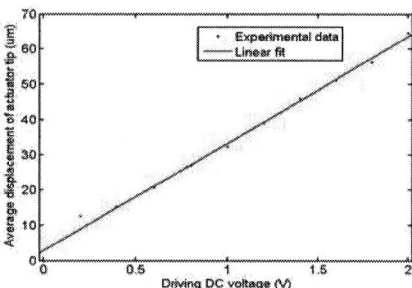

Fig. 3. Measured average displacement of cantilever tips versus dc driving voltage applied to the top electrodes of all ten cantilevers.

EXPERIMENTAL RESULTS

The ten cantilevers were biased simultaneously at the same dc driving voltage while the displacements of the ten cantilever tips were observed under the optical microscope. The measured displacements of ten cantilevers were averaged and repeated for various dc driving voltages. The results were plotted in Fig. 3. A linear fit for the experimental data was also plotted, and based on the approach discussed in [5], we derived d_{31} of the PZT thin films to be 113 ± 3 pmV^{-1}.

Fig. 4(a) shows the biasing configuration to induce torsional mode. A potential divider was implemented to split the dc power supply into four equal potential at the potential nodes between each resistors. As such, each of the cantilevers in both bias cases A and B will have a different dc bias value as evident from the look-up table in Fig. 4(b). It results in largest and zero cantilever displacement introduced at the mirror edges and center respectively. More importantly, the generated displacements for the 2 sets of cantilevers are towards opposite direction, resulting in torsional rotation of the mirror.

Prior to deriving the attenuation characteristic, the packaged device is first mounted on an x-y-z-θy-θz stage and the relative position of the mirror to the collimator is adjusted by moving and tilting the stage such that minimum insertion loss is reached. A 1550nm light source is used in this experiment. The red curve in Fig. 5 shows the measured attenuation curve versus dc voltage of the power supply. A dynamic attenuation range of 40 dB was achieved at a bias of 1.8V. The curve follows weakly that of a Gaussian distribution, similar to that of a Gaussian beam.

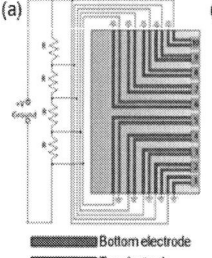

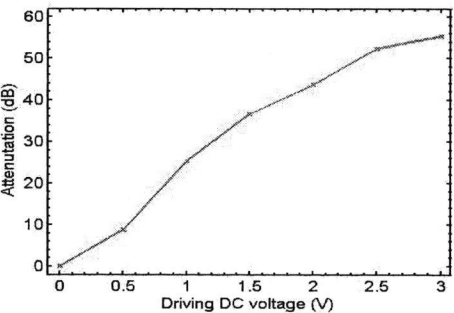

Power Supply	1V	2V	3V
Actuators	Relative bias of top to bottom electrode /V		
1	-1.00	-2.0	-3.00
2	-0.75	-1.5	-2.25
3	-0.50	-1.0	-1.50
4	-0.25	-0.5	-0.75
5	0.00	0.0	0.00
6	0.00	0.0	0.00
7	0.25	0.5	0.75
8	0.50	1.0	1.50
9	0.75	1.5	2.25
10	1.00	2.0	3.00

Bottom electrode
Top electrode

Fig. 4(a). Schematic diagram illustrating the electrical connections of the top and bottom electrodes of each cantilever to the dc power supply. (b) A look-up table showing the individual dc bias driving each cantilever for a dc power supply voltage.

Fig. 5. Measured attenuation curves versus dc driving voltage of power supply for torsional mode.

CONCLUSION

In this study, a novel piezoelectric driven 3-D VOA using mechanical supporting beam integrated with multiple cantilever actuators was explored and characterized in a 3-D torsional attenuation scheme. The rotation of a torsional mirror based on the relative difference in displacements under dc bias for the ten piezoelectric PZT cantilever actuators is demonstrated. The transverse piezoelectric constant, d_{31}, has been estimated to be $113 pmV^{-1}$. A dynamic attenuation range of 40dB was achieved at 1.8V.

REFERENCES

[1] C. Marxer, P. Griss, and N. F. de Rooij, "A variable optical attenuator based on silicon micromechanics," *IEEE Photon. Technol. Lett.*, vol. 11, no. 2, pp. 233–235, Feb. 1999.
[2]] H. Cai, X. M. Zhang, C. Lu, A. Q. Liu and E. H. Khoo, "Linear MEMS variable optical attenuator using reflective elliptical mirror," IEEE Photon. Technol. Lett., vol. 17, pp. 402-404. Feb. 2005.
[3] K. Isamoto, K. Katom A. Morosawa, C. Chong, H. Fujita and H. Toshiyoshi, "A 5-V operated MEMS variable optical attenuator by SOI bulk micromaching," IEEE J. Sel. Top. Quantum Electron., vol. 10, pp. 570-578, May 2004.
[4] C. Lee, F-.L. Hsiao, T. Kobayashi, K. H. Koh, P. V. Ramana, W. Xiang, B. Yang, C. W. Tan and D. Pinjala, "A 1-V operated MEMS variable optical attenuator using piezoelectric PZT thin-film actuators.," IEEE J. Sel. Top. Quantum Electron., vol. 15, Sep. 2009.
[5] T. Kobayashi, M. Ichiki, T. Noguchi, K. Nakamura and R. Maeda, "Fabrication of piezoelectric microcantilevers using LaNiO₃ buffered Pb(Zr,Ti)O₃ thin film," J. Micromech. Microeng., vol. 18, pp. 035007-1-035007-5, 2008.

DESIGN, FABRICATION, AND PACKAGE OF MEMS-BASED IMAGE STABILIZER FOR PHOTOGRAPHIC CELL PHONE APPLICATIONS

Jin Chern Chiou[1,2], Chen-Chun Hung[1] and Chun-Ying Lin[1]
[1]Institution of Electrical and Control Engineering, National Chiao Tung University, Taiwan
[2] School of Medicine, China Medical University, Taiwan, R.O.C.

ABSTRACT

This work presents a MEMS-based image stabilizer applied for anti-shaking function in photographic cell phones. The proposed stabilizer is designed as a two axis decoupling XY stage and strong to suspend an image sensor. Based on the special designs of a hollow handle layer and a corresponding wire bonding assisted holder, electrical signals of the suspended image sensor can be sent out with signal springs. The longest traveling distance of the stabilizer is 25μm with 51V applied voltage and sufficient to resolve the anti-shaking problem in a three megapixel image sensor and with excellent decoupling capability.

INTRODUCTION

The increasing number of camera pixels has exacerbated undesirable image blurring, as caused by hand shaking. Image stabilization function is a popular solution to resolve this problem. Among the familiar elements of image stabilization include lens shifting, [1] CCD shifting, and signal processing. Given the demand for device miniaturization, the lens shifting anti-shaking approach is insufficient since adding a movable lens causes nonlinearity in control that must be compensated for by a complex control algorithm. Despite requiring an actuating system associated with the image sensor, image sensor shifting is less disruptive of miniaturization; in addition, slimming of the system is better than when using the lens shifting method. [2]

This work describes a decoupling actuator to function as an image stabilizer. The design of an MEMS based image stabilizer that is suitable for a cell phone with a three-megapixel camera requires an actuator that can move by at least 25μm, [2] with a sufficiently strong structure to withstand the load of a 6.36 × 6.24× 0.1 mm³ image sensor, from which it must be de-coupled in two dimensions when driven.

DESIGN AND SIMULATION

Figure 1 schematically depicts the device design. One main decoupling beam and three pairs of folded springs of various sizes are designed in each direction to satisfy decoupling requirements. Designs of inner

comb-finger pairs can miniaturize the device size to comply with requirements for embedding in a photographic cell phone. Outer comb-driven finger pairs are designed as assisted-driving components to reduce the driving voltage to be suit for circuit integration in a cell phone module.

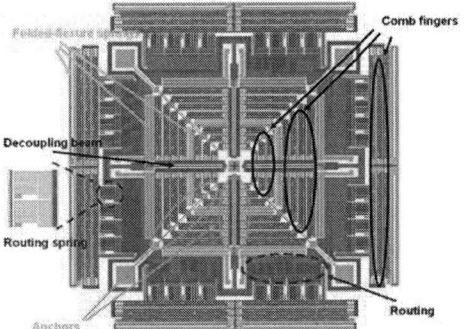

Figure 1. Structure design of XY stage.

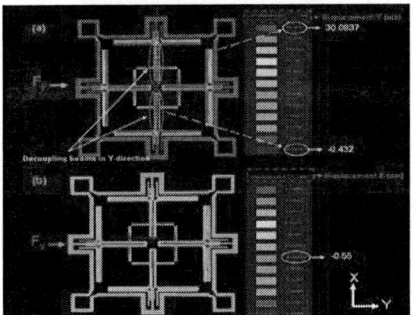

Figure 2. Simulation of decoupling XY stage.

The decoupling effect is simulated using the simplified decoupling composition, as shown in Fig. 2. During the simulation, a specific force F_y is set up in the Y-direction, resulting in a 30μm displacement in the Y-direction. Simulation results indicate that only 0.55μm moving displacement occurs in the X-direction. The simulation decoupling ratio of y displacement to x displacement displays an excellent decoupling effect in the design of the movable springs, which is significantly greater than the decoupling requirement.

FABRICATION AND PACKAGING

The fabrication process is illustrated in Fig.3. Firstly, the structure, routing and anchor patterns are prepared by double side lithography in device wafer. Then, we start with a 250μm-thick silicon wafer for structure holder. Fig.3(a) shows the Si_B for device holder. Then, Al is deposited and patterned as ICP hark mask. Fig.3(d) shows wafer punching through by ICP etching. Then, Cu is deposited for heat sinking. Fig.3(f)(g) shows the device combination by flip-chip bonder. device formation by 3rd ICP etching (i) device release by HF vapor.

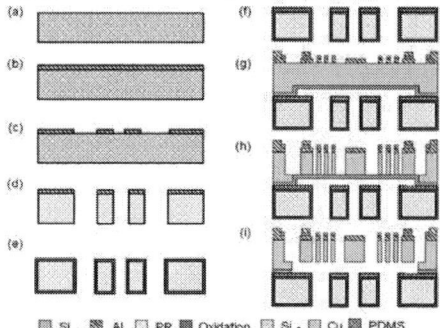

Figure 3. Fabrication process.

The wire-bonding assisted holder is designed for electric connection. Figure.4. shows the wire bonding process. The wire-bonding assisted holder is filled into the holes of substrate holder and sustains the suspended structure during wire bonding process. Figure.5. also shows the image stabilizer attached with an image sensor. The electrical connection is achieved by 32 bonding pads and routing springs.

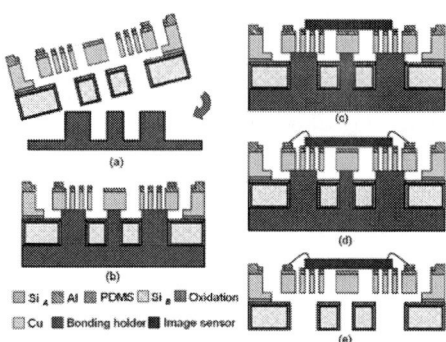

Figure 4. Packaging and wire bonding process.

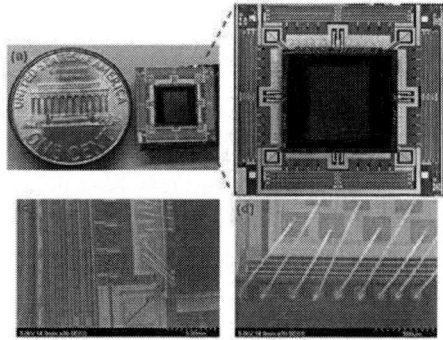

Figure 5. Pictures of decoupling XY stage.

EXPERIMENT RESULTS

To demonstrate the performance of the fabricated device, static characteristic of this 2-D decouple image stabilizer has been conducted. During the static driving test, the actuator was driven by DC voltages. The displacement of the actuator was measured by WYKO. As shown in Fig. 6, the displacement has achieved 25μm when applied 51V driven voltage and only allowed 0.42μm variation in vertical axis direction

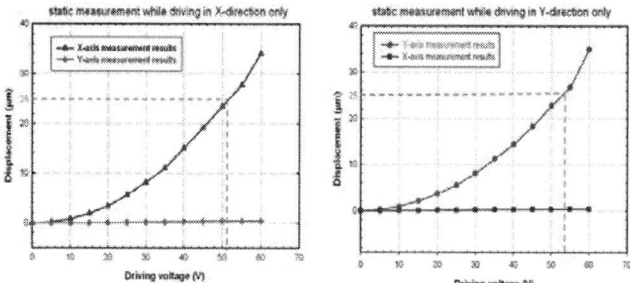

Figure 6. Static measurement of image stabilizer.

CONCLUSIONS

MEMS based image stabilizer has designed, fabricated and packaged. By the special design of wire-bonding assisted holder, we successfully package the image sensor upon the image stabilizer and complete the wire bonding for signal output. Experimental results indicate that a driving voltage of 51 V can cause incur a displacement of 25μm in the driving direction and a displacement of 0.42μm in the vertical direction, which are consistent with the anti-shaking purpose.

ACKNOWLEDGMENT

This work was supported in part by Ministry of Economic Affairs, Taiwan, R. O. C. under contract No. 97-EC-17-A-07-S1-011, and the National Science Council, Taiwan, R. O. C. under contract no. NSC-98-2220-E-009-014, NSC-98-2220-E-009-032, and NSC-98-2218-E-039-001. It was also supported in part by Taiwan Department of Health Clinical Trial and Research Center of Excellence under contracts No. DOH99-TD-B-111-004 and DOH99-TD-C-111-005.

REFERENCES

[1] K. Sato, S. Ishizuka, A. Nikami, and M. Sato 2005 Control techniques for optical image stabilizing system IEEE Trans. Consum. Electron. 39 461-66

[2] Yasuhiro Okamoto and Ryuichi Yoshida 1998 Development of linear actuators using piezoelectric elements Electron. Commun. Jpn. Pt. III 81 11-7

Development of A 2x2 Optical Switch
Using Bi-stable Solenoid-based Actuators

Bonnie Tingting Chia, Cheng-Wen Ma, Bo-Ting Liao, Sun-Chih Shih, and Yao-Joe Yang
Department of Mechanical Engineering, National Taiwan University, Taipei, Taiwan (R.O.C.)

ABSTRACT

This paper presents the development of a 2x2 optical switch which consists of a MEMS-based silicon micro-mirror structure and two bi-stable solenoid-based actuators. The silicon micro-mirror structure is realized by using a proposed simple single-step anisotropic silicon etching process. Bi-stable solenoid-based actuators are designed and developed. The proposed device, which adapts the split-cross bar (SCB) design as the optical-path configuration, has many advantages such as low power consumption, easy fiber alignment, simple manufacturing process, and simple actuation scheme. The measured insertion loss of the device is about -0.9 ~ -1.1 dB. The long-term reliability test shows that the deviations of the insertion losses are less than 0.03 dB after 10,000 switching cycles. Also, the measured cross-talk is about -60 dB, and the measured switching time is less than 10 ms.

INTRODUCTION

Optical switches are essential components in optical communication network. Low-port-count switches, such as 1x2 or 2x2 configurations, are useful for many applications, including network protection and optical add-drop multiplexing. Many research groups have demonstrated several types of 2x2 switches realized by using microelectromechanical systems (MEMS) technologies [1-3]. In this work, we propose a simple fabrication process and a reliable actuation method to realize a 2x2 optical switch using the *split cross-bar* (SCB) design [4]. When compared with the cross-bar switch [1,2], the SCB design does not requires very double-sided thin micro-mirrors which are quite difficult to be fabricated. When compared with mirror-array switches [3], the SCB switch requires fewer movable micro-mirrors so that the system is less complex. Also, the micro-mirror structure of the SCB switch can be monolithically created by using a proposed single-step silicon etching process. Furthermore, bi-stable solenoid-based actuators, which can effectively reduce the system power consumption, are also proposed for actuating micromirrors.

DESIGN

The design and the operational principle of the SCB switch are illustrated in Fig. 1. The bi-stable solenoid-based actuators are used to switch the movable mirrors between State 1 and State 2. As shown in Fig. 1(a), when the actuators do not contact the cantilevers (State 1), the optical signal is reflected by the movable mirror. When the cantilevers are pushed up by the actuators (State 2), the optical signals pass under the movable mirrors and are reflected by the fixed mirrors to the other pair of output channels, as shown in Fig. 1(b).

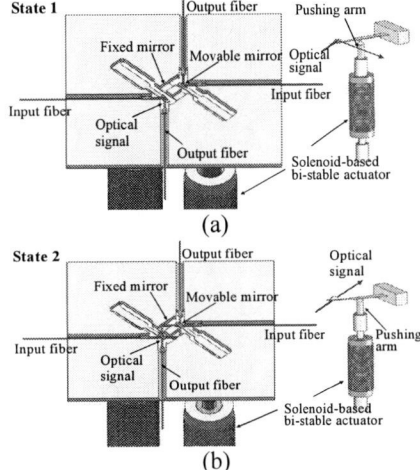

Fig. 1 The operational principle of the SCB switch.

FABRICATION

The front-side and the back-side mask layouts are shown in Fig. 2(a) and 2(b). The schematic drawing of the fabricated silicon micro-mirror structure is plotted in Fig. 2(c). Fig. 3 shows the proposed fabrication process, which is a single-step anisotropic silicon etching process with KOH etchant. The starting material is a double-side polished (100) silicon wafer (Step (i)). Vertical mirrors are formed on plane (100) during this KOH etching process under the conditions of high KOH concentration (55 wt%) and high etching temperature (75℃) (Step (ii)-(iii)) [5]. In addition, KOH solution is added with isopropyl alcohol (IPA) for reducing the roughness of the etched surfaces. After removing nitride layer (Step (iv)), a 2000 Å gold layer of about is deposited on the micro-mirror structure for improving the optical reflectivity of the mirrors (Step (v)). The fabricated micro-mirror structure of the SCB switch is shown in Fig. 4(a) and 4(b). The assembled system is shown in Fig. 5(a). Fig. 5(b) is the proposed solenoid bi-stable actuator. The exploded view of the system is illustrated in Fig. 5(c).

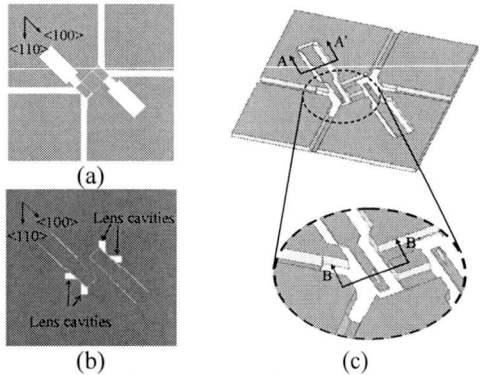

(a)

(b)   (c)

Fig. 2   Mask layouts and schematic drawing of the micro-mirror structure.

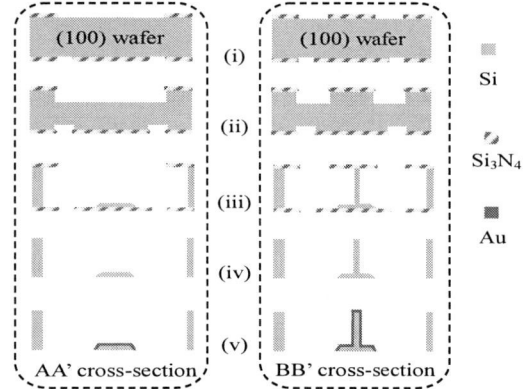

Fig. 3   Fabrication process of the micro-mirror structure.   The AA' and BB' cross-sections are indicated in Fig. 2(c).

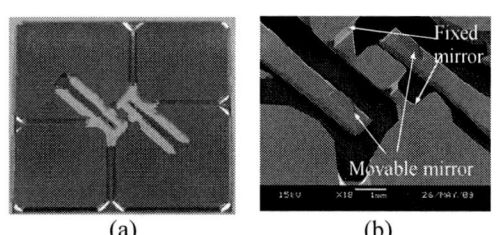

(a)   (b)

Fig. 4 Photos of the fabricated micro-mirror structure.

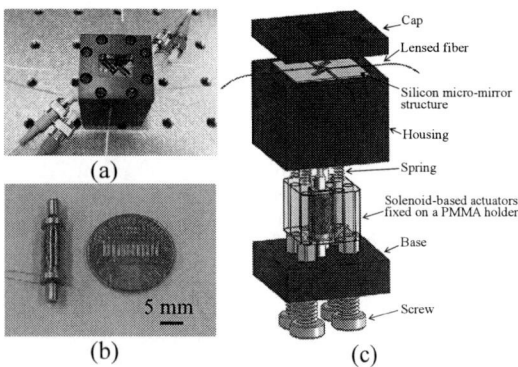

(a)   (b)   (c)

Fig. 5 (a) The assembled SCB switch; (b) the proposed solenoid bi-stable actuator; (c) the exploded view of the system.

MEASUREMENT RESULTS

The measured switching time is about 7 ms when the mirror switches from State 1 to State 2 (Fig. 6(a)). Also, the switching time is about 10ms for switching from State 2 to State 1 (Fig. 6(b)). The measured insertion losses for 100 switching cycles are shown in Fig 7(a).   The average values for State 1 and State 2 are -1.1dB and -0.9dB, respectively.   Fig. 7(b) shows the long-term reliability test result which is performed by continuously actuating the device with 10,000 switching cycles.   Each data point is the average of 100 measured data.   The error bars indicate a 95% confidence interval.   The deviations of the measured insertion losses are less than 0.03dB after 10,000 switching cycles.   In addition, the measured cross-talk between each channel is about -60 dB.   Theses optical performance of the switch is measured at the wavelength of 1550nm.

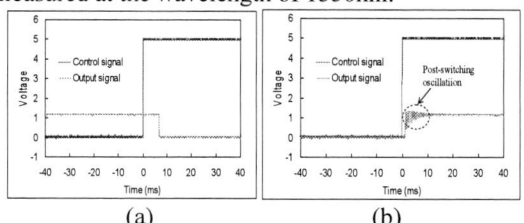

(a)   (b)

Fig. 6   Measured dynamic responses of the switch.

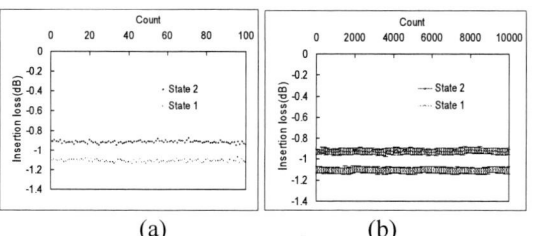

(a)   (b)

Fig. 7   Measured insertion losses of the switch.

REFERENCES

[1] C. Lee and C. Y. Wu, "Study of electrothermal V-beam actuators and latched mechanism for optical switch," *J. Micromech. Microeng.* vol.15, pp.11-19, 2005.

[2] W. C. Chen, C. Lee, C. Y. Wu, and W. L. Fang, "A new latched 2 x 2 optical switch using bi-directional movable electrothermal H-beam actuators," *Sens. Actuators A-Phys.* vol.123-124, pp.563-569, 2005.

[3] Z. L. Huang and J. Shen, "Latching micromagnetic optical switch," *J. Microelectromech. Syst.* vol.15, pp.16-23, 2006.

[4] Y. J. Yang, B. T. Liao, and W. C. Kuo, "A novel 2x2 MEMS optical switch using the split cross-bar design", *J. Micromech. Microeng.* vol.17, pp.875-882, 2007.

[5] O. Powell and H. B. Harrison, "Anisotropic etching of {100} and {110} planes in (100) silicon", *J. Micromech. Microeng.* vol.11, pp.217-220, 2001.

FABRICATION AND EVALUATION OF PIEZOELECTRIC DRIVE TYPE 2-AXIS TILT CONTROL DEVICE USING EPITAXIAL PZT THIN FILM

Katsuya Ozaki, Daisuke Akai, Kazuaki Sawada and Makoto Ishida
Toyohashi University of Technology, Japan

ABSTRACT

Piezoelectric drive type 2-axis tilt control device using epitaxial $Pb(Zr_{0.52},Ti_{0.48})O_3$ (PZT) thin film on epitaxial γ-Al_2O_3/Si substrate have been fabricated. A 500 nm-thick epitaxial PZT(111) thin film was sol-gel deposited on epitaxial $SrRuO_3$(111)/Pt(111)/γ-Al_2O_3(111)/Si(111) substrates, and hollow structure under actuate area was formed by XeF_2 gas etching. The fabricated device showed polarization-electric field (P-E) hysteresis loop of piezoelectric PZT thin film and the deflection. Therefore, the realization of piezoelectric drive type deformable mirrors (DMs) using epitaxial PZT thin film can be expected.

INTRODUCTION

Growing demands of adaptive optics (AO) for the industrial and medical applications have propelled the development of small-size and low-cost DMs. To achieve small and low-cost DMs, micro electro mechanical systems (MEMS) DMs have successfully been developed and manufactured as commercial products. The MEMS DMs may be classified in two groups: electrostatic drive type and piezoelectric drive type. Piezoelectric drive type can generate a large force, and its driving voltage is low. Recently, the DMs actuated by PZT thin films have been reported [1, 2]. However, these DMs are used poly crystalline PZT thin films. The deflection characteristics of piezoelectric DMs depend on crystalline of piezoelectric film. Our laboratory has studied about epitaxial γ-Al_2O_3 films as an insulating material on Si substrates and their ferroelectric device application [3-5]. Using crystalline insulator on Si substrates, it is useful for controlling orientation of piezoelectric film. In this work, piezoelectric drive type 2-axis tilt control device using epitaxial PZT thin films on epitaxial γ-Al_2O_3/Si substrate have been fabricated for application to AO.

FABRICATION

The process flow of the fabricated piezoelectric drive type 2-axis tilt control device is shown in Fig. 1. (a) γ-Al_2O_3 thin film was epitaxially grown on Si(111) substrate by Molecular Beam Epitaxy (MBE). The thickness of γ-Al_2O_3 is 10nm. Bottom electrode $SrRuO_3$/Pt was epitaxially grown on the γ-Al_2O_3 film by rf-sputtering. A 500 nm-thick Piezoelectric thin film PZT (with Zr:Ti ratio=52:48) was epitaxially grown on the substrate by sol-gel method. Top electrode Pt was deposited on the PZT film by rf-sputtering. (b) Top electrode, piezoelectric thin film and bottom electrode were patterned by Inductive Coupled Plasma-Reactive Ion Etching (ICP-RIE). (c) Interlayer dielectric SiO_2 was deposited and aluminum wiring was formed. (d) The etching hole for forming hollow structure was opened by ICP-RIE. (e) Hollow structure was formed by etching silicon under actuate area using XeF_2 gas. Fig. 2 show a view of fabricated piezoelectric drive type 2-axis tilt control device.

EVALUATION

The P-E hysteresis loop of the PZT thin film on epitaxial $SrRuO_3$/Pt/γ-Al_2O_3/Si substrate is shown in Fig. 3. The remnant polarization Pr and spontaneous polarization Ps are $9.6\mu C/cm^2$ and $28.7\mu C/cm^2$, respectively. The coercive electric field Ec was 18kV/cm. When the PZT thin film was applied DC voltage $\pm$ 20V, the deflection of fabricated device is shown in Fig. 4. This figure shows that the form of fabricated device deforms by applying voltage.

CONCLUSION

Piezoelectric drive type 2-axis tilt control device using epitaxial PZT thin film on epitaxial γ-Al_2O_3/Si substrate have been fabricated. The fabricated device showed P-E hysteresis loop of piezoelectric PZT thin film and the deflection. Therefore, the realization of piezoelectric drive type DMs using epitaxial PZT thin film can be expected.

REFERENCES

[1] I. Kanno et al, IEEE J. Sel. Topics Quantum Electron, 13 (2007), pp. 155-161
[2] T. Kunisawa et al, J. Soc. Mech. Eng, (2005), pp. 117-118
[3] D. Akai et al, Appl. Phys. Lett., 86 (2005) 202916
[4] D. Akai et al, Dig. of Tech. Papers Transducers '05, (2005), pp. 307-310
[5] M. Ito et al, Dig. of Tech. Papers Transducers '07, (2007), pp. 1275-1278

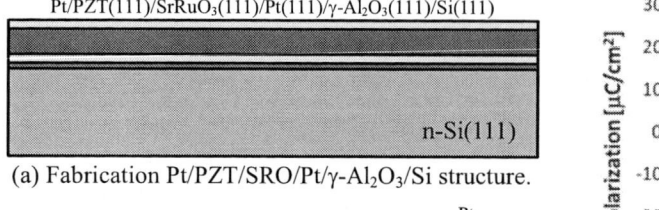

Pt/PZT(111)/SrRuO₃(111)/Pt(111)/γ-Al₂O₃(111)/Si(111)

(a) Fabrication Pt/PZT/SRO/Pt/γ-Al₂O₃/Si structure.

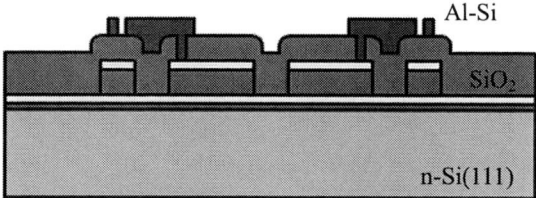

(b) Patterning Top Pt/PZT and bottom SRO/Pt.

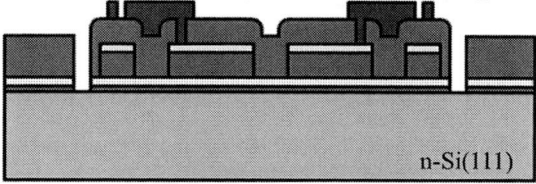

(c) Formation interlayer dielectric and wiring.

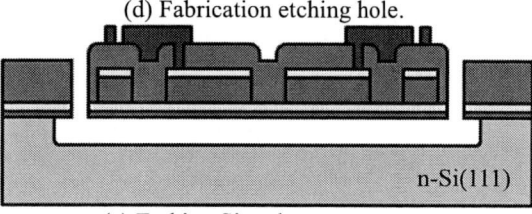

(d) Fabrication etching hole.

(e) Etching Si under actuate area

Fig. 1 Process flow of the fabricated piezoelectric drive type 2-axis tilt control device using epitaxial PZT/SrRuO₃/Pt/γ-Al₂O₃/Si substrate.

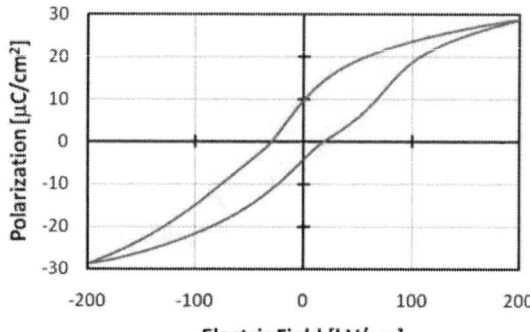

Fig. 3 The P-E hysteresis loop of epitaxial PZT thin film on epitaxial SrRuO₃/Pt/γ-Al₂O₃/Si substrate.

(a) Before voltage is applied.

(b) After DC 20V is applied.
Fig. 4 the deflection of fabricated device.

Fig. 2 Photograph of fabricated device. Chip size is 10mm × 10mm.

978-1-4244-8926-8/10 $26.00 © 2010 IEEE

COMPLIANT SCANNING MICROMIRROR ACTUATED WITH A DISPLACEMENT AMPLIFICATION MECHANISM

Tzung-Ming Chen[1,2], Florian Schneider[1], Ulrike Wallrabe[1]

[1]University of Freiburg, Germany

[2]Changhua University of Education, Taiwan

ABSTRACT

We present the performance of a piezo-actuated compliant micromirror made of single crystal silicon, which is assembled with a compliant displacement amplification mechanism made of copper. Both of those two parts are purely based on the elasticity and deformation of thin beams. The measurement results reveal that our optical device achieves the specified function as a linear optical scanner, and that the kinetic motion and out-of-plane bending are under control.

INTRODUCTION

Micromirrors are important devices, which have been well discussed in MEMS [1,2]. Normally, the specific rotational angle of the mirror, the motion response to the actuation source, and its application had to be included into the design considerations of a new micromirror. Our device is pushed by a piezo-electrical actuator in order to transfer a linear displacement to a rotational angle. It is to be used, as an optical scanner, e.g.. Our previous research [3,4] had discussed all of the considerations relying on complete design and analysis. In this paper the performance of this micromirror is ensured by means of static and kinetic measurements. There are three challenges in the fabrication and assembly procedure: firstly, an etch process with suitable parameters for multi-thin-beam structures, secondly, a simplified fabrication process for a suspended structure, and finally, an amplified input displacement for compensating the limited extension of the piezoelectrical actuator in order to get the maximum output rotational angle of the mirror.

FABRICATION

The purpose of this stage is to fabricate a 11 mm × 11 mm square chip with 0.5 mm thickness from (100) single crystal silicon wafer, in which a compliant micromirror is suspended, see Figure 1a. The dimension of the suspended compliant structure is 2.2 mm × 2.6 mm, and the length of the mirror part is 1.2 mm. The minimum beam width is 20 μm and the thickness of the suspended structure is 0.15 mm.

The STS Multiplex ICP is used as the processing machine to shape multi-thin-beam structures.

Photoresist AZ9260 and AZ4533 are used as the mask during etching, and SiO_2 is used as the stop layer. Besides, the handle wafer is adhered on the backside of the etched wafer with photoresist to avoid breakage and overcome thermal problems. Firstly, front side etching is performed in order to shape thin beam structures. Secondly, a SiO_2 layer is deposited to cover and protect sidewalls of etched silicon structures. Then, the backside etching is done to open the cavity of the silicon wafer and suspend the mirror structure. Furthermore, the SiO_2 layer is removed in a HF solution followed by drying the wafer and getting the mirror structure. The fabricated mirror structure is shown in Figure 1b.

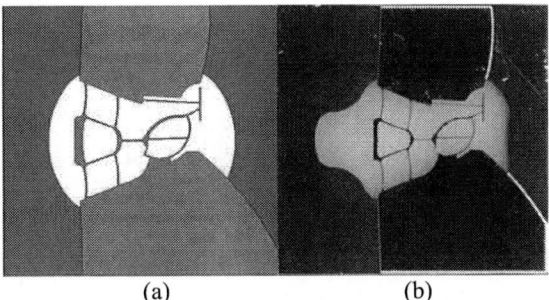

(a)   (b)

Figure 1 (a) The sketch of the micromirror structure. (b) The fabricated chip, in which a compliant mechanism is suspended.

Figure 2 The micro device is assembled of three parts. The maximum extension of the piezo (A) is 13 μm, which is amplified by the copper mechanism (B) to a 40 μm horizontal displacement, in order to push the compliant mechanism (C).

ASSEMBLY

This optical device includes three parts, the silicon chip, the copper amplification mechanism, and the piezoelectrical actuator, see Figure 2. All of those parts are assembled and fixed with glue on a base. The dimension of the commercially available piezo-electrical actuator, which is located in the center space of the copper mechanism, is 9 mm × 2 mm × 3 mm. The stroke of the piezoelectrical actuator is 13 μm at 120 V excitation voltage. In order to compensate for the difference between the required 40 μm of the input displacement and the stroke of the actuator, a copper amplification mechanism with an amplification ratio of 3.18 had been designed and fabricated. Considering the strength of the copper, the thickness of the amplification mechanism is 0.5 mm.

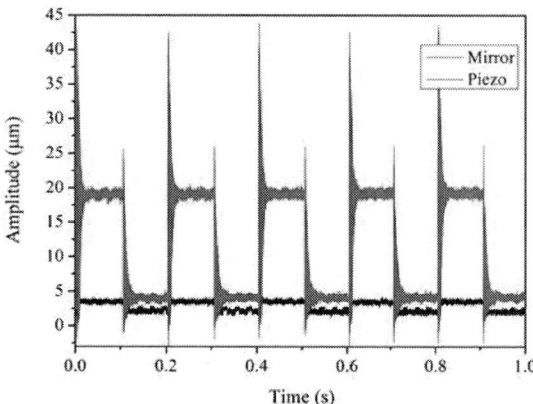

Figure 3 The kinetic behavior of the micromirror is tested with a 5 Hz square wave excitation voltage. The abnormal oscillation caused by the signal spike is eliminated in 30 ms.

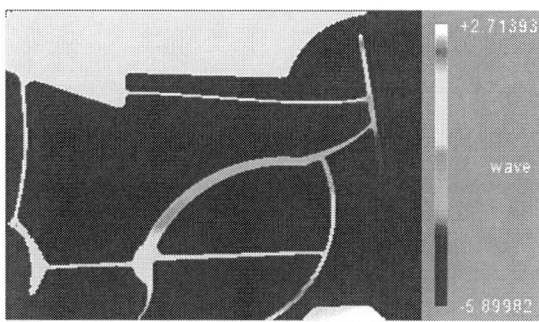

Figure 4 Scanning white-light interferometry is used to measure the lateral movement of the beam structure. The out-of-plane bending is less than 10 μm.

MEASUREMENT AND APPLICATION

In a kinetic measurement, the response of the micro-mirror and the actuator are compared. Figure 3 shows the oscillation of the compliant mechanism with a 5 Hz square wave excitation voltage. The overshooting of

the mirror caused by the square wave volatge is eliminated in 30 ms. Figure 4 displays that the micro-mirror in an actuated situation when measured with a scanning white-light interferometer. The depicted out-of-plane bending is less than 10 μm. Finally, the reflection of a laser beam proves that the micromirror can be used as a linear optical scanner, see Figure 5.

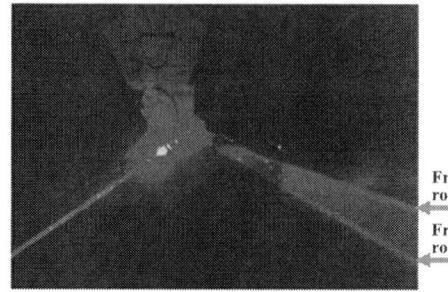

Figure 5 An optical view with low speed photography shows the reflection of a laser beam from the 0° and 6° rotational angle of the miror, separately.

CONCLUSION

From above results we conclude here that our fabricated micromirror, which is made of single crystal silicon, achieves the specified function as a linear optical scanner, and both of kinetic motion and out-of-plane bending are under control.

ACKNOWLEDGMENT

We appreciate the fabrication support from IMTEK cleanroom staff. We are also grateful to Mr. Jens Brunne, Mr. Jan Draheim and Mr. Stefan Krausse for the measurement support.

REFERENCES

[1] L. Zhou, J. Kahn and K. Pister, "Scanning micromirrors fabricated by an SOI/SOI wafer bonding process" Journal of microelectromechanical systems, vol. 15, pp. 24-32, 2006

[2] K. Wang, K. Wei, M. Sinclair and K. Boehringer, "Micro-optical components for a MEMS integrated display" The 12th International Workshop on The Physics of Semiconductor Devices (IWPSD), Dec. 16-20, Chennai, India, 2003, invited paper

[3] T. Chen, Z. Liu, J. G. Korvink, S. Krausse and U. Wallrabe, "Topology Optimization for Micro Rotational Mirror Design and Safe Manufacturing" Proc. of IEEE MEMS Conference, Sorrento, Italy, Jan. 25-29, 2009, pp. 1019-1022

[4] T. Chen, S. Krausse, J. G. Korvink and U. Wallrabe, "12° Design Rule for Single Crystal Silicon Curved Beam Compliant Mechanisms with Large Deformation" Proc. of IEEE MEMS Conference, Hong Kong, China, Jan. 24-28, 2010, pp. 552-555

978-1-4244-8926-8/10 $26.00 © 2010 IEEE

MULTILAYER PIEZOELECTRIC CERAMIC ACTUATOR FOR LASER SCANNER

Jae-Sung Song, In-Sung Kim, Soon-Jong Jeong, Min-Soo Kim

Korea Electrotechnology Research institute, Korea

ABSTRACT

This work studies the lead magnesium niobate-lead zirconium titanate (PMN-PZT) multilayer ceramic actuators (MLCA) for laser scanner applications. PMN-PZT MLCA was fabricated. $5\times5\times30$ mm^3 size MLCA was fabricated by conventional tape casting method. The displacement of the MLCA with 30 mm thickness was ~30 μm at 150 V. Tilt stage consisting of multilayer actuator and two hinge type lever mechanisms was investigated. Circular hinge type mechanism showed translation from linear motion to rotation motion along with stroke amplification. The hinge-based mechanical platform showed one-to-one linear translation with strain amplifier, which is suitable for high precision control system.

INTRODUCTION

Piezoelectric materials for actuators are widely used in applications requiring precision displacement control or high generative force [1]. In particular, multilayer piezoelectric ceramic actuators (MLCA) have been widely studied because the assets of MLCA are a rapid operation, low power consumption, high precision control and no noise compared with the electromagnetic actuators [2]. The most widely used piezoelectric ceramics are lead based materials because of their superior piezoelectric properties [3,4]. $0.2Pb(Mg_{1/3}Nb_{2/3})O_3-0.8Pb(Zr_{0.475}Ti_{0.525})O_3$ (PMN-PZT) ceramic has not only high displacement, and force due to its high piezoelectric coefficient, electromechanical coupling factor of 0.65 near morphotropic phase boundary, along with high temperature stabilization. The characteristics of piezoelectric actuators using reverse piezoelectric effects have been studied [5]. Especially, non-linear piezoelectric characteristics with respect to domain wall motion have been intensively investigated under electric field and compressive stress loading.

Piezoelectric ceramics have been widely used in various actuation applications because of high accuracy as well as fast response. Since the actuator got disadvantage of low strain/stroke and its hysteresis nature, the actuator should be used in a connection of mechanical flexure and control of motion mechanism of actuator for precision moving system. The combination of actuator and mechanical flexure are employed to many moving systems.

In this study, PMN-PZT MLCA for laser scanner applications was investigated.

EXPERIMENTAL PROCEDURE

The composition chosen for the present study was $0.2Pb(Mg_{1/3}Nb_{2/3})O_3-0.8Pb(Zr_{0.475}Ti_{0.525})O_3$. Powder synthesis was carried out using the columbite method to eliminate pyrochlore phase in calcination process. For the MLCAs, conventional tape-casting method was employed to fabricate devices from the powder. The fabricated powder was mixed with organic additives. The ceramic slurry was made in a form of thin tape with a thickness of 100 μm using the doctor blade method. 70Ag-30Pd electrode was printed on the 100 μm-thick piezoelectric green sheet, and 10 layers of the sheets were then laminated. $5\times5\times30$ mm^3 MLCAs were fabricated by a sintering process at an elevated temperature 1100 °C for 2 hrs. All alternative electrodes were connected to the external electrode. An Ag-based external lead was applied through a screen printing method. To measure their displacement characteristics, the curve of displacement versus electric field was made by the laser vibrometer (Graphtec Demodulator AT 3700). To measure resonance frequency, HP 4194A impedance/phase analyzer was employed. And resonance frequency was determined in the impedance-frequency curves taken from the impedance/phase analyzer.

RESULTS AND DISCUSSION

$5\times5\times30$ mm^3 size PMN-PZT MLCA was fabricated by conventional tape casting method (Fig.1). Figure 2 shows the longitudinal displacement of the MLCA as a function of AC voltage at 0.2 Hz at room temperature. Its displacements are 35 μm at 150 V.

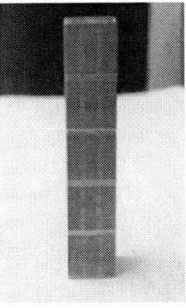

Figure 1. $0.2Pb(Mg_{1/3}Nb_{2/3})O_3-0.8Pb(Zr_{0.475}Ti_{0.525})O_3$ multilayer ceramic actuator with $5\times5\times30$ mm^3 size.

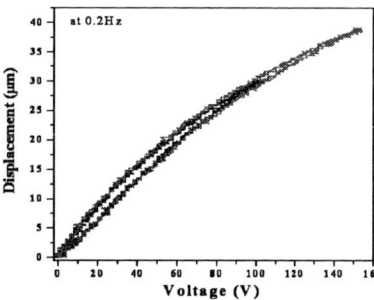

Figure 2. Longitudinal displacement as a function of the AC voltage at 0.2Hz for $0.2Pb(Mg_{1/3}Nb_{2/3})O_3$-$0.8Pb(Zr_{0.475}Ti_{0.525})O_3$ multilayer ceramic actuator

For laser scanner application of MLCAs, tilt stage consisting of MLCA and two hinge type lever mechanisms were designed by ANSYS. Circular hinge type mechanism showed translation from linear motion to rotation motion along with stroke amplification. On the motion translation, the displacement of 318 μm and rotation angle of 6 degree were obtained with a resolution of 0.005 degree. The hinge-based mechanical platform showed one-to-one linear translation with strain amplifier, which is suitable for high precision control system.

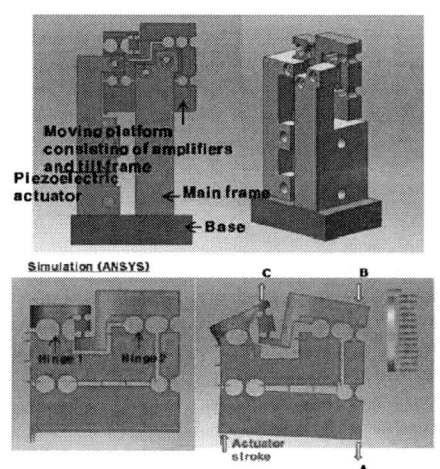

Figure 3. Schematic diagram of tilt stage and simulation result by ANSYS

Laser scanner system using tilt stages with MLCA was fabricated. Figure 4 shows the design and picture of the laser scanner system using two tilt stages with MLCAs. This system shows that the effective scanning area is 30x30 mm^2 and the scanning speed was upto 1 m/s. These properties indicate that this system is potentially good candidate for a wide range of optical applications.

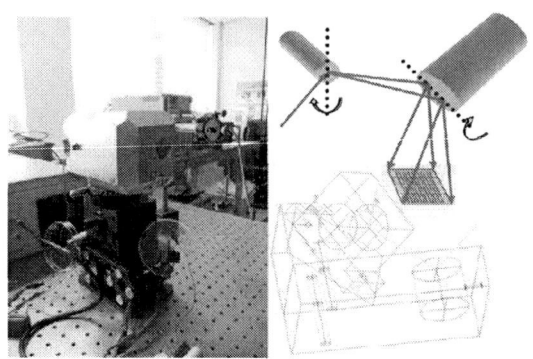

Figure 4. Laser scanner system using two tilt stages with MLCA.

CONCLUSION

This work studies the lead magnesium niobate-lead zirconium titanate (PMN-PZT) multilayer ceramic actuators (MLCA) for laser scanner applications. PMN-PZT MLCA was fabricated. 5×5×30 mm^3 size MLCA was fabricated by conventional tape casting method. The displacement of the MLCA with 30 mm thickness was ~30 μm at 150 V. Tilt stage consisting of multilayer actuator and two hinge type lever mechanisms was investigated. Circular hinge type mechanism showed translation from linear motion to rotation motion along with stroke amplification. The hinge-based mechanical platform showed one-to-one linear translation with strain amplifier, which is suitable for high precision control system. These MLCAs excellent piezoelectric and electro-mechanical properties indicate that this system is potentially good candidate for a wide range of electro-mechanical transducer applications.

REFERENCES

[1] K. Uchino, Expansion from IT/Robotics to Ecological/Energy Applications: Actuator2006, ed. by H. Borgmann (HVG, Bremen, 2006), pp. 48
[2] M. Suga, M. Tsuzuki, "Improved Drop Ejection Characteristics through Use of Micro-Valves in Ink Jet Head" Jpn. J. Appl. Phys., vol. 23, pp. 765-773, 1984
[3] B. Jaffe, R.S. Roth, S. Marzullo, "Piezoelectric Properties of Lead Zirconate-Lead Titanate Solid-Solution Ceramics" J. Appl. Phys., vol. 25, pp. 809-810, 1954
[4] Hirofumi Yamaguchi, "Behavior of Electric-Field-Induced Strain in PT-PZ-PMN Ceramics" J. Am. Ceram. Soc. vol. 82, pp. 1459-1462, 1999
[5] R. Newnham and G. R. Ruschau, "Smart Electroceramics" J. Am. Cer. Soc. vol.74, pp. 463-480, 1991

DROPLET-BASED LATERAL TUNABLE OPTOFLUIDIC MICROLENS ARRAY WITH PNEUMATIC CONTROL

Ye Liu and Hongrui Jiang
University of Wisconsin at Madison, USA

ABSTRACT

We report on a droplet-based, *in situ* formed tunable cylindrical microlens array in microchannels. The deionized water microlenses are formed via liquid-air interfaces of liquid droplets at T-shaped junctions of octadecyltrichlorosilane (OTS) treated polymerized isobornyl acrylate [poly(IBA)] microchannels, and their focal length can be tuned individually by pneumatic manipulation. These microlenses possess low spherical aberrations. Gravity, which affects the shape of the liquid microlenses, is considered in the design. Fluorescence enhancement for potential lab-on-a-chip applications is also demonstrated using such microlens array.

1 INTRODUCTION

Lateral optofluidic microlenses *in situ* formed on a microfluidic chip with tunable focal length are critical in efficient illumination and collection of light from limited areas in the channel [1, 2, 3], and would benefit many lab-on-a-chip applications, such as single cell analysis and single molecule sensing [4, 5].

Compared to fluid-dynamics-based microlenses [2, 3], droplet-based microlenses consume less chip area, thus more conducive to formation of multiple lenses with independent control, and do not require continuous fluidic flow for their operation. We previously demonstrated a simple method to fabricate lateral liquid (deionized, "DI" water) microlenses *in situ* and tune their focal lengths, all using pneumatic manipulation of droplets in microchannels [1]. Here, we extended the fabrication technology to realize lens arrays within microfluidics, where each lens can be tuned in focal length individually. We also studied the optical properties (aberrations) of the lenses, and the effect of gravity on their shape. Enhancement of fluorescence emission in a microchannel utilizing our microlenses was also demonstrated.

2 FORMATION AND TUNING OF THE MICROLENS ARRAY

The fabrication process of the microchannel network is similar to that described in [1], through liquid-phase photopolymerization of photosensitive polymers. The process requires only a single photomask film. The formation process of a 2-lens microlens array is shown in Fig. 1. Liquid droplets were pneumatically segmented from a static fluid and then guided along the designated channels to a desired junction by air pressures. The inner surfaces of the channels were chemically treated to form a hydrophobic OTS monolayer so that the curved

liquid–air interfaces of the droplet are perpendicular to both the top and bottom surfaces, thus a cylindrical lens [Fig. 1(a)], if gravitational effect is negligible.

The shape of the microlens is determined through Young-Laplace equation:

$$\Delta P = \gamma(1/R_1 + 1/R_2) \qquad (1)$$

where the radii of curvature $1/R_1$ and $1/R_2$ of the two lens facets of the droplet are determined by the surface tension γ and the pressure difference ΔP across the droplet. Therefore, the shapes and thus the focal-lengths of the microlenses can be easily tuned by adjusting the air pressure difference across their two faces, as shown in Fig. 2.

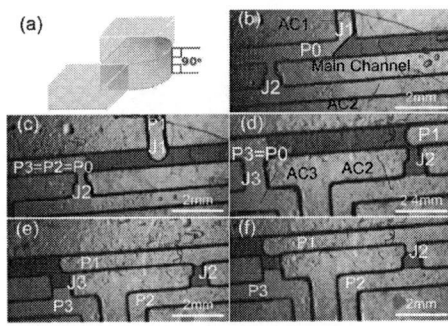

Fig. 1 (a) Schematic of a cylindrical lens. (b-f) *In situ* formation process of a 2-lens liquid microlens array within a microchannel network. (b) DI-water stream is flowed into the main channel, and an air plug starts to be injected through an air conduit (AC1). (c) Water droplet is cut out of the main stream by air pressure P1 at junction J1. (d) Lens droplet is split into the lens channel at junction J2, and air pressure in AC2 can be tuned to pin the lens droplet fraction at J2. The remaining droplet in the main channel is driven further towards junction J3 by P1. (e) P1 continues to increase in AC1, droplet splits at Junction J3, and air pressure in AC3, P3, is

978-1-4244-8926-8/10 $26.00 © 2010 IEEE 131

monitored. (f) The droplet is about to completely split at junction J3 and be pinned at the edges of the junction, and the microlens droplet at J2 remains pinned at J2. Notice the shape shift of the liquid lens at J2 is due to the pressure change of P1. P0 denotes atmospheric pressure.

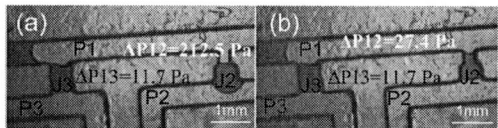

Fig. 2 Individual tuning of microlenses in the two-microlens array. The microlens at J2 is tuned in a wide range by changing P2, while the microlens at J3 remains its shape since ΔP13 is held at a constant value.

4 DESIGN CONSIDERATIONS AND FLUORENSENCE ENHANCEMENT

The effect of gravity is critical for a liquid microlens to maintain a cylindrical shape. The lateral surfaces (two lens facets) become asymmetrically deformed if the mass of the lens material is large enough. Simulation using Surface Evolver® suggests that the critical thickness of the microlenses where gravity causes obvious asymmetry for the lens surface profile is 0.732 mm for DI water (Fig. 3). In our experiments, the thickness of the lenses was ~400μm to ensure the cylindrical shape.

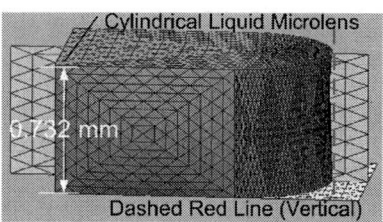

Fig. 3 Simulation results of a cylindrical liquid microlens. A gray vertical "wall" is placed to clearly display the cross-section view of the lateral profile. No Observable asymmetry occurs until the thickness of the liquid lens reaches 0.732mm.

We also calculated the focal lengths and aberrations of our lenses using ray-tracing simulation (Zemax®). For example, the l at J2 as shown in Fig. 2(a) has a spherical aberration coefficient of 0.0207, and a paraxial focal length of 1.03 mm.

The tunable microlenses were applied to enhance the fluorescence emission in a microchannel, as shown in Fig. 4. By tuning the pressure difference across the lens, excitation light can be focused into selected sections of a channel containing fluorescence sodium dye (0.3 mM/L) to increase the intensity of the fluorescence emission (Fig. 4).

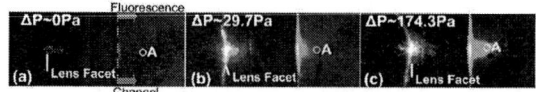

Fig. 4 Fluorescence enhancement using one lateral tunable microlens. Excitation light: blue; fluorescence: green. High light intensity was observed at interfaces due to scattering and reflection. Left interface (blue): the air-liquid interface of the microlens. As the lens becomes more convex (higher focusing power) from (a) to (c), fluorescence emission in the fluorescence channel is increased. For example, the fluorescence intensity at Spot A, a selected spot in the fluorescence channel, is increased by a factor of 37.

CONCLUSION

We demonstrated droplet-based lateral tunable cylindrical liquid microlens array in microfluidics. These lenses had low spherical aberrations, and their focal length could be tuned individually. The gravity effect on the cylindrical shape was found to be negligible when the thickness of the DI-water microlens does not exceed 0.732mm. Using such microlenses to focus excitation light into microchannels, fluorescence emission could be enhanced by as much as 37 times, showing promise of such lenses in lab-on-chip applications.

ACKNOWLEDGEMENT

This work was supported by the U.S. Department of Homeland Security, through a grant awarded to the National Center for Food Protection and Defense at the University of Minnesota (DHS-2007-ST-061-000003), the U.S. National Science Foundation (ECCS 0622202) and the Wisconsin Institutes for Discovery.

REFERENCE

[1] L. Dong and H. Jiang, *J. Microelectromech. S.*, vol. 17, no. 2, pp. 381 - 392, 2008
[2] S. Tang, C. A. Stan and G. M. Whitesides, *Lab Chip*, vol 8, pp. 395 – 401, 2008.
[3] X. Mao, J. R. Waldeisen, B. K. Juluri and T. J. Huang, *Lab Chip*, vol. 7, pp. 1303-1308, 2007.
[4] B. Huang, H. Wu, D. Bhaya, A. Grossman, S. Granier, B. K. Kobilka and R. N. Zare, *Science*, vol. 315, pp. 81-84, 2007.
[5] T.-H. Huang, Y. Peng, C. Zhang, P. K. Wong and C.-M. Ho, *J. Am. Chem. Soc.*, vol. 127, pp. 5354-5359, 2005.

Excellent fault tolerance of a MEMS optically differential reconfigurable gate array

Hironobu Morita and Minoru Watanabe

Electrical and Electronic Engineering, Shizuoka University

3-5-1 Johoku, Hamamatsu, Shizuoka, 432-8561, Japan

Email:tmwatan@ipc.shizuoka.ac.jp

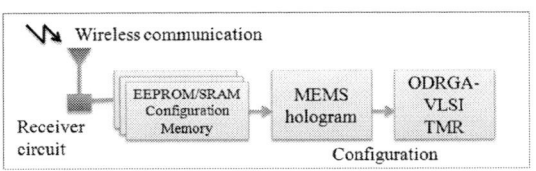

Fig. 1. Block diagram of a reconfigurable system for use in space.

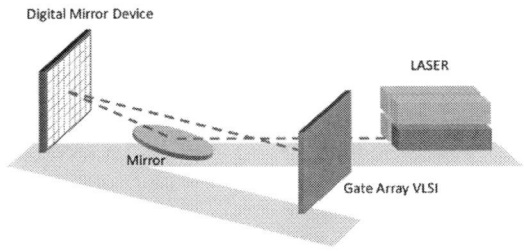

Fig. 2. Experimental system.

Abstract —This paper presents a four-context MEMS optically differential reconfigurable gate array that is useful in a space radiation environment. The technique enables rapid recovery of a programmable device that has been damaged by high-energy charged particles. It can use incorrect configuration data including some error bits resulting from damage by particles. This paper also clarifies the fault tolerance of the MEMS optically differential reconfigurable gate array.

I. INTRODUCTION

Demand of FPGA use for embedded systems in spacecraft, satellites, and space stations is increasing in order to realize hardware remote-update and remote-repair capabilities [1][2]. However, space systems are vulnerable to the effects of high-energy charged particles. Therefore, such FPGAs present the shortcoming that the circuit itself on the gate array is not robust because the configuration context on a configuration SRAM also suffers from single-event upsets (SEUs) and single-event latch-ups (SELs) [3]. To date, FPGAs have never been used in main components of space embedded systems.

This paper therefore proposes a four-context MEMS optically differential reconfigurable gate array that can be reconfigured based on incorrect configuration data that have been damaged by high-energy charged particles in a space-radiation environment. The technique enables rapid recovery of a programmable device even if the configuration data are damaged. This paper clarifies the fault tolerance of the MEMS optically differential reconfigurable gate array.

II. RAPIDLY REPARABLE HARDWARE SYSTEM

A rapidly reparable hardware system using an MEMS is presented in Fig. 1. This system comprises a wireless communication circuit, an EEPROM/SRAM, an MEMS device, and an optically reconfigurable gate array VLSI (ORGA-VLSI) [4]. In the system, configuration contexts are sent by wireless communication. However, the configuration context might

include error bits created during their transfer by wireless communication. Then, the configuration contexts are retained on the EEPROM/SRAM as they are received. In this situation, the quantity of error bits included on the configuration contexts is increased by the high-energy charged particles. Therefore, configuration context errors must be accepted. Using this method, the configuration context information is not communicated directly and is not stored directly on an EEPROM/SRAM. In advance, the configuration context information is converted to the corresponding holographic memory information. Then, the holographic memory information is sent by wireless communication and is stored on an EEPROM/SRAM. The stored holographic memory information is transferred cyclically to the MEMS device and is used for configuring a gate array at the necessary time.

In this system, a holographic configuration context stored on an MEMS device can be programmed onto an ORGA-VLSI in an extremely short time by exploiting two-dimensional optical connections between the MEMS device and ORGA-VLSI when an SEU occurs. In a holographic memory, each bit of a reconfiguration context can be generated from the entire area of its holographic memory pattern on the MEMS device. An optical majority voting operation is executed automatically for each configuration bit. Consequently, this architecture enables the use of incorrect configuration data.

III. EXPERIMENTAL SYSTEM

Figure 2 portrays a MEMS optically differential reconfigurable gate array that was constructed using a 532 nm, 300 mW laser (torus 532; Laser Quantum), a MEMS holographic memory, and an ORGA-VLSI. In this experiment, MEMS was controlled using a personal computer. An EEPROM/SRAM

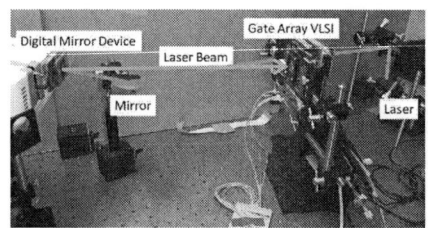

Fig. 3. Photographs of an experimental system and a digital micromirror device.

and wireless communication system are assumed to be implemented in the computer. Figures 3(a) and 3(b) respectively present photographs of the MEMS optically reconfigurable gate array and a MEMS device. The MEMS or digital micromirror device was provided by Texas Instruments Inc. [5]. The MEMS device chip is a type of spatial light modulator. It consists of $1,024 \times 768$ mirrors, each of $10.8 \times 10.8\ \mu m^2$. The switching speed of each mirror is as high as several thousand times per second. Such a device is useful in an electrically rewritable holographic memory. Calculated holographic memory patterns were displayed on the MEMS device. The ORGA-VLSI was placed at 300 mm distance from the MEMS device. The experiment described here uses an optically differential reconfigurable gate array VLSI (ODRGA-VLSI) [4] as an ORGA-VLSI. The ODRGA-VLSI was fabricated using a $0.35\ \mu$m triple-metal CMOS process. The photodiodes were constructed between the N-Well layer and the P-substrate. The photodiode size and distance between photodiodes were designed as $25.5 \times 25.5\ \mu m^2$ and as $90\ \mu$m to facilitate the optical alignment. The gate array structure is fundamentally identical to that of typical FPGAs. The ORGA-VLSI chip includes 4 logic blocks, 5 switching matrices, and 12 I/O bits. To program the gate array, 340 photodiodes are used. The ODRGA architecture can reduce the number of binary state Highs or bright bits so that the fault tolerance can be increased over that of conventional ORGA architectures. The high fault tolerance was analyzed.

IV. EXPERIMENTAL RESULTS

In this experiment, an AND circuit and an XOR circuit were implemented. To confirm fault tolerance for configuration data, up to 72,000 impulse noise were applied to each circuit's holographic memory patterns on a MEMS device. The impulse noise assumes high-energy charged particle incidents. The plots are shown in Fig. 4. When impulse noise is not applied, the configuration times of conventional ORGA architecture and ODRGA architecture were, respectively, 80 ns and 8 ns. Since optical reconfiguration time is inversely proportional to the number of bright bits in a configuration context and ODRGA architecture can reduce the number of bright bits in a configuration context, the ODRGA reconfiguration time is always faster than that of conventional systems. Moreover, results confirmed that the ODRGA architecture presents advantages in terms of the high fault tolerance of configuration contexts. Although a conventional ORGA is reconfigurable

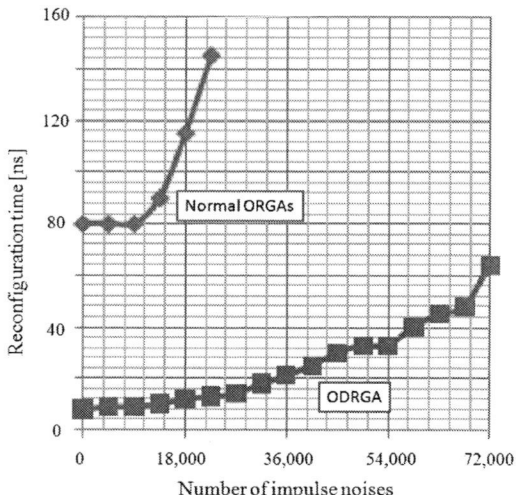

Fig. 4. Experimental result.

using a configuration context with up to 22,500 impulse noises, the ODRGA architecture can be reconfigured using a configuration context with 72,000 impulse noises. The ODRGA architecture can sustain much greater damage than conventional architectures. Therefore, results demonstrated that the MEMS ODRGA is a very rapidly reconfigurable device, making it a very robust device for use in radiation-rich space environments.

V. CONCLUSION

This paper has presented a four-context MEMS ODRGA, along with demonstration of its nanosecond-order reconfiguration capability. Results of these experiments demonstrated that, even if configuration data are damaged by high-energy charged particles, high-speed configurations taking less than 64 ns can be executed correctly. This paper therefore clarifies that the MEMS DORGA has high fault tolerance.

VI. ACKNOWLEDGMENTS

This research was supported by the Ministry of Education, Science, Sports and Culture, Grant-in-Aid for Scientific Research on Innovative Areas, No. 20200027. The VLSI chip in this study was fabricated in the chip fabrication program of VLSI Design and Education Center (VDEC), the University of Tokyo in collaboration with Rohm Co. Ltd. and Toppan Printing Co. Ltd.

REFERENCES

[1] Altera Corporation, "Altera Devices," http://www. altera.com.
[2] Xilinx Inc., "Xilinx Product Data Sheets," http://www. xilinx.com.
[3] S. Redant, R. Marec, L. Baguena, E. Liegeon, J. Soucarre, B. Van Thielen, G. Beeckman, P. Ribeiro, A. Fernandez-Leon, B. Glass, "Radiation Test Results on First Silicon in the Design Against Radiation Effects (DARE) Library," IEEE Trans. on Nuclear Science, vol. 52, no. 5, pp. 1550-1554, 2005.
[4] M. Nakajima, M. Watanabe, "A Four-Context Optically Differential Reconfigurable Gate Array," IEEE/OSA Journal of Lightwave Technology, Vol. 27, No 20, pp. 4460-4470, 2009.
[5] Texas Instruments, "DLP," http://www.ti.com/

BATCH FABRICATION OF FLOWABLE COLORIMETRIC PRESSURE SENSING PARTICLES VIA SURFACE MICROMACHINING

S. Chalasani, Y. Xie, Y. Zeng and C. H. Mastrangelo

University of Utah, Salt Lake City, UT, USA

ABSTRACT

The batch fabrication and test of artificial optical resonator slab-type micro particles (14 µm diameter, 0.7 µm gap) is presented as a means to map absolute pressure within microscopic environments. The pressure-sensing particles consist of a semi-transparent elastic polysilicon shell enclosing a reference vacuum cavity. The optical resonance frequency and the corresponding external pressure can hence be interrogated optically via reflectivity measurements. We demonstrate the measurement of internal pressures between 0-20 psi within microfluidic environments.

INTRODUCTION

The velocity field within a microfluidic environments can be readily mapped via particle image velocimetry (PIV). Unlike PIV there is no equivalent technique for the complementary pressure field. Pressure fields have been indirectly inferred via observations of channel deformation [1] and fluorescence of O_2-sensitive pressure sensitive paints [2-4]. In this paper we report the direct measurement of internal chip pressures via colorimetric pressure-sensing microparticles implementing a particle imaging manometry (PIM) system.

PRESURE SENSING PARTICLES

Each slab-type microparticle consists of a semi-transparent elastic shell enclosing a reference vacuum cavity as shown in Fig. 1. The thickness of the shell is designed such that the cavity gap and corresponding optical frequency is dependent on the

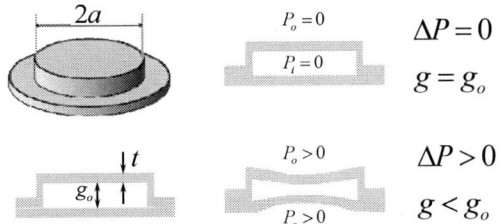

Fig. 1. (a) Schematic of slab-type spectroscopic pressure sensing hollow particle. (right) Note how the particle diaphragms deflect under a pressure differential thus changing the cavity gap.

external absolute pressure. The particle diaphragms thus form a Fabry-Perot resonator or etalon [5] of characteristic gap $g(\Delta P)$. In a simple Fabry-Perot resonator, at normal incidence the optical reflectance as a function of wavelength λ is

$$R(\lambda) = \cfrac{1}{1 + \cfrac{T_d^2}{4R_d} \cdot \csc^2(\cfrac{2\pi g}{\lambda})} \qquad (1)$$

Where T_d and R_d are the diaphragm transmission and reflection coefficients, and $g(P)$ is the pressure dependent gap. The reflectance is zero when the argument of the csc() is 2π hence $\lambda_{min} = g(\Delta P)$. The minimum reflectance wavelength shift is hence related to the external pressure change.

PARTICLE FABRICATION

Microparticles are batch fabricated on silicon wafers using the process shown in Fig. 2. The process flow

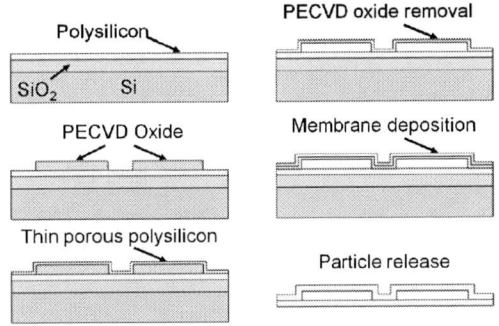

Fig. 2. Simplified pressure-sensing microparticle process flow. The particles can be released and stored in a methanol suspension.

requires just two lithography step. The first step is used to define the initial cavity gap, and the second is used to release the particle from its carrier substrate. The process starts with the growth of 0.6 µm of thermal oxide on silicon followed by 0.15 µm of undoped polysilicon. The particle cavity is next formed by the deposition of 0.7 µm of PEVD oxide and wet 6:1 BHF etching. Next the cavity oxide spacer is sealed with a 0.1 µm of porous polysilicon [6]. This material has small pores that permit the

978-1-4244-8926-8/10 $26.00 © 2010 IEEE

sacrificial etch of the spacer oxide in concentrated HF. Next we deposit a .05 µm of regular polysilicon to seal the cavity at the deposition pressure of the polysilicon sealing film (~ 200 mT). In the final step, the periphery of the particle is lithographically defined and the polysilicon is etched down to the underlying oxide. Next the particles are released by sacrificial etching of the bottom oxide in concentrated HF. The particles are collected via a series of gradual dilutions in de-ionized H_2O. The particle density is slightly lower than the density of H_2O; hence a final dilution in methanol produces microparticles in solution. Fig. 3 shows SEM photographs of a slab-type microparticle array on a

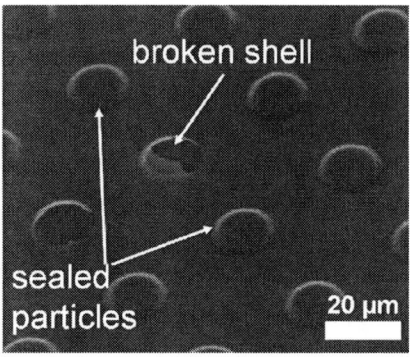

Fig. 3. SEM of slab-type 12 µm-diameter spectroscopic pressure-sensor sensor particles before their release.

carrier silicon substrate with a density of 310,000 particles per square centimeter.

EXPERIMENTS

Fig. 4 shows an optical photograph of released 0.7 µm-gap, 14 µm-diameter slab microparticles on a glass substrate, and the corresponding optical reflectance at atmospheric pressure measured using an Ocean Optics spectrometer attached to a microscope. The particle reflectance shows characteristic dips at 0.52 and 0.65 µm. Fig. 5 shows the pressure dependence of a microparticle reflectivity. The spectral dip change can be correlated to the pressure-dependent compression of the optical cavity.

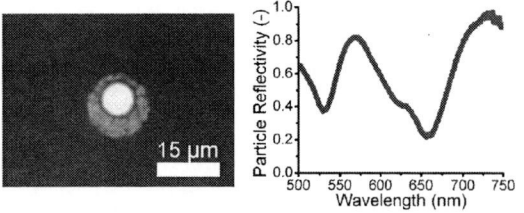

Fig. 4. Optical photograph of released 14 µm-diameter spectroscopic slab-type particles in a water suspension. (b) Measured microparticle reflectivity at normal incidence in air at atmospheric conditions.

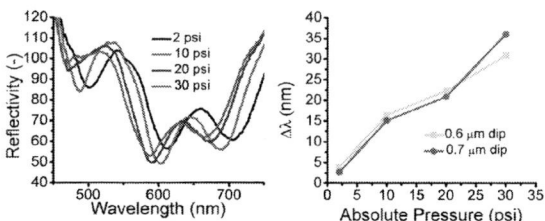

Fig. 5. (a) Measured particle reflectivity vs. pressure inside a liquid (H_2O) chamber. (b) Wavelength shift vs. chamber pressure for two different reflectivity dips.

In order to demonstrate the utilization of these devices we first introduced a large number of microparticles inside a test PDMS microfluidic chip with 100x25 µm² cross section which adhere to the capillary walls. The particle reflectivity was measured using the setup shown in Fig. 6(left) under

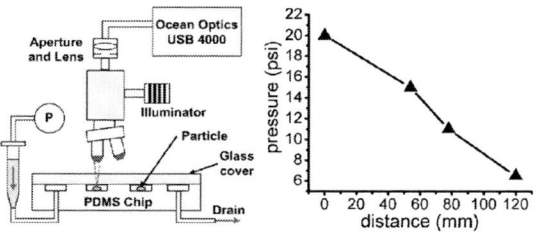

Fig. 6. (left) Experimental setup used for the measurement of internal pressure within a microfluidic chip. (right) Microparticle measured pressure vs. distance from chip inlet.

constant pressure driven flow of 3 cm/s. Fig. 6(right) shows the microparticle measured pressure drop vs. distance from the inlet. The measurements indicate an approximate linear pressure drop vs. distance.

REFERENCES

[1] H. Lee, H. Lu, *Lab Chip.* 2009, vol.9 ,pp. 3345-3353.
[2] J. W. Gregory, K. Asai, M. Kameda, T. Liu, and J. P. Sullivan," *Proc. IMechE* vol. 222 Part G: *J. Aerosp. Eng.*, review Paper 249, 2008.
[3] T. Liu and J. P. Sullivan. *Pressure and Temperature Sensitive Paints*, 2005 (Springer).
[4] S. H. Ima, G. E. Khalil, J. Callis, B. H. Ahnb, M. Gouterman, Y. Xia, *Talanta,*, vol. 67 (2005) 492–497.
[5] Miller, M. F.; Allen, M. G.; Arkilic, E. Breuer, K. S.; Schmidt, M. A., *Transducers' 97*, vol. 2, pp.1469-1472, vol.2, 16-19 Jun 1997
[6] G. M. Dougherty, T. D. Sands, and A. P. Pisano, *IEEE JMEMS*, vol. 12, No.4, pp. 418-424, 2003.

Enhanced Contrast of Wavelength Selective Mid-IR Detector Stable against Temperature Change

Katsuya Masuno, Shinya Kumagai, Kohji Tashiro, [1]Masaru Hori and Minoru Sasaki

Dept. of Advanced Science and Technology, Toyota Technological Institute,
2-12-1 Hisakata, Tenpaku-ku, Nagoya 468-8511, Japan
[1]Dept. of Electrical Engineering and Computer Science, Nagoya University,
Furo-cho, Chikusa-ku, Nagoya 464-8603, Japan
E-mail: sd08504@toyota-ti.ac.jp

Abstract

New methods to integrate wavelength selective filter into a thermopile detector using absorbance spectrum of a polymer material and to deposit the absorber selectively onto hot junctions of the thermopile detector are proposed. Hydrophilic absorber solution is selectively deposited on hydrophilic region on the detector. The fabricated detector shows +100% increases from baseline at λ_{peak} of the absorber. Use of molecular absorption is considered effective to accomplish stable SNR against temperature change.

Keywords: Infrared detector, thermopile detector, wavelength selective filter, atmospheric pressure plasma

1. INTRODUCTION

Recently, with the growing needs for detecting the amount of the substances, Mid-IR (MIR) sensor attracts attention. Using the nature that the substances have unique absorptions in MIR regions, optical detections are characterized as robust and long-time stable sensing methods. Further downsizing of the sensor will expand the application fields of the sensor. For portable sensors, thermopile detectors are preferably used as it operates at room temperature and has broad spectral responses. Since the spectral response is flat[1], optical bandpass filters (BPFs) are combined. BPFs generally have dependencies on incident angle and temperature [2,3]. From sensor performances, longer optical path for light absorption provides higher accuracy while size of an optical cell dominates the sensor dimension. Multiple reflections inside the optical cell may be effective to minimize the sensor size. Smaller detector dimension and independencies on incident angle will contribute to downsize the sensor.

In this study, a new method to integrate wavelength selective filter into a thermopile detector using an absorption spectrum of polymer material is proposed.

2. PRINCIPLE

The operating principle of the thermal IR detector is based on the temperature increase at hot junctions due to IR incidence. Certain wavelength selectivity can be integrated by selecting an absorber which has the spectral emissivity with acute peak at desired wavelength. Here, polyacrylonitrile (PAN) which has a peak at λ_{peak}=4458nm with FWHM of 25nm, induced by stretching of C≡N bond, is used. It's center wavelength is stable against temperature change [4]. PAN is chosen as an example. Absorption peaks of materials are generally expected to accomplish the same stability. As another material, use of ferric ferrocyanide,

having an isolated peak at 4790 nm and FWHM of 45nm, as an absorber material has a possibility of CO detection. Table 1 lists other molecular structures, center wavelengths (CWLs) and its detectable gases.

Table 1 : Molecular structures and possible detectable gases.

Molecular Structure	Vibration (Stretch)	CWL of Molecule $\lambda_{p\ molec,}$	Detectable Gas	CWL of Gas $\lambda_{p\ gas}$	ROI of Gases ROI_{gas}	$\left[\frac{\|\lambda_{p\ molec,} - \lambda_{p\ gas}\|}{(ROI_{gas})}\right]$ x100 [%]
CH₃N₃	N=N=N	4753	O₃	4750	110	2.7
Aliphatic Nitriles	C≡N	4462	N₂O	4496	100	34.0
Alkyl Thiocyanates	C≡N	4655	CO	4666	220	5.0
5-membered ring cyclic anhydrides	C=O	5390	NO	5333	270	21.1
Cyclopentene	C=C	6196	NO₂	6184	190	6.3

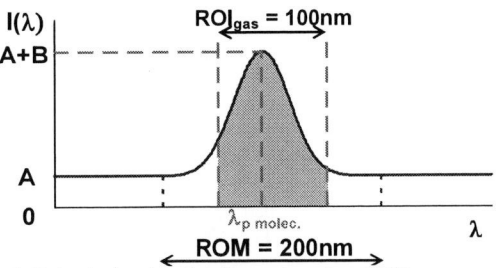

Fig. 1 Calculation model for estimating SNR.

Fig. 1 shows calculation model for estimating spectral SNR when molecular absorption is used as wavelength selective filter. The measuring spectral region is expressed as ROM, and region of interest as ROI. Supposing spectral shape to be Gaussian with flat baseline, expressed as follows

$$I(\lambda) = A + B\exp(-\frac{(\lambda - \lambda_p)^2}{2\omega^2}) \qquad (1)$$

where $I(\lambda)$ is the absorption intensity, A, B are magnitudes at the baseline and at the peak, respectively, λ is wavelength, λ_p is wavelength of absorption peak, and ω is width. Supposing signal is the area within the region of ROI and noise is the area within ROM subtracted by the signal, SNR is expressed as follows:

$$SNR = \frac{[\int_{ROI} I(\lambda)d\lambda]}{[\int_{ROM} I(\lambda)d\lambda - \int_{ROI} I(\lambda)d\lambda]} \quad (2)$$

According to the data of PAN, the width ω is

$$\omega = \frac{FWHM_{PAN}}{\sqrt{8\ln 2}} \approx \frac{25}{2.35} = 10.6 \quad (3)$$

Suppose ROM of 200nm and ROI of 100nm. Cases for spectral contrasts $(B+A)/A$ are 1 and 5, which is maximum spectral contrast from intrinsic PAN absorption data [5]. Positions of λ_p are set at the center of ROI and 10% inside from shorter (longer) edge of ROI. Table 2 shows estimated spectral signal to noise ratios (SNRs) based on absorption data ω of PAN. SNR is expected to be improved significantly.

Table 2 : SNR calculations based on PAN data.

SNR	$\lambda_{p\ molec.}$ is at center of ROI$_{gas}$	$\lambda_{p\ molec.}$ is 10% inside from shorter (longer) edge of ROI$_{gas}$
When (B+A)/A is 5	2.331	1.707 (1.707)
When (B+A)/A is 1	1.266	1.166 (1.166)

Fig. 2 shows the schematic of the thermopile detector. Chip size is 2mm × 2mm and consist of 64 thermocouples. The central part of the detector chip is a diaphragm, fabricated by anisotropic etching of silicon. The thermopile, comprised of platinum and highly doped p-type silicon, has hot junctions on the diaphragm and cold junctions on the substrate which acts as a heat sink. As the thermopile sensor generates signal proportional to the temperature difference between junctions, it is essential that the absorber material is selectively deposited onto hot junctions. Dimethylacetamide (DMA) is a solvent for PAN, and exhibits hydrophilicity. By creating hydrophilic / hydrophobic regions, the droplet of the solution is expected to stay inside the hydrophilic region.

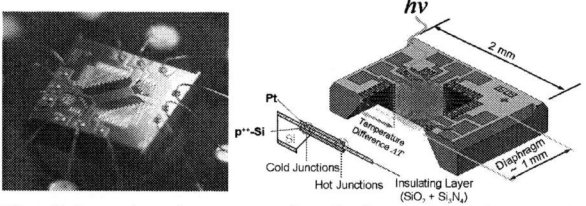

Fig. 2 Actual and cross sectional view of the thermopile detector used.

3. EXPERIMENTS

Fig. 3 shows fabrication sequence of the detector. The detector is hydrophobically modified using plasma polymerization of C_4F_8 (Step 1). Localized hydrophilic treatment over the diaphragm and hot junctions is carried out using a stencil mask and atmospheric pressure Ar / O$_2$ plasma (Step 2). Solution of the absorber is added dropwise (Step 3). Solvent is removed by heating the detector in vacuum chamber (Step 4). PAN is deposited onto the diaphragm and hot junctions of the detector.

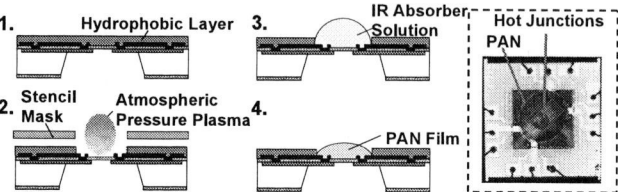

Fig. 3 Fabrication sequence of the detector. Inset shows fabricated detector.

4. RESULTS

Fig. 4 shows spectral response from the fabricated detector. The output signal showed +0.25µV or +100% increase from the baseline at λ=4450nm. CWL and FWHM of ~33nm show good correlation with the intrinsic absorption of PAN.

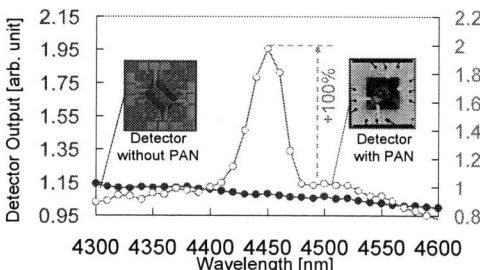

Fig. 4 Spectral responses of detector with and without PAN

Temperature stability of center wavelength of PAN is +0.042nm/°C [5]. Assuming cases of temperatures range from -20 to 60°C, estimated SNR deterioration is less than 0.036 for cases listed in table 2. Stable SNR against temperature change can be accomplished.

Part of this research is supported by the MEXT, "High-Tech Research Center" Project for Private Universities from 2007.

REFERENCES

[1] Laqua, B Schrader, G. G. Hoffmann, D.S. Moore, T. Vo-Dinh, Pure Appl. Chem. Vol. 67, p. 1751 (1995)
[2] Jiwang Yan, Katsuo Syoji, Tsunemoto Kuriyagawa, Kenji Tanaka, Hirofumi Suzuki, ASPE Proc. (2000)
[3] Irving H Blifford, Jr., Appl. Opt. Vol. 5, p.105, 1966
[4] A.Z. Gadzhiev, S.A. Kirillov: J. App. Spectrosc. Vol. 21, p. 1566 (1974)
[5] K. Masuno, K. Tashiro, M. Hori, S. Kumagai, and M. Sasaki : Jpn. J. Appl. Phys Vol. 49, 04DL18 (2010)

FABRICATION AND VERIFICATION FOR THE MICRO HOLOGRAPHIC OPTICAL PICKUP

Jin Chern Chiou[1,2], Kuan Chou Hou[1]

[1]Institution of Electrical and Control Engineering, National Chiao Tung University, Taiwan
[2]School of Medicine, China Medical University, Taiwan

ABSTRACT

In this paper, we describe an optical configuration with a holographic optical element (HOE) for the small-form-factor (SFF) pickup head. This system adopts a finite-conjugate objective lens for both red and blue wavelength. A holographic optical element is used for simplifying the optical configuration which provides a better approach for alignment. In addition to demonstrating this OPH by using existing discrete components, flip chip bonder with high accuracy alignment was to integrate it. Fabrication process of the micro holographic optical pickup has been successfully verified. The experimental results verify the optical performance and the system have been demonstrated.

INTRODUCTION

In high-density optical disc systems with high-NA objective lenses in optical system for small form factor optical drive, it is important to reduce the size of optical module. Miniaturization of optical system has many advantages to achieve light in weight and small in size. There are many miniature way to reduce size, for example, high NA of lens [1,2] and reflective light route [3]. A kind of small optical system was designed and simulated by Shih [4] but the study was lacked of systematic fabrication process and entire measurement results. Fabrication process of small optical was designed and implemented to demonstrate light path in expected way. In order to implement the structure and verify optical light route in the small optical system, flip chip bonder is the main technique which has high bonding precision to accomplish the FES requirement.

Construction of Fabrication Process

Fig. 1 shows the schematic diagram of the pickup head including a substrate, a laser diode (LD), a photodetector (PD), a mirror, two micro prisms (MP1, MP2), a holographic optical element (HOE), a lens module, silicon substrate with metal interconnection and a piece of PCB substrate. An edge-emitting laser diode chip (LD) is the main light source in the optical system. A PD with quadrant photo detector is used to create FES and RF signals. All of mirror and prisms work as reflective optical component in order to reduce dimension in this framework. HOE is used to be optical device to create the first order diffraction.

A lens modulus is consisted of finite-conjugate objective lens and lens holder and the purpose is to achieve light spot focusing. Photography of completed optical system unit is shown in Fig. 2. Fabrication process of small optical system is shown in Fig. 3. Firstly, a LD and a PD were bonded on the substrate that has patterned metal interconnection and then wire bonding was implemented to control driving current and acquiring signal. Secondly, mirror and micro prisms were also bonded on silicon substrate to reflect light path in order to reduce the dimension. Thirdly, glass substrate with HOE was bonding on both of prism and it also can calibrate the light path with virtual method. The last step of fabrication of optical system is micro lens bonding on the surface of HOE.

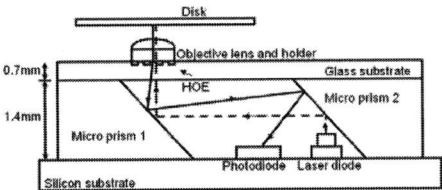

Fig.1. Structure of micro optical module.

Fig.2. Photography of micro optical pickup module.

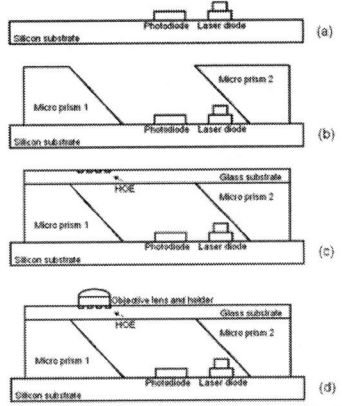

Fig.3. Bonding processes (a) Laser diode and photodiode. (b) Mirror and prisms. (c) HOE glass substrate (d) Lens modulus.

978-1-4244-8926-8/10 $26.00 © 2010 IEEE

Calibration of Optical Path

Alignment procedure can be utilized for adjusting the relative placement of the photodetector and HOE. The conceptual drawing is shown in Fig. 4. The virtual image of the laser is obtained by extending the diffracted beam to the virtual focal point. This point must be coincided with the location of the photodetector image.

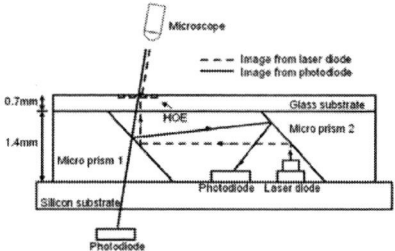

Fig.4. Virtual Image Method is used to adjust the optical path.

In order to accomplish the task that the reflecting light can be precisely located at the center of the quadrant PD, both MP2 and HOE that work to calibrate the two axes planar and circular location. The effort is that each small optical system will be ensured the precision of light path that begins from LD and reflect back to PD.

After calibration of virtual image method, lens module is bonded on the HOE substrate and HOE pattern is just beneath the lens. The laser beam came from LD will pass through lens. The central beam of these laser spots is zero-order diffraction spot and others are the diffractive light spots which are diffracted by HOE pattern. Diffraction of light is displayed on the surface of lens module is very important because meaning of fabrication process of lens module is that achievement with this phenomenon occurred.

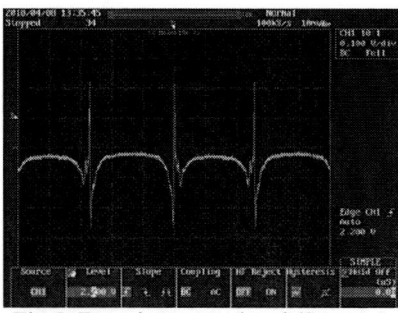

Fig.5. Focusing error signal (S-curve).

Experimental Results

Focusing error signal FES (S-curve) is shown in

Fig. 5. Working frequency of actuator for measurement is 5Hz and amplitude is 140mV. The horizontal axis is defocus position, and the vertical axis is optical input power. Output signal is calculated with (A+C)-(B+D) that output signal is from quadrant detectors. As the result, the adjustment state of an optical axis can be demonstrated confirmed by the experimental signals.

Conclusions

Light route in a small optical system has successfully verified and demonstrated. Finite-conjugate objective lens is adopted in the system. Small optical system reduces dimension and size of component is $1 \times 3 \times 5$ mm^3 for height, width and length. In presented study, a HOE is used to simplify the optical configuration and minimize scale. We implemented the structure and verified the system successful in the study.

Acknowledgment

This work was supported in part by Ministry of Economic Affairs, Taiwan, R. O. C. under contract No. 97-EC-17-A-07-S1-011, and the National Science Council, Taiwan, R. O. C. under contract no. NSC-98-2220-E-009-014, NSC-98-2220-E-009-032, and NSC-98-2218-E-039-001. It was also supported in part by Taiwan Department of Health Clinical Trial and Research Center of Excellence under contracts No. DOH99-TD-B-111-004 and DOH99-TD-C-111-005.

REFERENCES

[1] Hsi-Fu Shih, Yuan-Chin Lee, Yi Chiu, David Wei-Chung Chao, Gung-Ding Lin, Chun-Shin Lu, and Jin-Chern Chiou," Micro objective lens with NA 0.65 for the bluelight small-form-factor optical pickup head", OPTICS EXPRESS, vol. 16, no. 17, 2008

[2] Jin-Seung SOHN, Eun-Hyoung CHO, MyungBok LEE, Hae-Sung KIM, MeeSuk JUNG, Sung-Dong SUH, Wan-Chin KIM1, No-Cheol PARK and Young-Pil PARK," Development of Microlens for High-Density Small-Form-Factor Optical Pickup", Japanese Journal of Applied Physics vol. 45, no. 2B, pp. 1144–1151, 2006

[3] K.-S. Jung A H.-M. Kim A S.-J. Lee A N.-C. Park S.-I. Kang A Y.-P. Park," Design of optical path of pickup for small form factor optical disk drive", Microsystem Technologies, vol.11, no. 8-10, pp. 1041-1047, 2005

[4] Hsi-Fu Shih, Chi-Lone Chang, Kuei-Jen Lee, and Chi-Shen Chang," Design of Optical Head With Holographic Optical Element for Small Form Factor Drive Systems", IEEE TRANSACTIONS ON MAGNETICS, vol. 41, no. 2, pp. 1058-1060, 2005

A Novel Fabrication Method of the Micro Cube Beam-Splitter with Optical Surface Roughness

Kuo-Yung Hung[1] Ying-Chuan Chen[1] Shih-Hao Huang[2] Yun-Ju Chuang[3]

[1]Institute of Mechanical and Electrical Engineering, Ming Chi University of Technology, Taiwan.

[2]Department of Mechanical and Mechatronic Engineering, Taiwan Ocean University

[3]Department of BioMedical Engineering, Ming Chuan University, Taiwan

E-mail: kuoyung@mail.mcut.edu.tw, TEL: 886-2-29089899 ext: 4514, FAX: 886-2-29063269

ABSTRACT

This paper successfully used inclined exposure technology to fabricate 50/50 3D micro cube beam splitter with 45° optical grade surface (~1.4 mm thick). This paper tests the effect of solvent loss percentage of the polymer material (optimum surface roughness: 85.45% solvent remove) on the surface roughness of inclined surfaces. The smallest surface roughness achieved in the experiments using SU-8 material with a thickness of 1.4 mm was less than 70 nm ($400\mu m \times 400\mu m$ area). This type of micro-mirror can be used as a key component in Michelson Interferometer, bio detection systems, data storage system and linear encoder.

INTRODUCTION

A beam splitter (PBS) is an optical element that can separate a light beam into two orthogonally beams. It can be widely used in optical information processing systems, such as magneto-optic data storage, free-space optical switching, etc. A conventional PBS is usually a birefringent prism or consists of multilayer dielectric coatings [1], which can exhibit different optical properties for differently waves. During the past few years, several PBS [2-6] have been proposed which satisfy the above requirement for practical use. It is always necessary that a PBS should provide a good extinction ratio, certain angle and wavelength range. This suggests that various optics separating or polarization selective elements may be designed by using 3D MEMS device.

DESIGN and FABRICATION PROCESS

Micro-optical system little to develop of manufacturing technology for integrating beam-splitter and other optical component. This paper applications the inclined exposure technology to fabricate the 50/50 cube beam-splitter devices (fig. 1). The inclined exposure technology used in

this paper was a previously developed technology and exposure system [7-8].

The process steps designed in the paper are shown in Fig. 2. Figure 2a shows the spin coating of the SU-8 thick-film photo-resist on glass substrate 1. Figure 2b-c shows the inclined and directed exposure step following the soft bake, and Fig. 2d shows post-exposure baking, the developing process and Aluminum metal evaporation. Figure 2e shows the UV release layer and SU-8 spin coating on the glass substrate 2. Figure 2f shows the inclined and directed exposure step following the soft bake, and Fig. 2g shows post-exposure baking and the developing process. Finally, substrate 1 & 2 were aligned and combined together to form the micro cube beam-splitter (fig. 2h).

RESULTS AND DISCUSSION

In this paper, different soft-bake parameters (95, 105°C and 10, 12 hours) were tested. The relationships of the solvent contents in the polymer material with structure surface roughness are important for optical applications. Fig. 3 shows the results of the remove percentage of the polymer solvent in different soft-bake temperature and time. The smallest surface roughness achieved in experiments using SU-8 with a thickness of 1.4 mm was less than 70 nm (area: 400 X 400 μm) (Table 1). And the applied soft-bake parameters is 105°C for 10 hours, solvent contents only remain about 14.55% compare to original. This paper experimentally verified that special inclined exposure technology could be used to fabricate polymer cube beam-splitter devices and therefore offers an alternative to conventional mechanical processing. This cube beam-splitter could be used as a key component in Blu-Ray optical pickup heads used in portable, high-density storage systems.

REFERENCES

[1] Li L and Dobrowolski J A, "Splitter Operating

at Angles Greater than the Critical Angle" Appl. Opt. Vol. 39, pp. 2754-2771, 2000.

[2] Tyan R-C, Sun P-C, Scherer A and Fainman Y, "Polarizing beam splitter based on the anisotropic spectral reflectivity characteristic of form-birefringent multilayer gratings" Opt. Lett, vol. .21, pp. 761-763, 1996.

[3] Tamada H, Doumuki T, Yamaguchi T and Matsumoto S, "Al wire-grid polarizer using the s-polarization resonance effect at the 0.8-μm-wavelength band" Opt. Lett., vol. 22, pp. 419-421, 1997.

[4] Yi D, Yan Y, Liu H, Lu S and Jin G, "Broadband polarizing beam splitter based on the form birefringence of a subwavelength grating in the quasi-static domain" Opt. Lett., vol. 29, pp. 754-756, 2004.

[5] Zhou L and Liu W, "Broadband polarizing beam splitter with an embedded metal-wire nanograting" Opt. Lett., vol. 30, pp. 1434-1436, 2005.

[6] Lopez A G and Craighead H G, "Wave-plate polarizing beam splitter based on a form-birefringent multiplayer grating" Opt. Lett., vol. 23, pp. 1627-1629, 1998.

[7] Hung K Y and Liao J C, "The Application of Fresnel Equations and Anti-Reflection Technology to Improve Inclined Exposure Interface Reflection and Develop a Key Component Needed for Blu-ray DVD--Micro-Mirrors" Journal of Micromechanics and Microengineering, vol. 18, no. 7, 2008.

[8] Hung K Y, Hu H T, and Tseng F G, "Application 3D Glycerol-Compensated Inclined-Exposure Technology to Integrated Optical Pick-Up Head" Journal of Micromechanics and Microengineering, vol. 14, pp. 975-83, 2004.

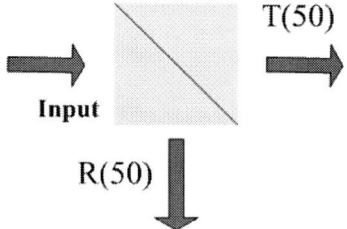

Figure 1. The concept diagram of the micro cube beam-splitter

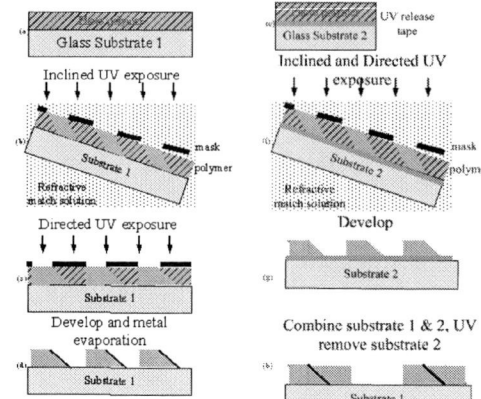

Figure 2. The process of the micro cube beam-splitter

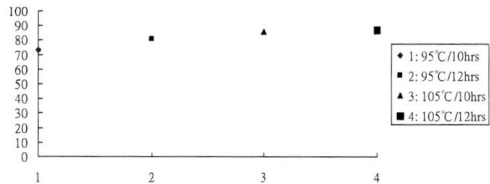

Figure 3. Solvent loss percentage of the polymer material by different soft-bake parameters

Table 1. Surface roughness measure results of the inclined surface by different soft-bake parameters in fig. 3.

95℃/10 hours	95℃/12hours
2D 400×400μm	
Roughness	
Surface Statistics:	Surface Statistics:
Ra: 124.25 nm	Ra: 210.11 nm
Rq: 150.96 nm	Rq: 261.59 nm
Rz: 813.87 nm	Rz: 1.38 um
Rt: 1.06 um	Rt: 1.56 um

105℃/10 hours	105℃/12hours
2D 400×400μm	
Roughness	
Surface Statistics:	Surface Statistics:
Ra: 66.76 nm	Ra: 191.32 nm
Rq: 83.31 nm	Rq: 235.89 nm
Rz: 523.75 nm	Rz: 1.31 um
Rt: 633.57 nm	Rt: 1.40 um

Dynamic trapping and release of
multiple particles in a polarized optical vortex

Baile Zhang and George Barbastathis

Singapore-MIT Alliance for Research and Technology Centre, Singapore 117543 and
Department of Mechanical Engineering, Massachusetts Institute of Technology, Cambridge,
MA 02139, US

ABSTRACT

We calculate rigorously the optical forces on two interacting particles trapped in an optical vortex with helical phase front and demonstrate the dynamic process of the two particles' evolution. We show that in this two-particle system, a particle trapped by an optical vortex at one moment can be released due to the interaction from the other trapped particle in the next moment. To our best knowledge, this work is the first theoretical description on dynamic evolution of multiple particles trapped by an optical vortex.

Optical forces from laser beams might be the most suitable tools for manipulating, analyzing and organizing mesoscopic objects [1-3]. For example, optical tweezers have already been widely used in cooling and trapping of atoms and operating DNA molecules [2]. Optical vortices with helical phase front have recently attracted a lot of attention because they can not only attract particles but also transfer orbital angular momentum, which might provide the potential solution to the actuators for microelectromechanical systems and mesoscopic pumps for microfluidic systems [2]. The realistic applications require deep understanding of the mechanism behind the interaction between mesoscopic objects and optical vortices. Although there have been a lot of experiments observing and studying the characteristics of an optical vortex, theoretical predictions are much fewer in the literature, especially for the objects with diameters roughly equal to one wavelength--a size range where many useful applications will fall in. Dynamical trapping of one particle in an optical vortex has been recently discussed in [3], but the scenario of multiple particles has yet to be examined. It is the purpose of this paper to study this problem.

We first describe an optical vortex that is linearly polarized along the x direction in the xy plane and is propagating along the z axis. For the sake of simplicity, we assume that E_x at $z = 0$ satisfies

$$E_x = e^{-(x^2+y^2)/w_0^2} \cdot \tanh(\sqrt{x^2 + y^2} / w_v)e^{-i\phi} \text{ [4], such}$$

that the optical vortex has a phase singularity with zero amplitude along the z axis. The optical vortex is first Fourier transformed into a spectrum of plane waves. These plane waves then are expanded into vector spherical waves with respect to each particle. By adopting the Lax-Foldy multiple scattering formulae, we are able to obtain rigorously the fields generated in this system. Then by numerically integrating the Maxwell-stress tensor over each particle's surface, we obtain the net force exerted to

each particle. A reasonable assumption, namely that the damping factor of the surrounding media is sufficiently large and the particles move slowly with negligible acceleration, can simplify the next treatment by letting the velocity of each particle be proportional to the force it is subject to. In our simulation, we adjust the position of each particle in each time step as $r^{k+1} = r^k + \alpha f^k$, where k is the time step index, r and f are the position and the net force of each particle in the k-th time step, and α is an arbitrarily chosen constant which satisfies the slow motion assumption of particles. While our model is three-dimensional, here we limit the discussion of the wave behavior and the particles' motion in the xy plane where the wave's phase front rotates clockwise.

The two particles we study have the same diameter of 800 nm and possess the same refractive index of 2.6. The parameters of the optical vortex are as follows: $w_0 = w_v = 3\lambda$, where $\lambda = 800$ nm is the wavelength of light in water. Fig. 1 shows the dynamical process of the two particles in the optical vortex. (Animation available at http://web.mit.edu/~bzhang/www/opticalmems2010/) The background is the absolute value of incident electric field E_x. The dark part at the center is due to the phase singularity of the optical vortex. The high field intensity ring is the orbit for the motion of the particles once they are trapped. Each portion of Fig. 1 has dimensions $10\lambda \times 10\lambda$.

The initial positions of these two particles are shown in Fig. 1 at time step 1, where one particle stays at the center and the other stays nearby. After 30 time steps, the first particle has been trapped and has started its orbital motion along the orbit with high field intensity. Due to the disturbance from the first particle, the second particle is pushed out toward the orbit with high field intensity. After 100 time steps, both particles have been trapped in the optical vortex, moving along the orbit. Due to their mutual

978-1-4244-8926-8/10 $26.00 © 2010 IEEE   143

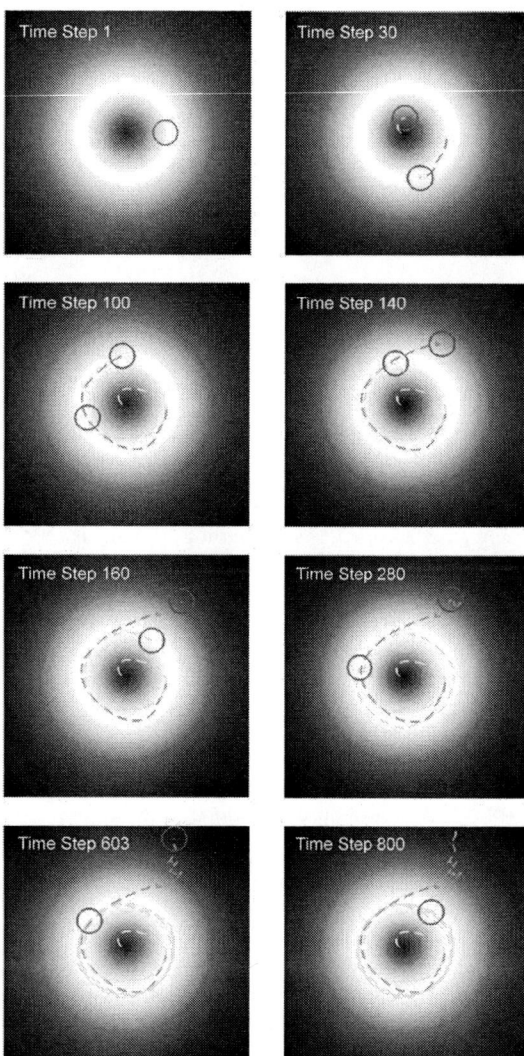

Fig. 1 Dynamic process of two particles evolving in an optical vortex polarized along the x direction in the xy plane at different time steps. The background indicates the amplitude of $|E_x|$. The red circles represent the positions of the particles. The green and yellow dotted lines indicate their loci, respectively.

interaction, the loci of their motion are not perfectly circular, but reminiscent more of a polygon. This is evidence of the oscillation of binding forces between these particles when their distance varies. From time step 100 to time step 140, the first particle has deviated from the orbit due to the repelling force from the second particle. Being subject to the attraction of the optical vortex to its orbit, the first particle slows down near the orbit's periphery. However, when the second particle gets closer to the first particle, a strong repelling force occurs and pushes the first particle farer away from the orbit, as shown in Fig. 1 at time step 160. After that, since the

field intensity around the first particle is very small, the first particle will be mainly subject to the force generated from the scattering of the second particle moving along the orbit. Generally, the scattering from a sphere is mainly distributed in the forward and backward directions. Therefore, when the second particle moves away from the first particle along its orbit, the first particle will move in the opposite direction due to the backward scattering of the second particle in the optical vortex. The motion of the first particle is almost along a tangent direction of the circular orbit, as shown at time step 280. We can expect that after time step 280, when the second particle approaches the first particle, its forward scattering will push the first particle away. Therefore, after the second particle rotates by several cycles, the locus of the first particle will be a zigzag curve, as shown at time step 603. When the first particle is sufficiently far away from the optical vortex, the scattering from the second particle is almost from the same direction, and therefore the first particle will move away along a smoother curve, as shown at time step 800. We can also see that the second particle is rotating along its orbit almost perfectly circularly.

Note that in our case, the release of the first particle is directly due to the interaction between multiple particles rather than the insufficient damping in the surrounding media [2]. In addition, in order to obtain sufficient interaction for this phenomenon to occur, the size of the particles cannot be much less than the wavelength. However, if the size of particles is much larger than the wavelength, then one particle has already occupied most of or the entire orbit ring, making the second particle difficult to be trapped in the beginning. Essentially, how many particles can be trapped in an optical vortex with a given size is a dynamic problem with certain trends of probability when the size of particles increases.

In summary, we have demonstrated the dynamic trapping process of two particles in an optical vortex and predicted that a particle trapped at one time can be eventually released due to the disturbance from the other trapped particle. This provides deeper understanding of the multiple-particle interaction at the mesoscopic scale in an optical vortex.

Reference

[1] A. Ashkin, J. M. Dziedzic, J. E. Bjorkholm and S. Chu, "Observation of a single-beam gradient force optical trap for dielectric particles," Opt. Lett. 11, 288-290 (1986).

[2] D. G. Grier, "A revolution in optical manipulation," Nature 424, 810-816 (2003).

[3] J. Ng, Z. Lin and C. T. Chan, "Theory of optical trapping by an optical vortex beam," Phys. Rev. Lett. 104, 103601 (2010).

[4] D. Rozas, C. T. Law and G. A. Swartzlander, Jr., "Propagation dynamics of optical vortices," J. Opt. Soc. Am. B 14, 3054-3065 (1997).

DESIGN AND FABRICATION OF LARGE FIBER-MODE-MATCHED THREE-DIMENSIONAL ADIABATIC TAPERED COUPLERS FOR INTEGRATED OPTICS

Chun-Wei Liao[1], Yao-Tsu Yang[1], Sheng-Wen Huang[1], and Ming-Chang M. Lee[1]
Pi-Yao Lin[2], Chao-Min Chou[2], and Jia-Ming Shieh[2]

[1]Institute of Photonics Technologies, National Tsing Hua University, Hsinchu, Taiwan R.O.C.
[2]National Nano Device Laboratories, Hsinchu, Taiwan R.O.C.

ABSTRACT

We presented the design and fabrication of fiber-mode-matched three-dimension (3D) tapered couplers for efficient coupling light from fibers to photonic integrated circuits (PICs). The 3D adiabatic taper was made by SU8 and the structure was designed by beam propagation method (BPM). Measuring waveguide transmission with this 3D taper, we observed that the coupling efficiency was improved by 12 dB. The misalignment tolerance can be around 5 μm.

INTRODUCTION

Sub-wavelength waveguides made by silicon or silicon oxynitride have received great attentions for highly integrated photonic circuits. Although the device footprint is significantly reduced, the waveguide core dimension becomes extremely small (typically less than 1 μm), causing severe mode-mismatching between waveguides and single-mode fibers (SMF). Such large mode-mismatching results in inefficient coupling and high insertion loss, which impedes practical applications. Many approaches, such as prism couplers, grating couplers, micro-lenses [1] and tapered optical fibers have been proposed to improve the coupling efficiency, but they induce other issues, for example, narrow band and stringent alignment tolerances. Three-dimension tapered couplers, sometimes called beam-spot converters, are very promising to solve this problem. However, to make a tapered structure especially in the vertical dimension is difficult. Several approaches such as anisotropic etching [2], grayscale masks [3] and tapered liquid waveguide [4] were reported to make this structure. But all these methods are critical and expensive in fabrication and incompatible with standard semiconductor fabrication process.

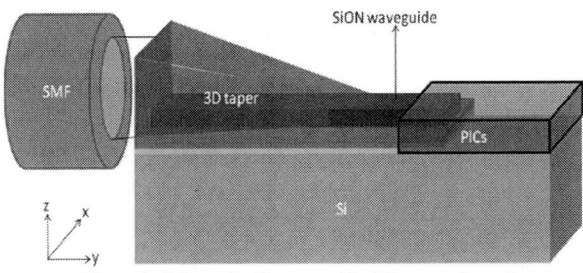

Figure 1: Schematic diagram of the 3D tapered coupler

Here, we propose a simple and low-cost method for fabricating a large fiber-mode-matched 3D adiabatic tapered coupler, which has a coupling area able to match the core dimension of SMF. The schematic drawing of the wedge-like 3D tapered structure is showed in Fig. 1. A negative photoresist SU-8-3010 (refractive index ~1.56) was used to make the tapered structure. SU8 has been successfully applied for high-aspect-ratio MEMS structures and optical waveguides [5] due to high optical transparency in the visible and near infrared spectral regions, as well as stable chemical and mechanical properties. Because the process is done in low temperature, this 3D taper can be easily added on as-fabricated devices. In this work, we show that this adiabatic tapered coupler is able to couple light from fibers to silicon oxynitride (SiON) waveguides and vice versa, where the waveguide index can be adjusted between 1.45 and 2.0 by plasma-enhanced chemical vapor deposition (PECVD) deposition system.

EXPERIMENTAL

BPM was used for calculating the coupling efficiency of the 3D tapered coupler. In this simulation, the SiON waveguides are on a silicon-on-insulator (SOI) substrate with 2 μm buried oxide and 0.15 μm Si device layer. Between the SOI and SiON waveguide, a 0.2 μm thick SiO_2 is inserted. This composite SOI and the SiO_2 layer perform as a bottom cladding for the SiON waveguide, preventing the light leaking to substrate. The refractive indices of SU8, Si and SiO_2 are assumed to be 1.56, 3.45 and 1.44, respectively. The width and height of the 3D SU8 taper decreases from a square of 10μm×10μm until directly connecting to the SiON waveguide, defined with a core dimension of 5 μm in width and 1 μm in height. The length of the 3D taper is 850 μm and the total length of the SiON waveguide is 1200 μm. The beam spot size of launched fiber modes was assumed to be 10 μm in diameter.

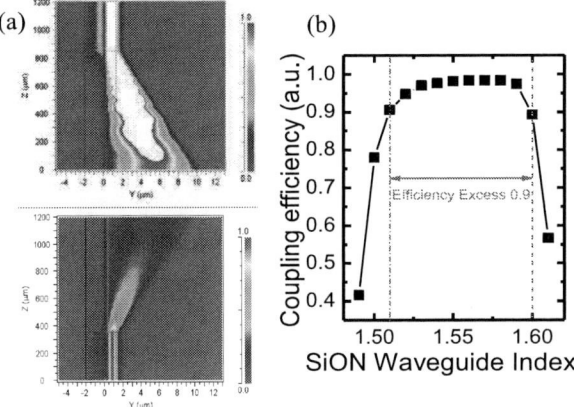

Figure 2: 3D BPM simulation: (a) wave propagation from taper to waveguide (top) and waveguide to taper (bottom), and (b) taper coupling efficiency versus different SiON waveguide index

Fig. 2 shows simulated wave propagation along the 3D taper and the coupling efficiency versus SiON index of refraction. The SiON waveguide index ranging between 1.51 and 1.60 exhibits a coupling efficiency more than 90% for both propagation directions. Outside this index range, the coupling efficiency decreases dramatically, which may be caused by index mismatch between the waveguide and the taper. The fabrication process is described as follow. The substrate is a Si wafer with 2 μm thick SiO_2 on the top surface. Next, a

978-1-4244-8926-8/10 $26.00 © 2010 IEEE   145

0.15-μm-thick Si and a 0.2-μm-thick SiO_2 layer was subsequently deposited by low-pressure chemical vapor deposition (LPCVD) system. To comply with the simulation result of the optimal waveguide refractive index, a 1-μm-thick SiON layer was deposited by PECVD with an appropriate process gas ratio and the waveguide structure was patterned via photolithography followed by dry etching process. Next, SU8-3010 was spin coated on the waveguide surface and a glass slide coated with Teflon was impressed onto the SU8 film obliquely, resulting in a wedge shape developed on the SU8 film. After removal of the glass slide, a photolithography process was used to define the lateral dimension of the 3D taper. Fig. 3 shows a 3D taper with increasing cross-sectional area from left to right.

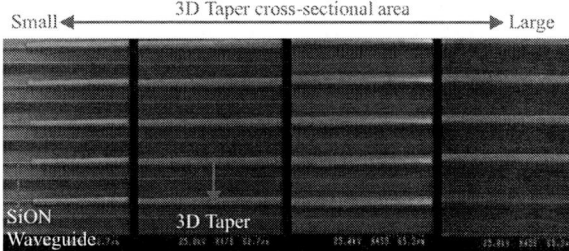

Figure 3: SEM images of different sections of 3D tapers

MEASUREMENT AND DISCUSSION

To examine the coupling efficiency of the 3D taper, we measured the insertion loss of SiON waveguides with one end covered with the SU8 tapered structure while the other left bare. Flat-cleaved single-mode fibers were aligned to the 3D tapers and conical lensed fibers were aligned to the bare waveguide ends. A tunable laser with scanning wavelength from 1290 nm to 1360 nm was launched at either one end. In our experiment, two pathways were defined according to the direction of wave propagation. One is from the taper to the bare waveguide (Pathway 1) and the other is the opposite way (Pathway 2). Fig. 4 shows transmission spectra of a waveguide with the 3D taper. In this figure, waveguide transmission without tapers was also calibrated. Apparently, with 3D tapered couplers, the coupling efficiency was improved by more than 12 dB for both pathways, attributed to a better mode match between the 3D taper and the fiber mode. Besides, the transmission spectrum is flat and shows the potential of broad-band applications.

Moreover, fiber misalignment tolerance was also investigated for the 3D tapered coupler. Fig. 5 shows the measured extra loss by intentionally varying the launched position of a fiber deviating from the centre of 3D taper (cross section) to ±5 μm both in x or z direction. If 3-dB loss is adopted as a reference value, the offset tolerance was ±4 μm in x direction and ±5 μm in z direction for both Pathway 1 and 2.

CONCLUSION

In this work, we proposed a new 3D tapered coupler for coupling light from single-mode fibers to sub-wavelength optical waveguides. The preliminary result shows that waveguides with the 3D tapers can have more than 12 dB improvement in coupling efficiency compared with those without 3D tapers and misalignment tolerance can be up to ±5 μm. This 3D taper is made by SU8, which can be easily fabricated by a simple and low-cost process.

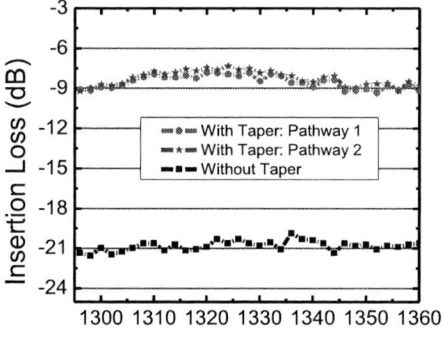

Figure 4: Insertion loss of waveguides with/without 3D tapered couplers

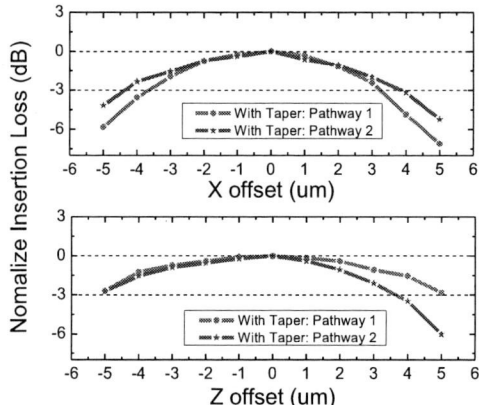

Figure 5: Normalized coupling loss versus alignment offset

ACKNOWLEDGEMENT

The work is supported by National Science Council (NSC98-2622-E-007-002-CC1) and National Nano Device Laboraories (NDL98-C02M3C-040) in Taiwan.

REFERENCE

[1]H. M. Presby, et al. "Near 100% efficient fiber microlenses," Electron. Lett. 28, (1992)
[2]R. Holly, et al. "Fabrication of silicon 3D taper structures for optical fibre to chip interface," Microelectronic engineering. Vol. 84, (2007)
[3]A. Sure, et al. "Fabrication and characterization of three-dimensional silicon tapers," Optics expresses. Vol. 11, (2003).
[4]Hisashi Terae, et al. "Tapered waveguide by liquid   for a coupler of optical fibers to mems devices," IEEE, pp 794-797, (2008).
[5]A. Borreman, et al. "Fabrication of Polymeric Multimode Waveguides and Devices in SU-8 Photoresist Using Selective Polymerization," IEEE, pp 84-86, (2002)

Fabrication of LED Based Ultra Slim Optical Pointing Device

Jae Young Joo[1], Do-Kyun Woo[1], Sun Sub Park[2] and Sun-Kyu Lee[1,#]

[1] School of Information and Mechatronics, Gwangju Institute of Science and Technology, Oryoung-dong, Buk-gu Gwangju, 500-712, South Korea

[2] Honam Technology Service Division, Korea Institute of Industrial Technology, Gwangju, Oryoung-dong, Buk-gu Gwangju, 500-480, South Korea

Abstract.

The Ultra Slim Optical Pointing Device (USOPD) is a slim optical mouse as an input device for the application of wireless portable personnel communication device like a smart phone. In this paper, we have fabricated optical components of GaN LED based USOPD. The USOPD consist of illumination optical components and imaging lens. LED beam shaping lens consisting of both aspheric lens and Fresnel facet successfully machined by Diamond Turning Machine (DTM). Additional V-shaped groove with refractive-reflective surfaces (VGRRS) for beam path banding was fabricated by bulk micromachining of silicon and the shadow effect in thermal evaporation. Fabrications of imaging lens, arrayed multilevel Fresnel lenses, were fabricated on electron beam lithography, FAB etching. The proposed optical components are extreme compactness as well as high optical efficiency, thereby applicable to the ultra slim optical system like USOPD.

INTRODUCTION

Recently, the size of an optical component in a mobile phone is becoming thinner and smaller and its functionality is required more various for entertaining. Thus necessity of input device in it is increasing for various types of application[1-2]. Under this technological requirement, Chip Scale Package (CSP) Light Emitting Diode (LED) offers portability for a mobile phone by achieving thin and lightweight design. In addition to LEDs' high brightness, the light output, coupled with optical components, plays an important role from the perspective of high illumination quality control and miniaturization of such devices[3-4]. Thus, there are increasing technological demands in ultra shim optical components[5-8].In this paper, we propose the Ultra Slim Optical Pointing Device (USOPD) as an input device like a slim optical mouse in a wireless portable personnel communication device.
We studied the fabrication of optical components of LED based Ultra Slim Optical Pointing Device (USOPD) in this paper.
As shown in Fig. 1, the USOPD consist of illumination optical components and imaging lens. LED beam shaping lens was successfully machined by Diamond Turning Machine

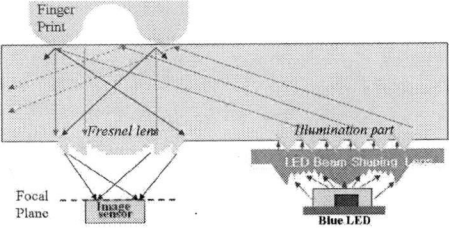

Fig. 1 Schematic of the USOPD

(DTM). Additional VGRRS for beam path bending was fabricated by bulk micromachining of silicon and the shadow effect in evaporation process. The proposed optical components are extreme compactness as well as high optical efficiency, thereby applicable to the ultra slim optical system like USOPD.

LED Beam Shaping Lens

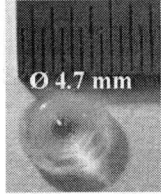

Fig. 2 Machining of LED beam shaping lens

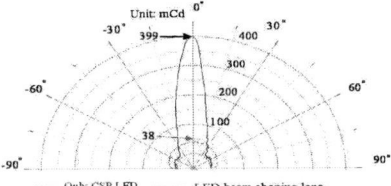

Fig. 3 Performance of the LED beam shaping lens

LED beam shaping lens consisting of both aspheric lens and Fresnel facet successfully machined by Diamond Turning Machine (DTM) as shown in Fig. 2.
Minimizing the tool alignment error of a sharp-edged natural diamond bite. The machined lens showed good surface quality on facets and negligible form error. The machined prototype of lens on poly methyl methacrylate (PMMA) plate reduced the viewing angle of the testing

978-1-4244-8926-8/10 $26.00 © 2010 IEEE

Chip Scale Packaged LED from 140° to 17.4° in FWHM as shown in Fig. 3. The thickness machined lens is about 500 μm with eleven facets. This miniaturized LED beam shaping lens is extremely compact and provides high collimation efficiency. Thus, it can be applicable for the USOPD.

Mold of V-shaped groove with refractive-reflective surfaces

Additional VGRRS for beam path bending was fabricated by bulk micromachining of silicon and the shadow effect in evaporation process. The VGRRS as the light bending optical component has been fabricated through a thermal oxidation process to the TMAH etching process. SiO2 with a thickness of 450 nm obtained by a thermal oxidation process was used for a mask in TMAH wet etching.

To successfully obtain the mold of VGRRS, it has to be considered that an anisotropy etching ratio of (100)/(111) in TMAH wet etching is 20 and the misalignment in photo-lithography is 0.5°. Designed pattern width and length for the V-groves was in 44 and 296 ͪm respectively. For that, a photoresist (AZ 1500 38cp) was coated with 2000 rpm for a resist thickness of 3 ͪm, SiO$_2$ was etched by etchant (HF:NH4F=9:100). Finally fabricated V-groves through TMAH anisotropy etching with conditions of an etching rate of 500 nm/min and etching temperature of 80° are shown in Fig. 4.

Fig. 4. Fabrication results of V-groves.

Imaging Lens

The mould of imaging lens, arrayed multilevel Fresnel lenses, were fabricated with an alignment method including three times repetitive processes of Electron-Beam Lithography(EBL) and fast atom beam (FAB) etching, as shown in Fig. 5.

EBL has the strong advantage that enables precise fabrication because of high resolution and the absence of a mask, and FAB etching with 1:1 selectivity of the EB resist and substrate can also help to fabricate precise patterns.

The eight-level lenses mould designed with a minimum pattern width of 240 nm and thickness of 807 nm have been successfully fabricated with the following fabrication results: an alignment error of approximately 50 nm and a minimum pattern width of 247 nm and thickness of 741 nm.

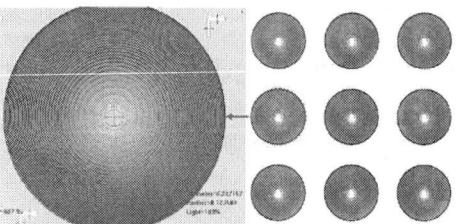

Fig. 5. 3 x 3 8 level Fresnel lens

Summary and Further Work

The proposed optical components are extreme compactness by precise fabrication technologies. These optical components are applicable to the ultra slim optical system like USOPD. For the further work, we will assemble each of these optical components including the LED in a single mould and fabricated USOPD system.

Acknowledgement

This work was supported by the Korea government (MEST, R0A-2008-000-10065-0) and partly by RISE

REFERENCES

[1] Korea patent application number 10-2008-0062799, Date 2008-06-30

[2]D. K. Woo, K. H. , S. C. Cho, S. K. Lee, "The development of an integral optics system for a slim optical mouse in a slim portable electric", J. Vac. Sci. Technol. B. Vo. 27 no. 3, pp. 1422-1427, 2009

[3] J. Y. Joo, C. S. Kang, S. S. Park, S. K. Lee, "LED beam shaping lens based on the near-field illumination" Opt. Express, Vol. 17, Iss. 2, pp. 23449-23458, 2009.

[4] J. Y. Joo, S. K. Lee, "Miniaturized TIR Fresnel lens for miniature optical LED applications" IJPEM, Vol.10 No.2, pp.137-140, 2009.

[5] D. K. Woo, K. H. , S. K. Lee, "High order diffraction grating using v-shaped groove with refractive and reflective surfaces " Opt. Express, Vo. 16, Iss. 25, pp. 21004-21011, 2008

[6] C. B. Lee, K. Hane, W. S. Kim, S. K. Lee. "Design of the retrodiffraction gratings for the polarization-insensitive and the polarization-sensitive characteristics by using Taguchi method" Appl. Opt. Vol. 47, pp. 3246-3253, 2008.

[7] D. K. Woo,Do-Kyun Woo, et al. J. Opt. A : Pure Appl. Opt. 10 (2008) , pp. 044001.

[8] D. K. Woo, K. H. , S. K. Lee, "Fabrication of a multi-level lens using independent-exposure lithography and FAB plasma etching" Journal of Optics A,Vol. 10, no.4, pp. 1-6, 2008.

978-1-4244-8926-8/10 $26.00 © 2010 IEEE

FAST ATOM BEAM-BASED FABRICATION OF HIGH-EFFICIENCT BLAZED GRATING USING SLANTING ANGLE CONTROL OF A SUBSTRATE

ChaBum Lee[1], Kazuhiro Hane[2], Sun-Kyu Lee,[*,1]

[1]Gwangju Institute of Science and Technology, Korea
[2]Tohoku University, Japan

ABSTRACT

This paper presents electromagnetic analysis of Si-based small period blazed reflection gratings for a period of 0.6 - 1.5 μm in terms of the rigorous coupled wave analysis (RCWA) and its fabrication using fast atom beam (FAB) etching method. The optimized blazed gratings were successfully fabricated on a slanted silicon substrate by FAB etching method, and diffraction efficiencies (DE) for four kinds of gratings were evaluated from optical testing and these results showed good agreement with these theoretical values, respectively.

INTRODUCTION

Recently, a great deal of research on surface relief has been performed. SRGs play an important role in diffractive optical elements (DOEs) such as spectrometer, monochromator and multiple imaging devices [1-6]. Keeping pace with these interests, a lot of scalar and vector analysis methods of diffractive gratings have been introduced. From the fabrication point of view, there exist hot issues such that how small period, how fine surface and how continuous profile of blazed gratings could be successfully fabricated because efficiency and characteristics of DOEs are highly dependent on their own surface profiles. So far, an excellent blazed structure with the period as small as the wavelength region has not been well introduced. Here blazed gratings (period 0.6 to 1.5 μm) with ideal profile are successfully fabricated on a slanted silicon substrate by the FAB etching method.

FABRICATION

Blazed gratings with a period of 0.6 - 1.5 μm and a depth of 0.32 μm have been successfully fabricated on a slanted silicon substrate by using both E-Beam direct writing and the FAB etching method. The used fabrication process was presented in Fig. 1.

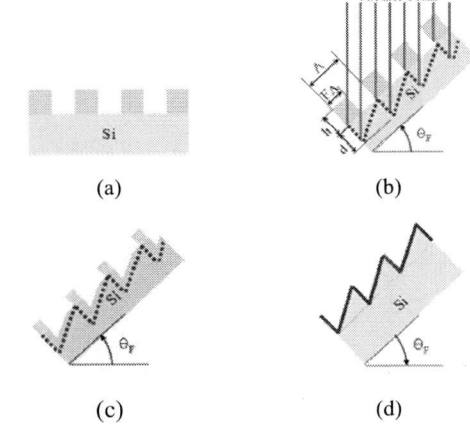

(a)   (b)

(c)   (d)

Fig. 1 (a) binary patterning, (b) slanting a substrate, (c) FAB etching; stage I and (d) FAB etching; stage II. The scanning electron microscope (SEM) and noncontact type stylus profiler, atomic force microscope (AFM), were used to measure and evaluate the geometry of blazed gratings. The fabricated blazed gratings measured by SEM and AFM were shown in Fig. 2 and Fig. 3, respectively. This measured peak-to-valley depth was slightly less than the designed one due to the edge-rounding effect occurring in stylus-profiling method. Furthermore, it has been known that the large blazed angle just as the fabricated blazed gratings in this paper may cause errors at steep blazed facets and it results in change in diffraction angle and efficiency [2, 5]

978-1-4244-8926-8/10 $26.00 © 2010 IEEE   149

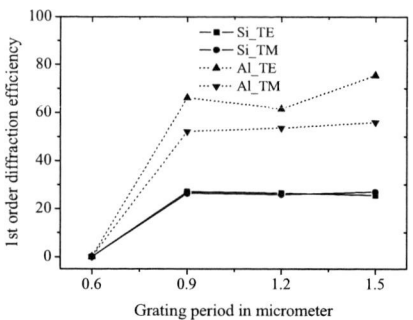

Fig. 2 SEM photographs of the fabricated gratings; periods (a) 0.6 μm, (b) 0.9 μm, (c) 1.2 μm, (d) 1.5 μm.

Fig. 4 Measurement of first order DE.

OPTICAL MEASUREMENT

A laser diode, λ=0.633 μm, was used for the light illumination. As shown in Fig. 8, the experimental results for Si-based blazed gratings were compared with the theoretical results; $(\eta_{r,1,TE,Si}, \eta_{r,1,TM,Si})$ = (0%, 0%), (21.2%, 21.2%), (24.7%, 24.7%) and (26.6%, 26.4%). It can be found that the experimental values are well matched with the theoretical ones as the period increases.

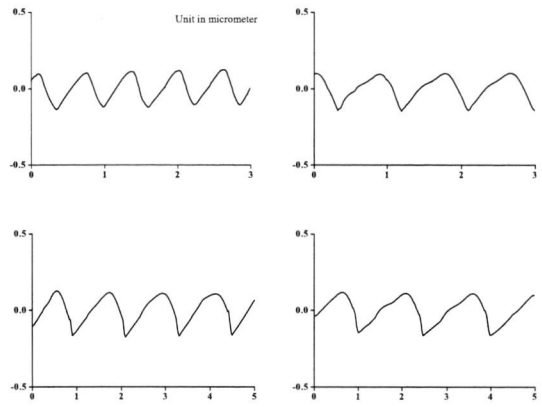

Fig. 3 Atomic force microscope profiles; period

(a) 0.6 μm, (b) 0.9 μm, (c) 1.2 μm and (d) 1.5 μm.

To enhance the reflectance Al is coated on the Si-based gratings. As expected, it was clearly seen that DE for TE and TM polarization increasingly rises compared to that of Si-based gratings due to facet materials, Si and Al as seen in Fig. 4.

CONCLUSION

Analysis, fabrication and measurement of blazed gratings were carried out. The optical and geometric diffraction characteristics of four kinds of blazed gratings with the periods 0.6-1.5μm were investigated for analysis and fabricated on a slanted silicon substrate using the FAB etching method. From optical measurement, it was observed that the experimental values are well matched with the theoretical ones as the period increases. This dry etching method for nanoscale patterning is expected to be utilized in company with high reflection coating technology.

This work was supported by the Korea government (MEST) (No.R0A-2008-000-10065-0) and RISE

REFERENCES

[1] N. F. Borrelli, Microoptics Technology, 2nd Ed., (Marcel Dekker, 2004).

[2] ChaBum Lee et al. J. Micromech. Microeng. 18 (2008) 045014.

[3] M. G. Moharam et al. J. Opt. Soc. Am. A, Vol. 12, No. 5 1068-1076 (1995).

[4] ChaBum Lee et al. Appl. Opt. Vol. 47 (2008) pp.3246-3253

[5] ChaBum Lee et al. J. Micromech. Microeng. 20, (2010) 055028

[6] JaeYoung Joo et al. Opt. Express, Vol. 17 (2009) 23449

978-1-4244-8926-8/10 $26.00 © 2010 IEEE

X-RAY IMAGING TEST FOR A SINGLE-STAGE MEMS X-RAY OPTICAL SYSTEM

Ikuyuki Mitsuishi[1], Yuichiro Ezoe[2], Kensuke Ishizu[2], Teppei Moriyama[2],
Yoshitomo Maeda[1], Takayuki Hayashi[2], Takuro Sato[2], Makoto Mita[1], N Y. Yamasaki[1], K. Mitsuda[1],
Mitsuhiro Horade[3], Susumu Sugiyama[3], Raul E. Riveros[4], Taylor Boggs[4], Hitomi Yamaguchi[4], Yoshiaki Kanamori[5],
Kohei Morishita[6], Kazuo Nakajima[6] and Ryutaro Maeda[7]

[1] Institute of Space and Astronautical Science (ISAS), Japan Aerospace Exploration Agency (JAXA), Japan,
[2] Tokyo Metropolitan University, Japan, [3] Ritsumeikan University, Japan, [4] University of Florida, US,
[5] Tohoku University, Japan, [6] Kyoto Univeristy, Japan, [7] AIST, Japan

ABSTRACT

An X-ray imaging test for an X-ray optical system based on MEMS technologies was conducted at the ISAS 30 m beamline. An X-ray reflection and focusing were successfully verified at Al K_α 1.49 keV for the first time. The image quality estimated as a half power diameter was $\sim$20 arcmin. This was consistent with the angular resolution estimated from the surface roughness of 200 nm rms at 100 μm scale. In this paper, the experimental setup and the result of X-ray imaging analysis are reported.

INTRODUCTION

The novel ultra-lightweight and high-resolution X-ray optics based on MEMS technologies were proposed and being developed for future missions [1, 2, 3] . The fabrication processes are divided into three steps. Firstly, curved micro pore structures are made by DRIE (Deep-Reactive Ion Etching). Secondly, the etched side walls are smoothed by annealing and magnetic field assisted finishing [4] to reflect X-rays (rms <1 nm). Finally, the wafer is deformed spherically to focus parallel X-rays [5]. Due to the micro pores (the width of 20 μm), the optical system fabricated in this way could be lightest among space X-ray telescopes. Thanks to the spherical deformation, high angular resolution can be achieved at the same time. Our goals are the area to mass ratio of 1000 cm^2 / kg and the angular resolution of 15 arcsec respectively. We fabricated a single stage optical system as shown in figure 1. It was annealed in Ar atmosphere for 4 hrs after DRIE and plastically deformed with a curvature of radius of 1000 mm. We already demonstrated a reflection using an optical light [6] and hence proceeded to evaluate X-ray imaging performance of this optic using an X-ray.

EXPERIMENTAL SETUP

Figure 2 shows the ISAS 30 m beamline. The beam line consists of an X-ray generator, 30 m beam line and a detector chamber. Al target was chosen to utilize a fluorescent emission line of Al K_α (1.49 keV). X-rays are collimated by molybdenum slits, hit side walls of our optic and are focused at a focal point in a detector chamber. The dispersion angle of X-rays under this setup is small (<0.02 degree). We utilized an X-ray CCD as a detector to take an image and the CCD was located in the detector chamber about 530 mm away from our optic on a focal plane. The effective area of the CCD is 24.6 $\times$ 24.6 mm^2 and 0.154 mm corresponds to 1 arcmin on the CCD in our system. To determine an actual focal length geometrically,

a 3-dimensional surface profile was taken by NH3 (Mitaka Kohki) as shown in figure 3. Then we extracted cross-sectional profiles and curvature of radius through fitting processes assuming a spherical function. Consequently, the curvature of radius was found as 1061 $\pm$ 1 mm. The focal length corresponds to half of the curvature of radii, i.e., 530 mm.

RESULT

We scanned a square pencil beam with various sizes, took images and extracted angular resolutions at the energy of 1.49 keV. As shown in figure 4, we verified the X-ray reflection and focusing by using MEMS X-ray optics for the first time and angular resolutions are about 20 arcmin. An example of the angular profile is shown in figure 5. This value was almost constant over the optical system. An example of CCD image when the beam size is 9.3 $\times$ 9.3 mm^2 is shown in figure 4.

This angular resolution is consistent with that of the expected one from mirror surface roughness shown in figure 6. However, it is two orders of magnitude worse than our goal. Thus a new process and further optimizations are required. The key task is to remove this kind of this roughness. According to SEM images and surface profiles on side walls, wavelike structures were found which was created during the DRIE process. Thus we will introduce an oxidation and BHF etching cycles after the DRIE. We expect an efficient removal of wavelike structures by this and optimized an annealing process in the near future.

References

[1] Y. Ezoe et al., " Ultra light-weight and high-resolution X-ray mirrors using DRIE and X-ray LIGA techniques for Space X-ray Telescopes, " Microsystem Tech., 2009, ISSN 0946-7076.

[2] I. Mitsuishi et al., " Novel ultra-lightweight and high-resolution mems x-ray optics, " in Society of Photo-Optical Instrumentation Engineers (SPIE) Conference Series, ser. Society of Photo-Optical Instrumentation Engineers (SPIE) Conference Series, vol. 70360, p. 7360C, 2009.

[3] Y. Ezoe, et al., in these proceedings.

[4] H. Yamaguchi, et al., " Magnetic field assisted finishing for micropore X-ray focusing mirrors fabricated by deep reactive ion etching " CIRP Annals - Manufacturing Technology, in press.

[5] K. Nakajima et al, "Shaped silicon crystal wafers obtained by plastic deformation and their application to silicon crystal lenses," Nature Materials, **4**, 47-50 (2005).

[6] I. Mitsuishi, et al., " Optical Image Analysis of the Novel Ultra-Lightweight and High-Resolution MEMS Optics " Proc. of Optical MEMS, Florida, US, Aug 17-20 2009, pp.123-124.

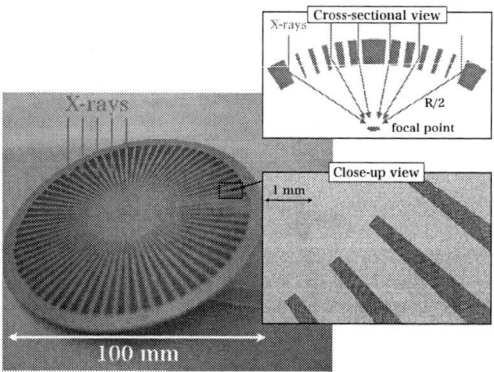

Figure 1: A sample of DRIE-processed X-ray optic with cross-sectional and close-up views.

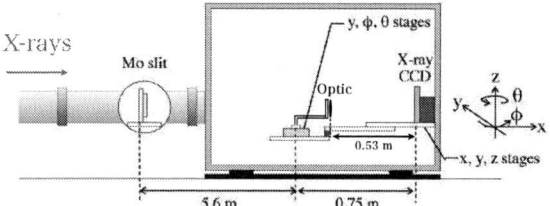

Figure 2: The experimental setup in the ISAS 30 m beam-line.

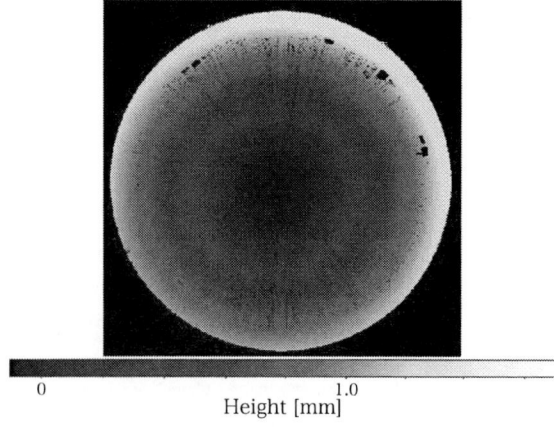

Figure 3: The three dimensional surface profile of the measured optic.

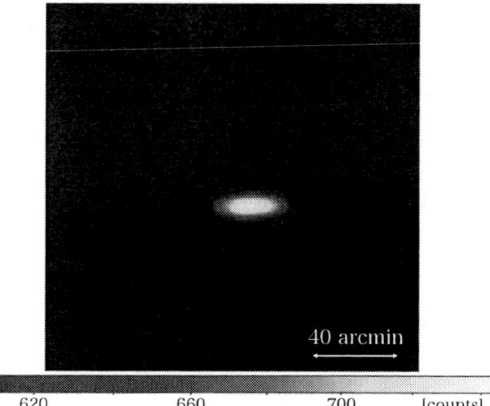

Figure 4: X-ray focused CCD image.

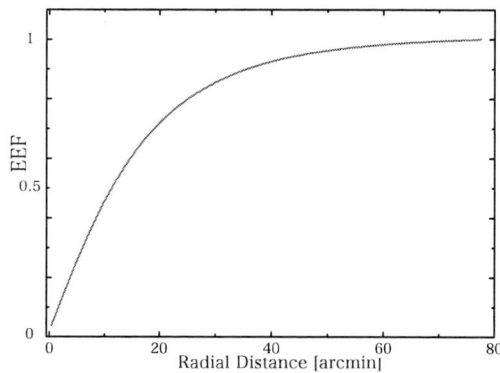

Figure 5: The EEF (Encircled Energy Function) profile. The EEF means a ratio of total counts within a radial distance to total counts. In this figure, the radial distance at 0.5 corresponds to a half of the HPD (Half Power Diameter). Within the HPD value, a half of total counts is included.

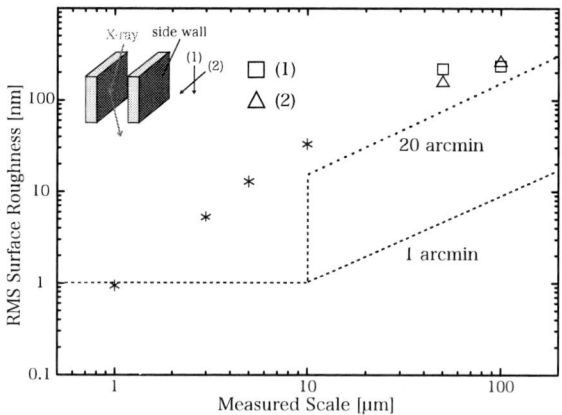

Figure 6: Average velues of rms surface roughness at various measured scales. Rms surface roughness at smaller scales ($\leq 10~\mu$m) are taken by the DFM (SII Nano Technology), the others are taken by the Dektak stylus-based profiler. Two dotted lines show angular resolutions of 1 (our goal) and 20 arcmin respectively.

Improvement of GaN crystalline quality on nanoscale patterned sapphire substrates

Yu-Sheng Lin*, J. Andrew Yeh

Institute of NanoEngineering and MicroSystems National Tsing Hua University, Hsinchu, Taiwan, R.O.C.

*d949205@oz.nthu.edu.tw

ABSTRACT

A method for the reduction of defect density in GaN epilayer using nanoscale patterned sapphire substrates (NPSS) was proposed. The sapphire substrates were patterned by natural lithography and inductively coupled plasma reactive ion etching (ICP-RIE). The undoped GaN films were grown on NPSS through metal organic chemical vapor deposition. The pits density was analyzed by atomic force microscope (AFM) and threading dislocation distribution was observed by scanning electron microscopy (SEM). The optical characteristics were measured from X-ray diffractometry and photoluminescence spectroscopy. These results indicate NPSS can improve crystalline quality by effectively reducing threading dislocations.

Keywords—**nanostructures, patterned sapphire, natural lithography, GaN**

INTRODUCTION

Gallium nitride (GaN) semiconductors attract much attention because of their applications in light-emitting diodes, laser diodes, transistors, detectors, and so forth. [1] The performance of GaN-based devices affected dramatically by inherent threading dislocation density (TDD) in the epitaxial GaN films. The TDD in the range of 10^9-10^{10} cm^{-2} results from the large mismatch of lattice constants and thermal expansion coefficients between the GaN films and the underneath sapphire substrates [2]. Therefore, the major effort in improving performance of GaN-based devices is how to grow GaN films with high epitaxial quality. Currently, many effective methods have been developed to minimize TDD by patterned sapphire substrates (PSS) [3-5]. The fabrication methods of PSS included photo lithography [3], electron-beam lithography [4] and laser holography [5]. Nevertheless, the above methods require complicated process and are time-consuming.

In this study, we presented nanoscale patterned sapphire substrates (NPSS) by natural lithography and inductively coupled plasma reactive ion etching (ICP-RIE) with high uniformity of nanostructures and wafer-level characteristics. The advantages of this method are simple and suitable for GaN epitaxy with low defect density on their surfaces.

FABRICATION

The poly-silicon thin film was deposited on 2-inch single side polished c-plane (0001) sapphire substrate by low pressure chemical vapor deposition at 600 °C. Then, the substrate was dipped in the Wright etching solution [6] for 30 minutes to texture poly-silicon surface. This etching solution reacted with silicon along grain boundary. Thus, it formed the individual

island (i.e. grain) on surface. Each island of the poly-silicon functions as hard mask for subsequent etching. The wafer was vertically etched with BCl_3/Cl_2 mixed gas of 90 sccm by ICP-RIE at 5 mtorr with RF power of 1900 W for 10 minutes. Finally, the hard mask was removed by KOH solution at 80 °C.

The morphology of NPSS was measured by atomic force microscopy (AFM). Fig. 1 shows the 2D and 3D AFM images of NPSS. The depth of nanostructures is 33 nm averagely and density of nanotrenches is 43%. After AFM measurement of NPSS, the undoped GaN films of 2.5 μm in thickness were grown on the NPSS (GaN-NPSS) and a reference flat sapphire substrate (GaN-FSS) by metal organic chemical vapor deposition at pressure of 300 torr. Two samples are loaded into the same chamber.

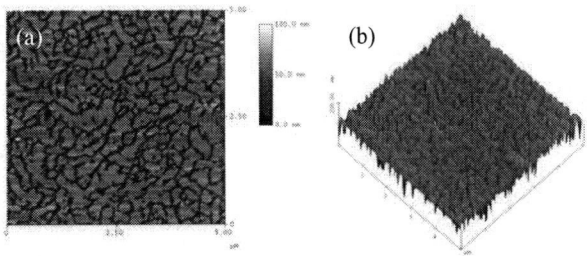

Fig. 1 – (a) 2D; (b) 3D AFM images of NPSS.

RESULTS AND DISCUSSIONS

The morphology and distribution of pits of GaN-FSS and GaN-NPSS were analyzed by AFM. The GaN films quality were measured from high resolution X-ray diffractometry (HRXRD) and photoluminescence (PL) spectroscopy at room temperature. The distribution of threading dislocations in GaN films were observed by scanning electron microscopy (SEM).

The AFM images of GaN-FSS and GaN-NPSS are shown in Fig. 2. The measured area is 5 μm x 5 μm. It is evidence that pits density of GaN-FSS is much more than the GaN-NPSS sample. The size of the pits is randomly distributed which attributed to the surface termination with threading dislocations. The estimated TDD in the GaN-FSS and GaN-NPSS is about 1.13 x 10^9 cm^{-2} and 2.20 x 10^8 cm^{-2}, respectively. It is clear that the GaN-NPSS shown fewer and smaller pits than GaN-FSS. The TDD was reduced by one order of magnitude estimated from the AFM results.

The results of HRXRD are shown in Fig. 3. The rocking curve of HRXRD shows the GaN peak (0002) reflections from GaN-FSS and GaN-NPSS examined at room temperature. The full-width at half maximum (FWHM) of the ω/2θ (0002) reflections was 248 arcsec for GaN-FSS and 215 arcsec for GaN-NPSS. The X-ray intensity of GaN-NPSS is 2 times higher than GaN-FSS. That indicated the NPSS can effectively improve the GaN crystalline quality.

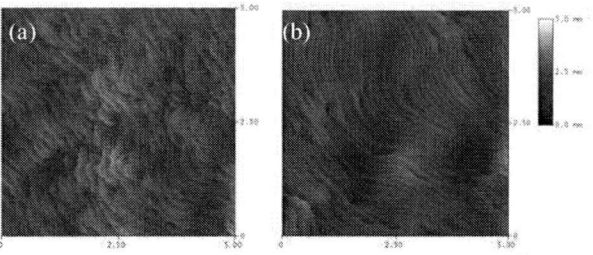

Fig. 2 - AFM images of (a) GaN-FSS; (b) GaN-NPSS.

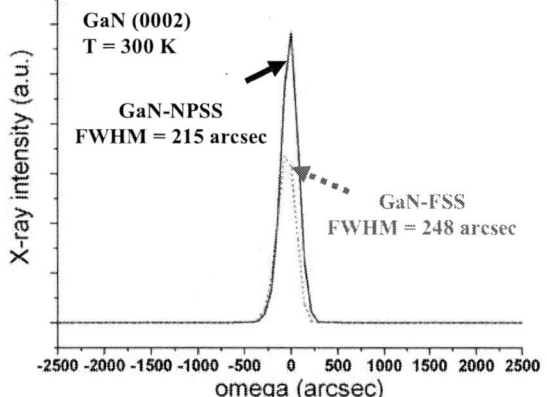

Fig. 3 – XRD results of (0002) reflections from GaN-FSS and GaN-NPSS.

The PL results of two samples measured at room temperature as shown in Fig. 4 (a). Obviously, five-fold enhancement of emission intensity was obtained on the GaN-NPSS which peak is the same in 362 nm. The enhancement of PL intensity mainly attributed to the effective reduction of TDD, which acted as nonradiative recombination centers. These centers can effectively localize carriers to avoid capture from dislocations. Fig. 4 (b) and (c) are the cross-sectional images of SEM performed to investigate the dislocation distribution in GaN films. Fig.4 (c) shows a large number of extended dislocations propagate throughout the GaN epilayer, originating from the strain related to the lattice mismatch between GaN/sapphire interfaces. In Fig. 4 (b), the GaN epilayer buried the nanotrench incompletely that threading dislocations were only generated from c-plane of sapphire surface. The threading dislocations are not existence in the nanotrenches, because free standing laterally grown GaN epilayer were achieved. This surface geometry of the NPSS is adequate for the relaxation of compressive strain during the GaN growth on sapphire. These results indicated that a significant reduction in the TDD was achieved via the lateral epitaxial overgrowth on the NPSS.

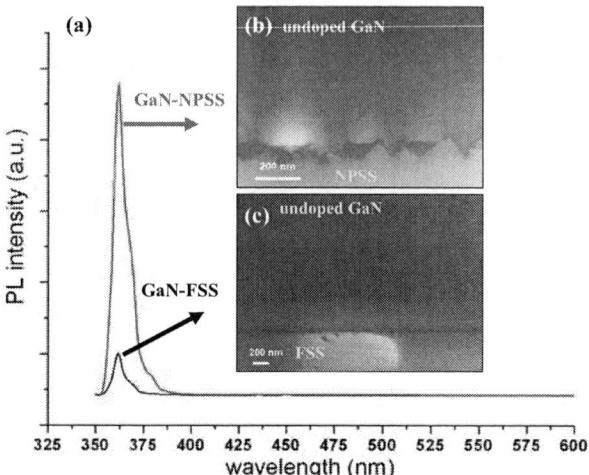

Fig. 4 – (a) PL spectra of GaN-FSS and GaN-NPSS. The SEM images of (b) GaN-NPSS; (c) GaN-FSS.

CONCLUSION

This paper demonstrated an effective approach to fabricate nanoscale patterned sapphire substrate and how it improves the crystalline quality of GaN epitaxy. The pits density on GaN surface improved one order of magnitude estimated by AFM. The FWHM of GaN-NPSS was found to be 33 arcsec narrower than GaN-FSS from HRXRD results. The PL intensity of GaN-NPSS was five-fold higher than GaN-FSS. The results show the NPSS can effectively improve crystalline quality in GaN films by reduction of the threading dislocation density.

REFERENCES

[1] E .F. Schubert and J. K. Kim, "Solid-State Light Sources Getting Smart" *Science*, vol.308, pp. 1274, 2005.

[2] S. D. Lester, F. A. Ponce, M. G. Craford, and D. A. Steigerwald, "High dislocation densities in high efficiency GaN-based light-emitting diodes" *Appl. Phys. Lett.*, vol.66, pp. 1249, 1995.

[3] H. Gao, F. Yan, Y. Zhang, J. Li, Y. Zeng, and G. Wang, "Enhancement of the light output power of InGaN/GaN light-emitting diodes grown on pyramidal patterned sapphire substrates in the micro- and nanoscale" *J. Appl. Phys.*, vol.103, pp. 014314, 2008.

[4] K. Nakano, M. Imura, G. Narita, T. Kitano, Y. Hirose, N. Fujimoto, N. Okada, T. Kawashima, K. Iida, K. Balakrishnan, M. Tsuda, M. Iwaya, S. Kamiyama, H. Amano, and I. Akasaki, "Epitaxial lateral overgrowth of AlN layers on patterned sapphire substrates" *phys. stat. sol. (a)*, vol.203, pp. 1632, 2006.

[5] J. Lee, S. Ahn, S. Kim, D. Kim, H. Jeon, S. Lee, and J. H. Baek, "GaN light-emitting diode with monolithically integrated photonic crystals and angled sidewall deflectors for efficient surface emission" *Appl. Phys. Lett.*, vol. 94, pp. 101105, 2009.

[6] R. D. Horning, A. Mirza, and R. R. Martin, "Characterization of Si-Si Bonding by Wright Etching" *J. Electrochem. Soc.*, vol.141, pp.796, 1994.

978-1-4244-8926-8/10 $26.00 © 2010 IEEE

THE MORPHOLOGICAL CONTROL OF MEH-PPV FILMS ON AN ITO ELECTRODE FOR HYBRID SOLAR CELL FABRICATION

Quynh Nhu Nguyen Truong [1], N. T. N. Truong [2], C. Park[3] and Jae Hak Jung[3*]

School of Display and Chemical Engineering, Yeungnam University,
214-1 Dea-dong, Gyeongsan, Gyeongbuk, Rep. of Korea
*Corresponding author: jhjung@ynu.ac.kr Tel.: (+82) 053-810-3513

ABSTRACT

The poly[2-methoxy-5-(2-ethylhexyloxy-p-phenylene -vinylene)] (MEH-PPV) thin films were deposited on indium tin oxide (ITO) coated glass by using spin-coating technique at room temperature. The effect of spin-coating speed on the electrical and morphological properties of MEH-PPV film was studied. The surface roughness would be controlled by varied spin-coating speeds. I-V characteristics of ITO/MEH-PPV/Al structure have measured by solar simulator at the optimum condition.

Keywords: spin-coating, roughness mean square, indium tin oxide.

INTRODUCTION

Recently, polymer solar cell has become one of the most imperative issues for many researchers because of its low cost, simple processing, and promising technology application [1-3]. The performances of the devices have been developed by the combination of inorganic semiconductors to form hybrid organic – inorganic solar cells [4, 5]. In this study, we investigated the MEH-PPV film characterizations which are optical, electrical, and morphological properties. Furthermore, the hybrid solar cell using ITO/MEH-PPV+ (PCBM or CdSe)/Al structure is considered for the fabrication.

EXPERIMENTALS

The ITO-glass substrate was cleaned by using the sequence of trichloroethylene (TCE), acetone and methanol in an ultrasonic bath in 10 minutes. The THF was used as the solvent to obtain the solution. The MEH-PPV films were spinning on a cleaned ITO glass with spin-coating speed in range of 1000-5000 rpm and then dried at 100° C in 10 minutes that is used for the drying oven. The morphological, electrical properties of films were measured by atomic force microscopy (AFM), Al was deposited on MEH-PPV coated ITO to form device structure as ITO/ MEH-PPV/Al. Device characteristics were measured by A. M 1.5 Keithley solar simulation.

RESULTS AND DISCUSSION

Figure 1 shows the AFM topography of MEH: PPV films with spin coating speed in range of 1000-5000 rpm. The surfaces of films that have high spin speed are smoother than those of low spin coating speed. The surface roughness of films was reduced, since the spin coating speed was increased. The surface morphology of films was observed by an optical microscopy. The film at 5000 rpm is uniform. On the basis of the result, it is indicated that the spin coating speed is one of the most important factor to control the morphology of MEH-PPV film.

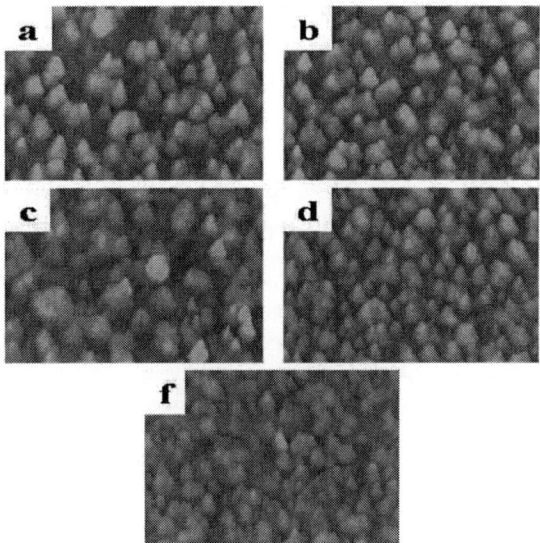

Fig. 1. The AFM topography images of MEH-PPV films, (a) 1000 rpm, (b) 2000 rpm, (c) 3000 rpm, (d) 4000 rpm, (f) 5000 rpm.

Figure 2 shows the surface roughness of samples in terms of the varied spin coating speeds. As one can be seen from the figure, the surface roughnesses (RMS and Ra) is reduced as the function of spin speed.

However, the surface roughness is still less than 10 nm.

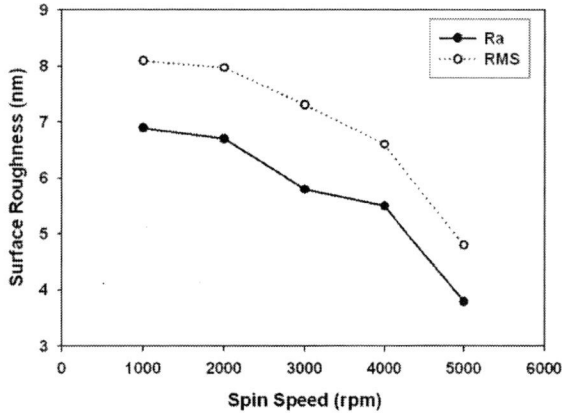

Fig. 2. The surface roughness of MEH-PPV films based on the spin speed.

I-V characteristic is measured by the fabrication of ITO/MEH-PPV/Al device. I-V curve of film obtained at 3000 rpm of the spin coating speed and drying at 100° C for 30 minutes is the highest value. Moreover, the fabrication of ITO/MEH-PPV + (PCBM, CdSe)/Al are considered to demonstrate the main effects of the morphological control on the hybrid solar cell performance.

CONCLUSIONS

The morphology control of MEH-PPV film is one of the most effectiveness methods to achieve optimization of device performance. The thin film of hybrid solar cells required optimum surface roughness to improve the transportation for the polymer/Al interfaces. The proposed method is clearly shown the transparent and systematic way to control the film surface morphology by considering various spin coating speeds. As a result, the device performance was optimized in the condition of 3000 rpm of spin coating speed.

ACKNOWLEDGEMENT

This work was supported by the BK21 Display Materials and Process Engineering Program.

REFERENCES

[1] M. L. Cantu, F. C. Krebs. "Hybrid solar cells based on MEH-PPV and thin film semiconductor oxides (TiO2, Nb2O5, ZnO, CeO2 and CeO2-Tio2) : performance improvement during long-time irradiation" Solar Energy Materials and Solar Cells. 90. pp. 2076-2086, 2006.

[2] H. Hoppe,, n. Arnold, D. Meissner, N. S. Sariciftci, "Modelling of optical absorption in conjugated polymer/fullerene bulk heterojunction plastic solar cells" Thin Solid Films, 451. pp. 589-592, 2004.

[3] X. Yang, J. Loos, S. C. Veenstra, W. J. H. Verhees, R. A. J. Janssen, " Nanoscale morphology of high performance polymer solar cells" Nano Letters, 5. pp. 579-583, 2005.

[4] W. U. Huynh, J. J. Dittmer, W. A. P. Alivisatos, "Hybrid nanorod-polymer solar cells" *Science* 295 pp.2425-2427,2002.

[5] W. U. Huynh, J. J. Dittmer, W. C. Libby, G. L. Whiting. A. P. Alivisatos, "Controlling the morphology of nanocrystal-polymer composites for solar cells" Advanced Functional Materials, 13. pp. 73-79, 2003.

Photonic MEMS Vibrating at X-band Rates (11 GHz)

Matthew Tomes, Tal Carmon

University of Michigan, USA

ABSTRACT

We report on an opto-mechanical resonator with vibration excited by compressive radiation pressure via Stimulated Brillouin Scattering [SBS]. We experimentally excite an 11 GHz mechanical whispering gallery mode [WGM] from an optical WGM. The mechanical vibration is detected via the red Doppler shifted light it scatters. We numerically solve the stress-strain equation to calculate the circumferentially circulating mechanical WGM and reveal mechanical WGMs with a variety of transverse shapes. The rate of vibration can be high irrespective of the fabrication tolerances as the driving forces are applied on an optical wavelength scale.

INTRODUCTION

Previous opto-mechanical resonators have been based on light radiation applying pressure on a cavity wall. Examples for the moving part include the wall of a WG cavity [1], a flexible mirror in a Fabry-Perot [2], and a flexible membrane [3]. Here, we experimentally demonstrate the use of optical electrostriction (Stimulated Brillouin Scattering) to excite mechanical vibration in WGM cavities by compressive pressure as in Fig. 1.

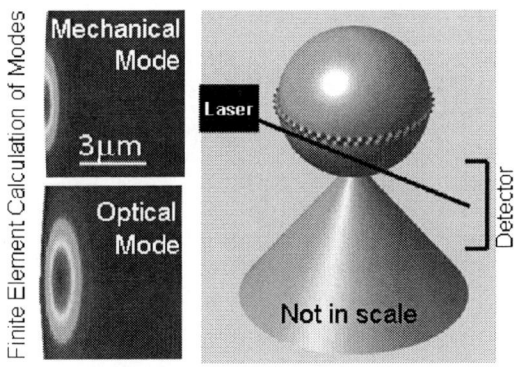

Figure 1. Finite element calculation of a mechanical WG mode with 600 acoustical wavelengths around the circumference where color represents mechanical strain. Similarly, an optical mode [4], with 300 wavelengths around the circumference where color represents electric field.

With this method, the acoustical wavelength is half of the optical-pump wavelength, and the vibration frequency is related to the time it takes sound (travelling at 5800 m/s) to cross half an optical wavelength. We are able to measure excitation of high order mechanical modes with vibration rates that scale inversely with optical wavelength, irrespective of the resonator size.

RESULTS

We experimentally observe mechanical vibration in a 100μm diameter SiO_2 sphere at rates of 11 GHz excited by a telecom compatible infrared pump (Fig. 2 inset). We measure a threshold of 26μW (Fig. 2) and slope efficiency of 90% for light scattered and Doppler shifted by the mechanical mode (Stokes). The oscillations are interogatable via the same fiber used for coupling into the cavity [5]. We also measure cascaded generation of stokes lines, frequency tunability, frequency scaling inversely with wavelength and high order transverse mechanical modes. In the future, using high Q diamond cavities, via currently available crystal polishing techniques [6], it will be possible to use a short wavelength UV source, and achieve mechanical frequencies above 200 GHz.

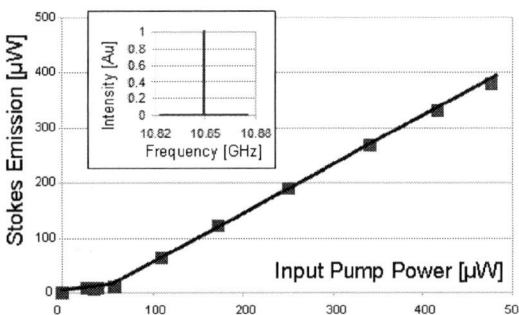

Figure 2. Stokes emission vs. input pump power for the process shows a threshold value of 26 μW. Inset is the mechanical vibration frequency as measured by an electrical spectrum analyzer.

REFERENCES

[1] T. Carmon, H. Rokhsari, L. Yang, T. Kippenberg, and K. Vahala, "Temporal behavior of radiation-pressure-induced vibrations of an optical microcavity phonon mode," *Physical review letters,* vol. 94, p. 223902, 2005.

[2] D. Kleckner and D. Bouwmeester, "Sub-kelvin optical cooling of a micromechanical resonator," *Nature,* vol. 444, pp. 75-78, 2006.

[3] J. Thompson, B. Zwickl, A. Jayich, F. Marquardt, S. Girvin, and J. Harris, "Strong dispersive coupling of a high-finesse cavity to a micromechanical membrane," *Nature,* vol. 452, pp. 72-75, 2008.

[4] M. Oxborrow, "Traceable 2-D finite-element simulation of the whispering-gallery modes of axisymmetric electromagnetic resonators," *IEEE Transactions on Microwave Theory and Techniques,* vol. 55, pp. 1209-1218, 2007.

[5] M. Cai, O. Painter, and K. Vahala, "Observation of critical coupling in a fiber taper to a silica-microsphere whispering-gallery mode system," *Physical review letters,* vol. 85, pp. 74-77, 2000.

[6] A. Savchenkov, A. Matsko, D. Strekalov, M. Mohageg, V. Ilchenko, and L. Maleki, "Low Threshold Optical Oscillations in a Whispering Gallery Mode CaF_ {2} Resonator," *Physical review letters,* vol. 93, p. 243905, 2004.

GYROSCOPIC OPTOMECHANICS
Xingyu Zhang, Matthew Tomes, Tal Carmon
University of Michigan, Ann Arbor, USA

ABSTRACT

Optically-induced forces can lead to mechanical deformation for micro- or nano-structures. Scattering optical forces and gradient optical forces, both originating from the linear momentum of light, have been widely demonstrated. Here we suggest the use of gyroscopic optical forces, originating from the angular momentum of light, to facilitate mechanical deformation. We calculated gyroscopic precession by light, arising from the angular momentum of a circularly polarized beam of light propagating inside a bent nano-waveguide. This gyroscopic optical force can deform the structure, with right-handed and left-handed circular polarization inducing opposite displacements.

INTRODUCTION

Optical forces have been found to cause significant mechanical displacement if light is confined to structures of micro- or nano-scale[1]. Scattering optical forces[2-4] and gradient optical forces[5, 6], both originating from the linear momentum of light, have been demonstrated. Here we exploit gyroscopic precession induced by the angular momentum of light to control MEMS deformation.

As shown in Figure 1a, when a turboprop airplane takes a turn with its propeller rotating, it will feel a gyroscopic "nose pitch" effect[7]. To maintain level flight, the pilot has to adjust the control surfaces and let aerodynamic forces exerted on the airframe counteract this gyroscopic moment. Similarly, in optics, each photon in a circularly polarized beam must carry spin angular momentum of $+\hbar$ or $-\hbar$ directed parallel to propagation, associated with the right-handed or left-handed circular polarization state of light [8]. Therefore, if we let a circularly polarized beam propagate along a curved trajectory in a bent nano-waveguide (Figure 1a), then distributed torque will arise on this waveguide. This torque is finally expected to deform the waveguide, as analyzed in Figure 1b.

CALCULATION

We derived the formula for optically-induced distributed torque (in units of $N \cdot m / m$) on the bent waveguide to be

$$\vec{\tau} = \pm \frac{P\lambda_0}{2c_0 \pi R} \hat{r} \qquad (1)$$

where P is the input power, λ_0 is the vacuum wavelength, c_0 is the vacuum velocity of light, R is the radius of curvature of the bent waveguide, and $\hat{r}$ is the unit vector along the radial direction. The right-handed and left-handed circularly polarized light will induce distributed torque along and opposite to the radial direction. This torque is present everywhere along the bent part of the nano-waveguide. With the ends fixed, the waveguide is expected to be deformed into a new twisted shape.

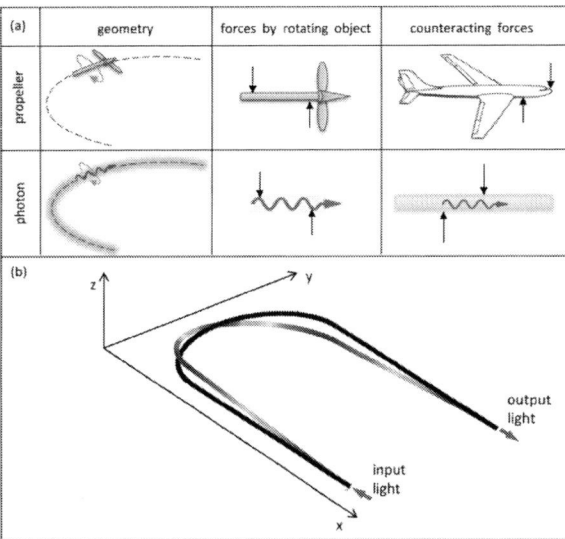

Figure 1: Gyroscopic precession by light. (a) Analogy between a propeller and photon. When a turboprop airplane takes a turn with its propeller rotating, gyroscopic moment will arise and cause the airplane to dip. Similarly, each photon in a circularly polarized beam carries angular momentum. When this beam of light propagates inside a bent nano-waveguide, gyroscopic moment will cause the waveguide to deflect from its original position. (b) A tapered waveguide bent into a U shape will exhibit a significant deflection when excited with an input light. The black curve indicates the original shape and the colored curve indicates the deflected shape with the amplitude of deformation greatly exaggerated for clarity.

RESULT

In order to get the deformation of a bent fiber as shown in Figure 1b, we solve the strain-stress tensorial equation where the only assumption is discretization in space. The boundary conditions on this structure include the gyroscopic forces by light calculated by Eq. 1.

Our major result is that a micron-scaled deformation is possible when a 1-Watt optical power is propagating at a typical taper that is 1 micron thick and bent as in Figure 2a. The deformation direction can be reversed by changing the polarization of light from right-hand to left-hand circular polarization (Figure 2c). As for competing forces, we were calculating that gravity and centrifugal radiation pressure have a negligible effect on the gyroscopic precession deformation seen in Fig 2c.

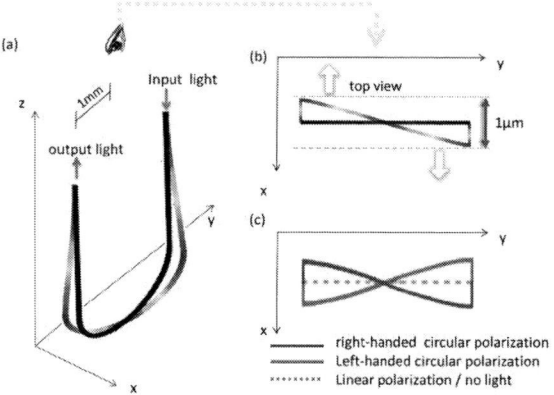

Figure 2: Proposed experiment. a. 3-D perspective. The colored curve indicates the deflected shape, and the black curve indicates the original U shape. The amplitudes of deformation are greatly exaggerated for clarity. b. Top-view of the deformation shows that the maximum deformation is greater than 1 μm. c. Comparison among deformation from right-handed circularly polarized light (blue curve) and left-handed circularly polarized light (red curve). There is no deformation from linearly polarized light (dashed line).

CONCLUSTION

We suggest gyroscopic precession by light to control the deformation of micro devices. These forces can be amplified by counting also on the orbital angular momentum of Laguerre-Gaussian type modes that were reported with angular momentum of more than $150\hbar$ per photon[9].

REFERENCES

[1] T. Carmon, H. Rokhsari, L. Yang *et al.*, "Temporal behavior of radiation-pressure-induced vibrations of an optical microcavity phonon mode," *Physical Review Letters,* vol. 94, no. 22, pp. 223902, 2005.

[2] A. Dorsel, J. McCullen, P. Meystre *et al.*, "Optical bistability and mirror confinement induced by radiation pressure," *Physical Review Letters,* vol. 51, no. 17, pp. 1550-1553, 1983.

[3] T. Kippenberg, H. Rokhsari, T. Carmon *et al.*, "Analysis of radiation-pressure induced mechanical oscillation of an optical microcavity," *Physical Review Letters,* vol. 95, no. 3, pp. 33901, 2005.

[4] H. Rokhsari, T. Kippenberg, T. Carmon *et al.*, "Radiation-pressure-driven micro-mechanical oscillator," *Optics Express,* vol. 13, no. 14, pp. 5293, 2005.

[5] M. Povinelli, S. Johnson, M. Lon ar *et al.*, "High-Q enhancement of attractive and repulsive optical forces between coupled whispering-gallery-mode resonators," *J. Opt. Soc. Am. B,* vol. 20, pp. 1967-1974, 2003.

[6] M. Povinelli, M. Loncar, M. Ibanescu *et al.*, "Evanescent-wave bonding between optical waveguides," *Optics Letters,* vol. 30, no. 22, pp. 3042-3044, 2005.

[7] L. E. G. W. H. Warner, *Dynamics*, pp. 441-442, Belmont: Wadsworth Publishing Company, Inc, 1961.

[8] R. Beth, "Mechanical detection and measurement of the angular momentum of light," *Physical Review,* vol. 50, no. 2, pp. 115-125, 1936.

[9] J. Curtis, and D. Grier, "Structure of optical vortices," *Physical Review Letters,* vol. 90, no. 13, pp. 133901, 2003.

MEASUREMENTS OF LIGHT FIELDS EMERGING FROM FINE AMPLITUDE GRATINGS

Myun-Sik Kim,[*] Toralf Scharf, and Hans Peter Herzig

Ecole Polytechnique Fédérale de Lausanne (EPFL), Switzerland
[*]myunsik.kim@epfl.ch

ABSTRACT

High resolution amplitude and phase of light fields emerging from a 2-μm-period amplitude grating are measured for different wavelengths. The amplitude gratings lead to highly periodic patterns caused by the Talbot effect. Such patterns reach periodicities of a fraction of the grating period. We discuss the effect of wavelengths and the number of diffraction orders participating in the imaging.

INTRODUCTION

Amplitude and phase of light fields emerging from periodic structures, although considered by many researchers, are still an interesting subject of research. Talbot observed the self-imaging effect of a periodic structure in 1836 [1]. Later, Lord Rayleigh showed that its origin is the interference of the diffracted beams and found that the regular repeat distance of self-images, the so-called Talbot Length, can be expressed as

$$z_T = 2 \, \Lambda^2 / \lambda, \qquad (1)$$

where Λ is the grating period and λ is the wavelength of the incident light [2]. Since smaller gratings have larger diffraction angles given by the diffraction formula as $\sin \theta = \lambda / \Lambda$, high numerical aperture (NA) and high resolution techniques are required to observe these effects for small gratings. This is especially important for grating with periods comparable to the operation wavelength. Moreover, phase measurements need often special techniques to be performed.

In this work, we demonstrate high resolution measurements of amplitude and phase of light fields diffracted by an amplitude grating with 2 μm period and using the high resolution interference microscope (HRIM) at several wavelengths. The Talbot effect is observed in the intensity and phase maps and results are discussed. Conditions were light field patterns smaller than the period of the diffraction grating can be observed are evaluated. The newly introduced phase measurements allow accurate evaluation of the Talbot length and provides deeper inside.

EXPERIMENTS

The HRIM is basically a microscope with objectives that have highest numerical aperture such as 100X / NA 0.9 combined with a Mach-Zehnder interferometer. Detailed information of system setup is given in references [3, 4]. In the object arm of the interferometer, the 2-μm-period grating is illuminated at normal incidence with a plane wave. In the reference arm a piezo-electrically driven mirror is mounted to change optical path lengths and the phase distribution of the wave-field is obtained by employing a classical five-frame algorithm [4]. The grating is mounted on a z-axis piezo stage with a scan range of

500 μm and an accuracy of 1 nm. The objects are scanned through the observation plane and three-dimensional (3D) intensity and phase maps are recorded during the measurement. While the amplitude measurement is diffraction limited and has finite spatial resolution, the phase measurement is, in principle, not subjected to resolution issues.

RESULTS AND DISCUSSIONS

Three different wavelengths, red at 642 nm, green at 532 nm and blue at 405 nm are considered and the diffraction angle and the Talbot length are studied. The corresponding x-z slices of the measured 3D intensity maps for three wavelengths are shown in Fig. 1. The propagation direction is z. One observes self-images at z_T and shifted pattern by half period at the half of the Talbot length ($z_T/2$) as expected from theory. The Talbot lengths are calculated by using Eq. (1) to be 12.5 μm for 642 nm, 15 μm for 532 nm, and 19.8 μm for 405 nm. The visibility of the smaller fractions of the Talbot images differs depending on the illuminating wavelength because the number of collected diffraction orders by the measurement objective varies for different wavelengths.

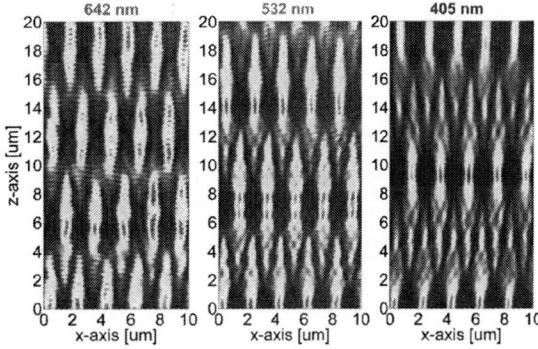

Fig. 1. The x-z slices of the measured 3D intensity map of 10 x 20 μm²: (a) 642 nm, (b) 532 nm, and (c) 405nm.

For a 2-μm-period grating at normal incidence one finds the angle for the first diffraction order as 18.7° for 642 nm, 15.4° for 532 nm, and 11.7° for 405 nm. The observing objective has a numerical aperture of NA = 0.9 corresponding to a maximum acceptance angle of 64°. Hence, diffraction from orders 0 to ± 3 can be collected by this objective at 642 nm illumination. For 532 nm and 405 nm, higher diffraction orders from 0 to ± 4 and 0 to ± 5 contribute, respectively. This can be verified by observing the back focal plane of the objective with the help of a telescope or Bertrand lens. Figure 2 shows images of the back focal plane of the objective and the bright spots are the diffraction orders. In practice, intensity falls off for large angles and less diffraction orders might be collected and certain experimental errors limit the NA of the objective to be smaller than 0.9. However, the number of diffraction orders gets fewer as the wavelength increases, as shown in Fig. 2. Better contrast and higher resolution are obtained for the intensity measurement at 405 nm thanks to the contributed higher orders.

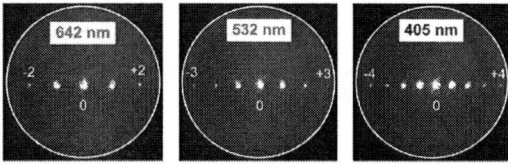

Fig. 2. Images of the back focal plane of the 100X / NA 0.9 objective lens. The grating is illuminated by a plane wave of (a) 642 nm, (b) 532 nm, (c) 405 nm. White circle shows the exit pupil and the 0 order as well as the highest diffraction order which are indicated.

To obtain more information the phase of the emerging fields are investigated by applying the aforementioned interferometric technique. Figures 3(a)-3(c) show the x-z slices of the measured 3D phase maps for the three wavelengths. At the multiple of the half of the Talbot length ($z_T/2$), planes with particular phase structures, such as phase singularities, are found. Because of the extremely low intensity near the plane of the singularities, measurement noise is amplified. Compared to the intensity measurements, the phase maps are more detailed. More precise evaluations of the Talbot length can be done by measuring the distance between the planes where phase singularities occur.

CONCLUSIONS

We present high resolution amplitude and phase measurements of light fields emerging from the 2-μm-period amplitude grating measured with a high resolution interference microscope. The number of the collected diffraction orders determines resolution and visibility of the smaller fractions of the Talbot images in the intensity and phase measurements. The phase maps allow more accurate measurements of the Talbot length by using the position of singularities because the phase field is not subjected to a resolution limit.

ACKNOWLEDGMENT

The research leading to these results has received funding from the European Community's Seventh Framework Programme FP7-ICT-2007-2 under grant agreement No. 224226.

REFERENCES

[1] F. Talbot, "Facts relating to optical science. IV" Philos. Mag. 9, 401-407, 1836
[2] L. Rayleigh, "On copying diffraction gratings and some phenomena connected therewith" Philos. Mag. 11, 196-205, 1881
[3] C. Rockstuhl, et al., "High Resolution Interference Microscopy: A Tool for Probing Optical Waves in the Far-Field on a Nanometric Length Scale" Current Nanoscience 2, 337-350 ,2006
[4] M.-S. Kim, et al., "Amplitude and Phase Measurements of Highly Focused Light in Optical Data Storage Systems" accepted for publication in Jpn. J. Appl. Phys., 2010
[5] J. Schwider, et al., "Digital wave-front measuring interferometry: some systematic error sources" Appl. Opt. 22, 3421-3432, 1983

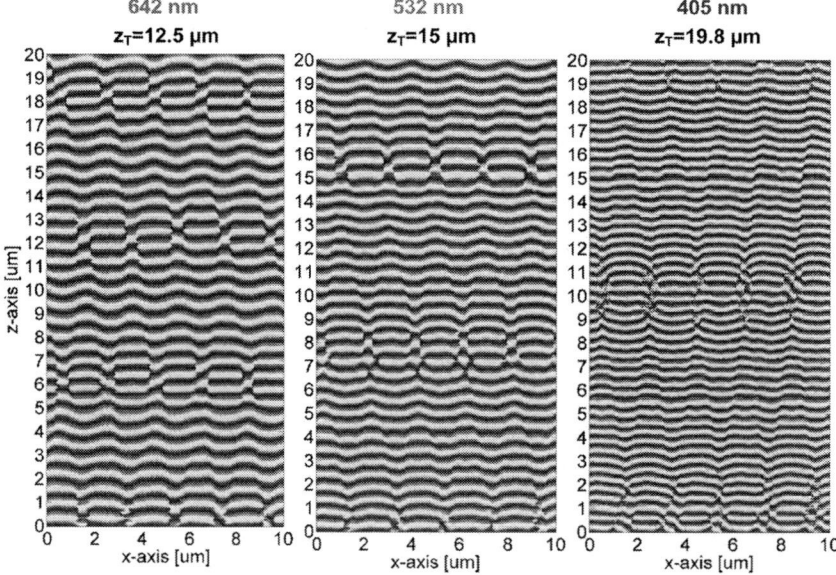

Fig. 3. The x-z slices of the measured 3D phase map of 10 x20 μm^2: (a) 642 nm, (b) 532 nm, and (c) 405 nm. At the multiple of the half Talbot length ($z_T/2$), the self-images of the plane wave surrounded by the phase singularities are observed.

Nanoscale Epitaxial Growth of GaN on Freestanding Circular GaN Grating

Yongjin Wang[*], Fangren Hu, and Kazuhiro Hane

Department of Nanomechanics, Tohoku University, Sendai, JAPAN

[*]wang@hane.mech.mech.tohoku.ac.jp

ABSTRACT

Freestanding circular GaN gratings are fabricated on a GaN-on-silicon substrate, and epitaxial growth of GaN is subsequently conducted on the prepared GaN template by molecular beam epitaxy growth. With the assistance of circular GaN gratings, the surface diffusion is improved and thus, the selective growth of GaN takes place. Epitaxial circular gratings with InGaN/GaN multiple quantum wells are generated with self-organized lateral facets and demonstrate the promising photoluminescence performances. This work provides a feasible approach to produce integrated GaN-Si devices by a combination of fast atom beam etching of GaN, silicon micromachining and epitaxial growth of GaN.

KEYWORDS

circular grating, GaN-on-silicon, fast atom beam etching, molecular beam epitaxy

INTRODUCTION

Recent progresses in the deposition of GaN thin films on silicon substrates have made high quality GaN-on-silicon substrates possible [1, 2], and bulk silicon micromachining is a mature technique [3, 4]. Additionally, the surface diffusion can be sufficiently enhanced with the assistance of nanostructures and thus, the selective growth of GaN is possible [5-8]. The resultant epitaxial structures with smooth self-organized facets are free of etching damage. These offer a great potential for producing novel GaN-based devices by a combination of epitaxial growth of GaN, etching of GaN and silicon micromachining [8, 9].

Here, we extend our research on the epitaxial growth of GaN on freestanding circular GaN gratings. Various freestanding GaN gratings are fabricated on a GaN-on-silicon substrate by combining electron beam (EB) lithography, fast atom beam (FAB) etching of GaN, and deep reactive ion etching (DRIE) of silicon. The epitaxial growth of GaN by molecular beam epitaxy (MBE) is performed on the prepared GaN template without introducing additional dielectric mask. Through the introduction of nanoscale GaN grating structures, the selective growth of GaN occurs and depends on the grating period P and the grating width W. The optical performances of the resultant epitaxial gratings are characterized in photoluminescence measurements.

THEORETICAL APPROACH

The epitaxial growth of GaN on freestanding GaN circular grating is implemented on a GaN-on-silicon substrate, which consists about 200nm GaN layer, 450nm $Al_xGa_{1-x}N$ layer (0.70~0.20 Al mole fraction), 200nm AlN buffer layer and 200μm (111) silicon substrate. Figure 1 schematically illustrated the fabrication procedure. A positive EB ZEP520A resist was first spin-coated onto the GaN device layer, and circular grating patterns were defined by EB lithography (steps a-b). The resist nanostructures were used as a mask for FAB etching of GaN with Cl_2 gas (step c). After removing the residual resist, the processed GaN circular gratings were protected by thick photoresist, and silicon substrate beneath the GaN grating region was subsequently patterned from the backside by photolithography and etched down to the AlN layer by DRIE of silicon (steps d-e). The freestanding GaN circular gratings were generated by removing the residual photoresist and cleaned for the epitaxial growth of GaN (step f). The epitaxial growth of GaN was conducted on the processed GaN template by MBE with radio frequency nitrogen plasma as gas source. The 6 pair 3nm InGaN/9nm GaN multiple quantum wells (MQWs) active layers were integrated at the upper part of ~420nm thick epitaxial films.

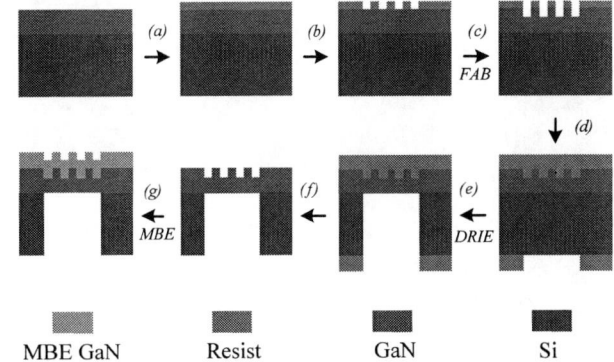

Figure 1: Schematic fabrication process of nanoscale epitaxial of GaN on freestanding circular GaN grating

RESULTS AND DISCUSSIONS

Figure 2(a) shows an optical micrograph of fabricated circular GaN grating on freestanding GaN slab. The size of freestanding GaN slab is about 300μm in diameter. Although the freestanding GaN slab has a slight deflection due to the residual stress, circular GaN gratings are well fabricated with various grating periods. Figure 2(b) and 2(c) illustrate scanning electron microscope (SEM) images of circular gratings, where the grating periods P are 600nm and 400nm, respectively. Figure 2(d) shows the tilt-view SEM image of 400nm period circular grating.

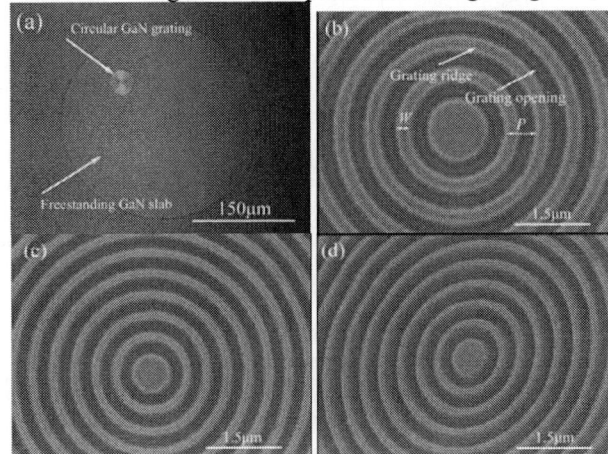

Figure 2: (a) optical micrograph of circular GaN grating on freestanding GaN slab; (b) and (c) 600nm and 400nm period circular GaN grating; (d) tilt-view of 400nm period GaN grating.

Figure 3(a) shows SEM image of the epitaxial circular GaN gratings. The freestanding GaN slab is easily fractured due to the large changes in residual stress during the epitaxial growth of

GaN. The coherent growth takes place on unpatterned GaN substrate, and the crack networks illustrated in the inset of Fig. 3(a) are generated in the epitaxial layers [1]. The 600nm period epitaxial grating is shown in Fig. 3(b), and the inset is the corresponding zoom-in view. With the assistance of circular gratings, the surface diffusion is sufficiently improved and thus, the selective growth of GaN takes places, resulting in the smooth self-organized lateral facets. The symmetrical cracks are also observed. Figure 3(c) shows SEM image of the epitaxial gratings with 400nm period. The lateral facets are smooth, and the coalescences of the lateral facets are realized. For a fixed grating period P, the grating width W has an important influence on the epitaxial growth of GaN. Figures 3(d)-3(f) illustrate the 700nm period epitaxial gratings with the grating widths W of 350nm, 300nm, and 250nm. As grating width W decreases, self-organized lateral facets tend to be fully coalesced and the selective growth of GaN also occurs in the grating openings.

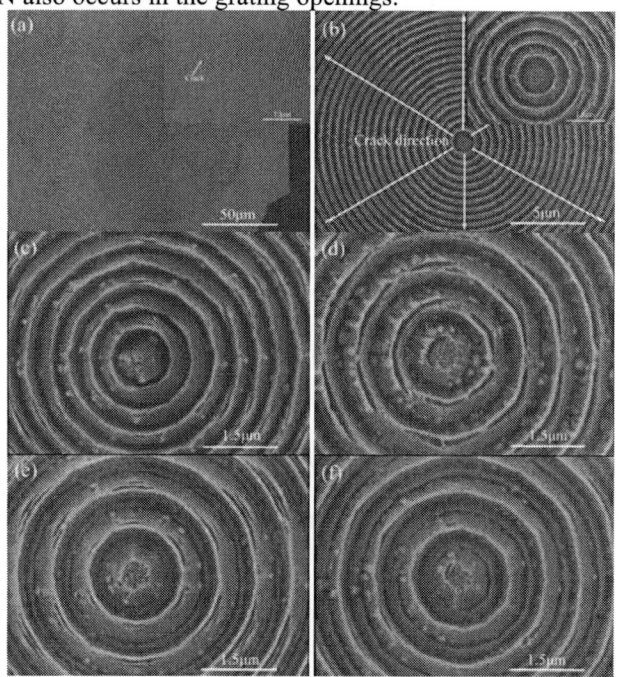

Figure 3: (a) Epitaxial circular GaN grating on freestanding GaN slab, and the inset is the zoom-in view; (b) 600nm period epitaxial grating and the inset is the zoom-in view; (c) 400nm period epitaxial grating; (d),(e) and (f) 700nm period epitaxial gratings with various grating widths.

The photoluminescence (PL) spectra of the epitaxial structures are measured at room temperature using a 325nm He-Cd laser source. The pump laser beam is focused onto the sample through a UV-compatible objective lens (numerical aperture: 0.36), and the emitted light is collected by the same objective lens and measured using a multichannel Hamamatsu analyzer system. Figure 4(a) shows the PL spectra of the epitaxial circular GaN gratings with various grating periods. The similar PL spectra are observed for all the epitaxial circular GaN gratings. The dominant PL peaks are found at about 434.1nm, which is associated with the excitation of InGaN/GaN MQWs. With decreasing the grating period P from 600nm to 400nm, the PL intensities are greatly increased. Additionally, coalescences of the lateral facets are easier to complete as the grating width W decreases. The PL intensities shown in Fig. 4(b) are thus increased.

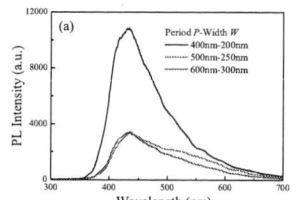 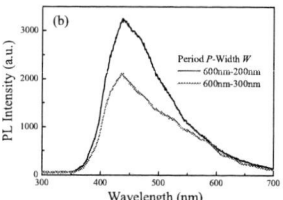

Figure 4: (a) PL spectra vs. grating period P; (b)PL spectra vs. grating width W.

CONCLUSIONS

Freestanding circular GaN gratings are fabricated on a GaN-on-silicon substrate by a combination of FAB etching of GaN and DRIE of silicon. Nanoscale epitaxial growth of GaN is subsequently conducted on the prepared GaN circular grating by MBE growth. Selective growth of GaN occurs and depends on the grating period P and the grating width W. Self-organized lateral facets are smooth and tend to be fully coalesced. The PL spectra demonstrate the promising optical performances for the epitaxial circular GaN gratings. This work opens the possibility for a large variety of integrated GaN optics devices.

ACKNOWLEDGEMENTS

This work was supported by the Research Project, Grant-In-Aid for Scientific Research (19106007). Yongjin Wang gratefully acknowledges JSPS for financial support.

REFERENCES

[1] J. -M. Bethoux, P. Vennéguès, F. Natali, E. Feltin, O. Tottereau, G. Nataf, P. De Mierry, and F. Semond, "Growth of high quality crack-free AlGaN films on GaN templates using plastic relaxation through buried cracks," J. Appl. Phys. **94**, 6499-6507(2003)

[2] F. Schulze, A. Dadgar, J. Bläsing, A. Diez, and A. Krost, "Metalorganic vapor phase epitaxy grown InGaN/GaN light-emitting diodes on Si(001) substrate," Appl. Phys. Lett. **88**, 121114 (2006)

[3] T. Zimmermann, M. Neuburger, P. Benkart, F. J. Hernández-Guillén, C. Pietzka, M. Kunze, I. Daumiller, A. Dadgar, A. Krost, and E. Kohn, "Piezoelectric GaN Sensor Structures," IEEE Electron Device Lett. 27, 309-312(2006)

[4] J. Lv, Z. C. Yang, G. Z. Yan, W. K. Lin, Y. Cai, B. S. Zhang, and K. J. Chen, "Fabrication of Large-Area Suspended MEMS Structures Using GaN-on-Si Platform," IEEE Electron Device Lett. **30**, 1045-1047(2009)

[5] K. Y. Zang, Y. D. Wang, S. J. Chua, and L. S. Wang, "Nanoscale lateral epitaxial overgrowth of GaN on Si (111)," Appl. Phys. Lett. **87**, 193106(2005)

[6] H. Matsubara, S. Yoshimoto, H. Saito, J. L. Yue, Y. Tanaka, and S. Noda, "GaN Photonic-Crystal Surface-Emitting Laser at Blue-Violet Wavelengths," Science **319**, 445-447(2008)

[7] K. Kishino, H. Sekiguchi, and A. Kikuchi, "Improved Ti-mask selective-area growth (SAG) by rf-plasma-assisted molecular beam epitaxy demonstrating extremely uniform GaN nanocolumn arrays," J. Cryst. Growth **311**, 2063-2068(2009)

[8] Y. J. Wang, F. R. Hu, and K. Hane, "Lateral epitaxial overgrowth of GaN on patterned GaN-on-silicon substrate by molecular beam epitaxy," submitted

[9] Y. J. Wang, F. R. Hu, H. Sameshima, and K. Hane, "Fabrication and characterization of freestanding circular GaN gratings," Opt. Express **18**, 773-779(2010)

Preparation of Anodic Aluminum Oxide Nano-Template using Al/Si Substrate for Large Area LED Applications

Lan Shen[1], Doogwook Kim[1], Bonggi Min[1], Zhengbin Gu[2] and Chinho Park[1]*

[1]Yeungnam University, Rep. of Korea, and [2]Nanjing University, China

*Corresponding author: chpark@ynu.ac.kr

ABSTRACT

Anodic aluminum oxide (AAO) nano-templates were prepared using the Al/Si substrates with the thickness of the aluminum films of about 500 nm. The pores of various sizes and depths were fabricated electrochemically through anodic oxidation. The optimum morphological structure for large area InGaN LED applications was formed by adjusting the applied potential, temperature, types of acid solution and their concentration. Scanning electron microscopy (SEM) investigations showed that hexagonal-close-packed nanopore arrays have smooth wall morphologies and well-defined diameters corresponding to the diameter of the applied template.

INTRODUCTION

Nanoporous anodic alumina has many favorable characteristics as a template material for nanoparticle array fabrication because the template formation process is very simple and results in a high density of parallel pores. The pore diameter can be tuned from ~10 to >100nm by varying the anodization conditions. Applying anodic aluminum oxide as an etching mask provides an effective non-lithographic and free-of-foreign-catalysts method to fabricate ordered and dense nitride nanostructures in the application of high efficiency GaN-based light emitting diodes. In addition, the nano-scale structures have attracted much attention in the field of fabrication of display devices and semiconductor devices due to their superior optical properties and unique electronic, magnetic properties suited for such novel devices [1-3].

EXPERIMENTAL

Aluminum films were deposited on silicon by RF magnetron sputtering. The electro-polishing process utilizing perchloric acid and ethanol mixture was first attempted to reduce the surface irregularities of the aluminum film by varying polishing times and conditions. The two or three steps anodization process was used to prepare an ordered porous alumina nano-template using a typical anodization method, of which the details could be found elsewhere [4]. The thin aluminum film was first anodized in a oxalic or sulfuric acid solution, forming irregular pores. After the first anodization, the formed AAO layer was etched by immersing the specimen in a mixture of chromic acid and phosphoric acid. The second anodization was then performed under the same process conditions as used in the first anodization. Finally the pore-widening process with phosphoric acid was applied to enlarge the diameter and depth of the AAO nano-template structure.

RESULTS&DISCUSSION

Figure 1 show the pores with a diameter of approximately 45~75nm were typically obtained after the second anodization. Depending on the conditions of synthetic process, pores of various sizes and depths could be reproducibly formed. As the anodization voltage and temperature changed from 40 to 60 V and 0 to 20 °C, respectively, the size of the pores was changed from 45 to 75 nm. However, this method required extremely accurate and stable control in handling the experiments, because the results could vary with even a small change in the voltage or temperature due to the remarkably thin aluminum film. The sensitivity of the results on the change in the process conditions was too high. A better way to increase the pore-size with controllable sensitivity is to use the phosphoric acid pore-widening step under 30 °C temperature and 0 V. And the increase of anodization time did not affect the size of pores significantly but caused the increase of pore depth. Since the starting aluminum thickness was very thin, over-time anodization resulted in the decrease of aluminum thickness drastically, resulting in the collapse of the pores.

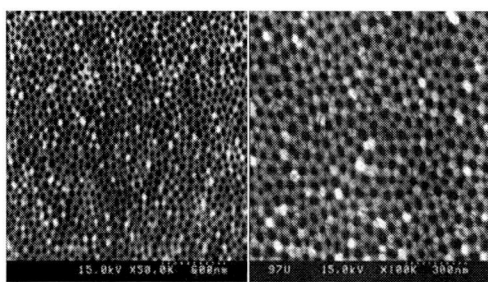

Fig. 1. SEM micrograph of surface view of a AAO film anodized at 25V and 15 °C using sulfuric acid and the average pore diameter is 45 ~75 nm.

After a two-step anodization, the resulting template has parallel pores with a fairly narrow size distribution and few defects, as shown in Figure 2. The SEM sample was prepared by first cross-sectional cutting the AAO film on Si substrate, the growth of straight parallel channels perpendicular to the substrate can be confirmed. It was combined with two parts: anodized, unanodized .

Fig 2. An Cross-sectional SEM image of the AAO template on Si substrate. The length of anodized is about 400nm.

Conclusion

Ordered nanoporous templates have been fabricated using an anodic aluminum oxide film. The results show that the nanopores have a relatively uniform size distribution with an average diameter of 45~75 nm. Our results indicate that the template approach is versatile and may be employed to fabricate various nanoparticle array systems on a range of substrates.

ACKNOWLEDGEMENTS

This work is supported by the National Center for Nanomaterials Technology of Pohang through Yeungnam University, and the researchers involved were partially supported by the 2nd phase of the BK21 Program.

REFERENCES

[1] Q. Li, S. M. Han, S. R. J. Brueck, S. Hersee, Y. B. Jiang and H. Xu, *Appl. Phys. Lett.*, 83, 5032 (2003).

[2] M. D. Austin, H. Ge, W. Wu, M. Li, Z. Yu, D. Wasserman, S. A. Lyon and S. Y. Chou, *Appl. Phys. Lett.*, 84, 5299 (2004).

[3] H. Chik, J. Liang, S. G. Cloutier, N. Kouklin and J. M. Xu. *Appl. Phys. Lett.*, 84, 3376 (2004).

[4] Soo-Hwan Jeong and Kun-Hong Lee, *Synth. Metals*, 139, 385-390 (2003).

Polarization Control of InAs Quantum Dot Semiconductor Laser using External Light Injection Technique

P. C. Peng [1], R. L. Lan [1], S. T. Hsu [1], H. H. Lu [1], G. Lin [2], H. C. Kuo [3], G. R. Lin [4], J. Y. Chi [5]

1. Department of Electro-Optical Engineering, National Taipei University of Technology, Taipei, Taiwan R.O.C.
2. Department of Electronics Engineering, National Chiao Tung University, Hsinchu, Taiwan, R.O.C.
3. Institute of Electro-Optical Engineering, National Chiao Tung University, Hsinchu, Taiwan, R.O.C.
4. Graduate Institute of Photonics and Optoelectronics and Department of Electrical Engineering, National Taiwan University, Taipei, Taiwan, R.O.C.
5. Institute of Opto-Electronic Engineering, National Dong Hwa University, Hualien, Taiwan R.O.C.
E-mail: pcpeng@ntut.edu.tw

Abstract

This work investigates experimentally the polarization control of InAs-InGaAs quantum dot (QD) semiconductor laser. The polarization characteristic of QD laser at different bias current is studied. The results of this study will be useful in the field of optical signal processing.

Summary

Optical signal processing systems using semiconductor lasers have become very appealing, because of their inherent compactness, simplicity of operation, low cost, and low power consumption [1]. Recently, an all optical inverter using a quantum well (QW) vertical-cavity surface-emitting laser (VCSEL) was reported [2]. Furthermore, all optical signal processing based on polarization switching in a QW VSCEL also has been explicated [3]. However, these systems generate the optical inverted and non-inverted data signals at the different wavelengths. That could limit the application in optical signal processing systems and polarization modulation systems. This work presents polarization control of quantum dot (QD) VCSEL based on a light injection technique. The results demonstrate its potential application in all optical signal processing systems.

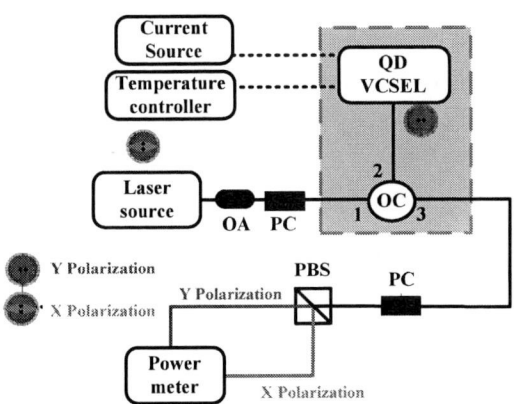

Fig. 1 Experimental setup for polarization control of QD VCSEL (PBS: polarization beam splitter, OA: optical attenuator, OC: optical circulator, PC: polarization controller)

Figure 1 shows the experimental setup for polarization control of QD VCSEL using an external light injection technique. The structure of QD VCSEL has been described in our earlier works [4]. The polarization of the QD laser output is Y polarization, and the QD laser exhibits no polarization switching. The light injection power is controlled by an optical attenuator (OA) at the output of the laser source, and the polarization of laser source is adjusted using a polarization controller (PC). The output light with X polarization from the laser source is guided through an optical circulator (OC). The polarization characteristic of QD laser is investigated using a polarization beam splitter (PBS) and an optical power meter. The external light injection caused polarization switching in the QD laser.

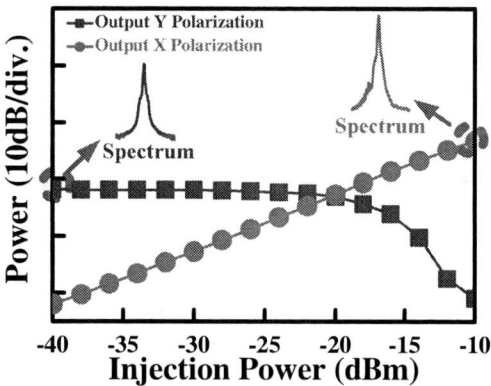

Fig. 2 Measured X and Y polarization power at 1.4 Ith.

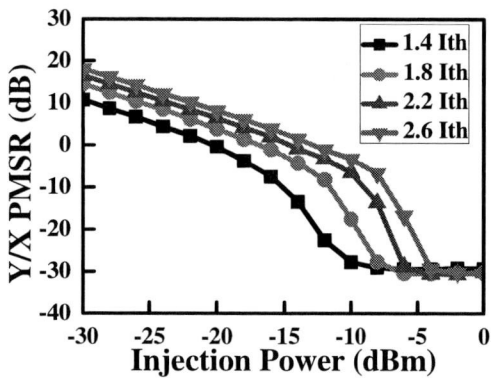

Fig. 3 Measurements of polarization mode suppression ratio (Y/X PMSR) at various bias currents.

Figure 2 shows the measured X and Y polarization power of QD VCSEL at 1.4 Ith (Ith: threshold current). In addition, the inset of Fig. 2 shows the

optical spectra of the QD VCSEL. When external light injected into the QD VCSEL, the output power of the X polarization will increases whereas that of the Y polarization will gradually decrease. We observed that varying the injection power causes polarization switching. Fig. 3 shows the measured Y/X polarization mode suppression ratio (PMSR) values as functions of the injection power. When the bias current of QD laser was increased, the injection power needed to be increased. The Y/X PMSR is less than -29 dB at the injection power of -4 dBm.

In conclusion, we experimentally demonstrate the polarization control of QD VCSEL using an external light injection technique. The switched PMSR of the QD VCSEL is achieved by adjusting the injection power. This result can be applied in novel signal processing systems.

References

[1] A. Kuramoto and S. Yamashita, "All-optical regeneration using a side-mode injection-locked semiconductor laser," IEEE Journal of Quantum Electronics, vol. 9, pp. 1283–1287, 2003.

[2] Y. Onishi, N. Nishiyama, C. Caneau, F. Koyama, C. E. Zah, "All-optical inverter based on long-wavelength vertical-cavity surface-emitting laser," IEEE Journal of Quantum Electronics, vol. 11, pp. 999–1005, 2005.

[3] K. H. Jeong, K. H. Kim, S. H. Lee, M. H. Lee, B. S. Yoo, K. A. Shore, "Polarization switching in a 1.5 um wavelength single-mode vertical cavity surface emitting laser under modulated optical beam injection control," Photonics in Switching, pp. 71 – 72, 2007.

[4] P. C. Peng, H. C. Kuo, W. K. Tsai, Y. H. Chang, C. T. Lin, S. Chi, S. C. Wang, G. Lin, H. P. Yang, K. F. Lin, H. C. Yu, and J. Y. Chi, "Dynamic characteristics of long-wavelength quantum dot vertical-cavity surface-emitting lasers with light injection," Optics Express, vol. 14, pp. 2944-2949, 2006.

Experimental Demonstration of the Vernier Effect using Series-Coupled Racetrack Resonators

Robi Boeck*, Nicolas A. F. Jaeger and Lukas Chrostowski
University of British Columbia, Vancouver, British Columbia, V6T-1Z4, Canada
* Email: rboeck@ieee.org

ABSTRACT

We demonstrate the design and performance of series-coupled racetrack resonators in silicon-on-insulator. The device was modeled and optimized by tuning various parameters such as power coupling factors and resonator dimensions. The Vernier effect is experimentally verified with an extended free spectral range of 36.2 nm and an interstitial resonance suppression of more than 11 dB. To the best of our knowledge, we have shown the closest match between experimental and theoretical results for the Vernier effect. Devices exhibiting similar performance to the one reported on here should be useful in applications such as wavelength division multiplexers and optical sensors.

INTRODUCTION

The advent of silicon-on-insulator (SOI) technology has helped to reduce the overall size of numerous integrated photonic devices down to the submicrometer scale. Ring and racetrack resonators have highlighted the benefit of SOI in applications such as wavelength division multiplexing and various types of sensors. In recent years, multiple identical rings either coupled in series or in parallel have been experimentally proven to be superior to single ring configurations [1]. However, one limitation that cannot be mitigated by using multiple identical rings is that the free spectral range (FSR) is inversely proportional to the total length of the resonator structure. To increase the FSR, the length of the resonator must decrease, thus increasing the bending losses. One solution to this problem is to use coupled ring resonators to create the Vernier effect and, thus, to significantly extend the FSR [1]. To the best of our knowledge, B. Timotijevic et al. [2] were the first to fabricate coupled SOI ring resonators to show the Vernier effect. Their devices showed relatively weak interstitial resonance suppression, as compared to the greater than 11 dB reported here. In this paper, the design and experimental performance of series-coupled racetrack resonators (shown in Fig. 1) showing the Vernier effect in single mode SOI strip waveguides is reported.

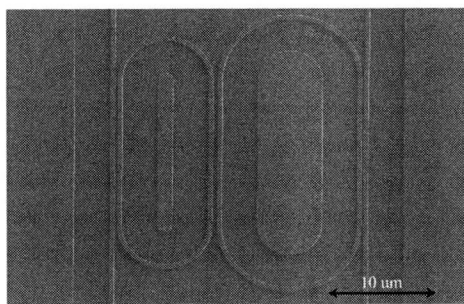

Fig. 1. SEM image of the fabricated series-coupled racetrack resonators.

DESIGN

The architecture commonly used to create the Vernier effect is to couple two racetrack resonators, R1 and R2, in series with one through port bus waveguide and one drop port bus waveguide [1]. The SEM shown in Fig. 1 is an example of this architecture. R1 and R2 have different FSRs created by choosing the length of R1 to be different from that of R2. The FSR of the device is significantly increased when $N \times FSR1$ equals $M \times FSR2$, where N and M are coprime integers [1]. Interstitial resonances will be present in the drop response due to one of the ring's output not being totally zero. The main goals in our design is to reduce all of the interstitial resonances and to have a large FSR comparable to the entire C-band. The fabricated SOI strip waveguides have a width of 420 nm (measured from SEM image) and a height of 220 nm. To determine the effective index and the group index of the waveguide, simulations are performed using a 2D mode solver [3]. The mode number and extended FSR are 23 and 37.7 nm, respectively. N and M are 5 and 4, respectively. The extended FSR, N, and M, determine the total length of each resonator. The choice of the length of the racetrack, ring radius, and gap distances depends on the optimization of the power coupling factors and reduction of losses. The symmetric power coupling factors to the bus waveguides are optimized to be 0.35. The inter-ring power coupling factor is 0.01. The straight track length for both rings is 15 μm. The radii of R1 and R2 are 6.795 μm and 4.475 μm, respectively. A small straight section of 50 nm is added in the middle of each half circle of the racetrack resonator. The gap distances to the bus waveguide for R1 and R2 is 150 nm. The inter-ring gap distance is 330 nm. The simulation of the drop port response is shown in Fig. 2.

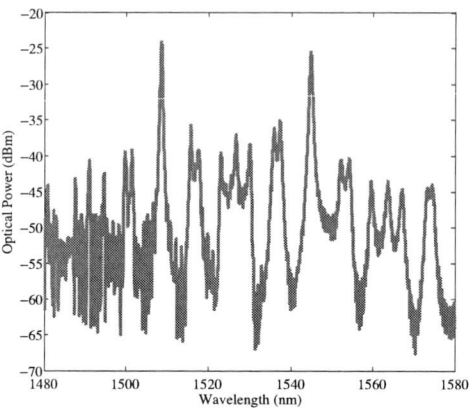

Fig. 3. Experimental drop port response.

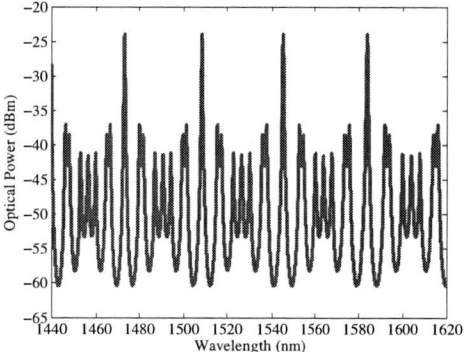

Fig. 2. Theoretical drop port response of series-coupled racetrack resonators.

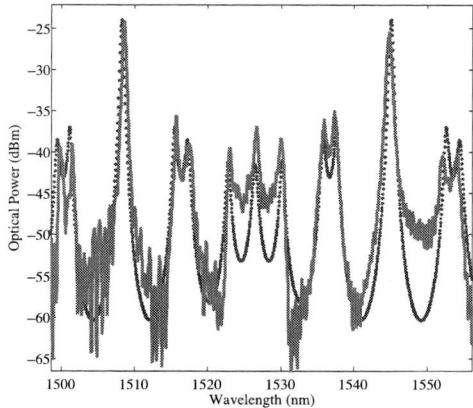

Fig. 4. Experimental (solid/red) and simulated (dotted/blue) drop port response.

EXPERIMENTAL RESULTS AND DISCUSSION

Series-coupled racetrack resonators, with the parameters mentioned above, were fabricated by ePIXfab at IMEC using 193 nm lithography. 1-D periodic grating couplers with tapers for reducing the width of the waveguide from 10 μm to 500 nm are used to couple light into and out of the waveguides. The length of the taper is 400 μm to ensure single mode propagation (TE at 1550 nm). Fig. 3 shows the measured TE-polarized drop response of the device shown in Fig. 1. The extended FSR is 36.2 nm, which is 4-5 times larger than the constituent FSRs. The possible reason that the theoretical FSR, 37.7 nm, does not match the experimental FSR is that the precise waveguide dimensions are not known due to fabrication tolerances. The interstitial resonance suppression is greater than 11 dB. The measured Q-factor is found to be 3260. The experimental data is also compared with the simulation results as shown in Fig. 4. The small discrepancies seen may, in part, be the result of neglecting second order dispersion in the theoretical model.

CONCLUSION

In summary, we demonstrated compact, high performance, series-coupled racetrack resonators exhibiting the Vernier effect. Our design goal was to minimize the interstitial resonances as well as to have a large FSR. The extended FSR of the device reported on was 36.2 nm. The interstitial resonance

suppression was greater than 11 dB. We therefore conclude that our design goals can be met.

ACKNOWLEDGMENTS

The authors would like to thank Raha Vafaei and Stephanie Flynn for obtaining SEM images and Dr. Nicolas Rouger and Nadia Makan for building the experimental set-up. We would also like to acknowledge CMC Microsystems, NSERC Canada (CGS-M) and Lumerical Solutions Inc. for supporting this project.

REFERENCES

[1] O. Schwelb,"The nature of spurious mode suppression in extended FSR microring multiplexers," *Optics Communications*, vol. 271, pp. 424–429, 2006.
[2] B. Timotijevic, et al., "Tailoring the spectral response of add/drop single and multiple resonators in silicon-on-insulator," *Chinese Optics Letters*, vol. 7, no. 4, pp. 291–295, 2009.
[3] A. Fallahkhair, K. Li, and T. Murphy, "Vector finite difference modesolver for anisotropic dielectric waveguides," *Journal of Lightwave Technology*, vol. 26, no. 11, pp. 1423–1431, 2008.

978-1-4244-8926-8/10 $26.00 © 2010 IEEE

POLARIZATION INDEPENDENT GRATING COUPLER FOR SILICON-ON-INSULATOR WAVEGUIDES

Chun-Chia Chiu and Ding-Wei Huang*

Institute of Photonics and Optoelectronics, National Taiwan University
No. 1, Sec. 4, Roosevelt Rd., Taipei, 10617, Taiwan, R.O.C.
Tel/Fax +886-2-33663664, *dwhuang@cc.ee.ntu.edu.tw

ABSTRACT

A polarization independent grating coupler for coupling the light from an optical fiber into a silicon-on-insulator (SOI) waveguide with a coupling efficiency independent of the polarization of the incident light was proposed. Such a polarization independent grating coupler can be achieved by superimposing the grating couplers which were originally designed for coupling solely the TE and the TM modes. Furthermore, it may have the advantage of being fabricated by a single etching step.

INTRODUCTION

The coupling of the light from an optical fiber into an optical waveguide is always an important issue for the design and fabrication of the integrated optical communication devices and it can be achieved by using the grating couplers with high coupling efficiencies . In addition, without additional steps in the fabrication of these devices, the grating couplers can also be designed to exhibit auxiliary functions such as focusing and splitting of the coupled light. The multifunctional capability of grating couplers may reduce or eliminate the need for additional optical waveguide components. An important limitation of these grating couplers is that they are designed solely for a specified polarization of the incident light. This may be a serious drawback that the polarization of the input light from an optical fiber is in most cases unknown and may even vary with time. Because the grating coupler is generally highly sensitive to the polarization, it will not work properly when input light having a polarization different from that given in the design specification, and thus it is impractical in a real optical fiber system. So we find it would clearly be desirable to have a grating coupler which has a coupling efficiency independent of the polarization of the incident light.

In this paper we present the couplers designed by simultaneous optimization for two orthogonal polarization states of the light incident on the two-dimensional subwavelength gratings (SWGs) [1]. The grating structure is designed where the diffractive grating is formed in the longitudinal direction, whereas the nondiffractive SWG structure in the lateral direction acts as effective medium of the grating [see Fig. 1]. The results show that the polarization response is highly flexible as the longitudinal and lateral grating structures vary. Polarization independence can be achieved when the coupler is designed to have the same response for coupling the two orthogonal polarizations of the incident light into the TE and TM modes, respectively, of the waveguide.

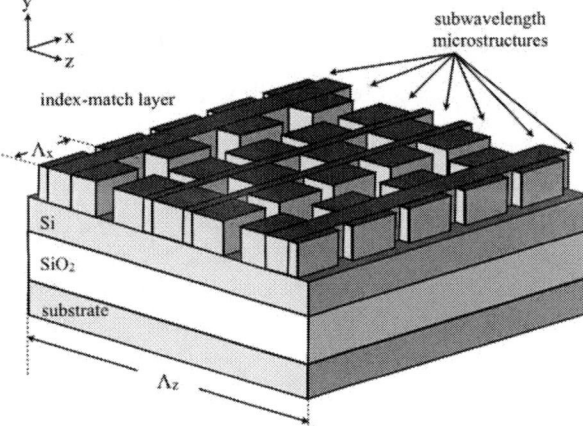

Fig.1 Grating coupler with SWG in lateral direction.

STRUCTURE AND SIMULATION METHOD

First we designed two different grating couplers for coupling solely the TE and the TM modes according to the methods described by the previous studies [2]. For TE polarization, the grating period was $\Lambda_{TE} = 0.51$ μm, filled factor was $ff_{TEz} = 0.5$ and refractive index was $n_{SWG,TE} = 2.458$. For the TM polarization, the parameters were $\Lambda_{TM} = 0.68$ μm, $ff_{TMz} = 0.5$ and $n_{SWG,TM} = 2.458$. Both of these grating couplers are 12.24 μm long. Then these two gratings were superimposed in a way that the refractive index of the intersection of the two merged grating couplers is equal to the refractive index silicon ($n_{Si} = 3.476$); the refractive index of the complement of two grating couplers is equal to the index-matching

978-1-4244-8926-8/10 $26.00 © 2010 IEEE   171

layer above the grating coupler (n_{iml} = 1.46); and the refractive indices of the rest parts were kept unchanged. Fig. 2 schematically illustrates the geometry of a polarization-independent grating coupler. After the merging of the two grating couplers, the resultant grating coupler has an equivalent grating period Λ_z = 2.04 μm which is the least common multiple of the two original grating couplers separately for TE and TM modes.

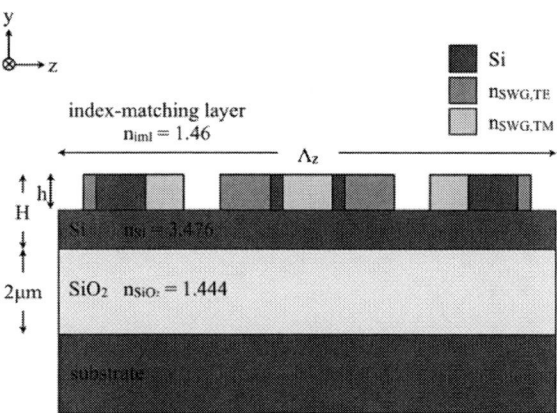

Fig.2 The vertical dimension showing the period Λ_z.

The SOI waveguide consists of the silicon core is H = 300 nm and the etch depth is h = 120 nm. The grating is etched into the silicon core layer and we introduce a SWG structure in the lateral (x) direction. For the proposed geometry, assuming that the pitch of the grating in the lateral direction is designed in a way that only the 0th order diffraction in the x-z plane is allowed, so

$$\Lambda_x < \lambda_0 / \max(n_{eff}) , \qquad (1)$$

where λ_0 = 1.55 μm is the free-space wavelength and $\max(n_{eff})$ is the maximum effective index encountered in the structure. For convenience, the three-dimensional grating structure was reduced to a two-dimensional (2D) equivalent grating structure [3] as the subwavelength regions can be equivalent into effective media. The end facet of the fiber was close to the grating, and the fiber was slightly tilted with respect to the vertical axis (θ = 10.18°) in order to avoid a large second order reflection. The entire structure is then analyzed numerically by using the finite-difference time-domain method (FDTD) with a simulation area of 21.62 × 4.84 μm^2 and a fundamental mode of the optical fiber incident toward the grating coupler and finally coupled into the SOI waveguide. Furthermore, we also used the micro genetic algorithm to optimize the coupling efficiency.

RESULTS AND CONCLUSION

The optimized parameters which allow the grating coupler to couple both the TE and TM polarizations with a high coupling efficiency of about 33% and 32% are listed in Table 1. The magnitude squared time average electric field and magnetic field of the optimized polarization-independent grating coupler is shown in Fig. 3 along with the simulation geometry.

Table 1 Optimal grating parameters of the polarization-independent grating coupler

H	h	Λ_{TE}	Λ_{TM}	θ
262 nm	114 nm	518 nm	690 nm	12.3°

ff_{TEz}	ff_{TMz}	$n_{SWG,TE}$	$n_{SWG,TM}$
0.401	0.535	2.609	2.473

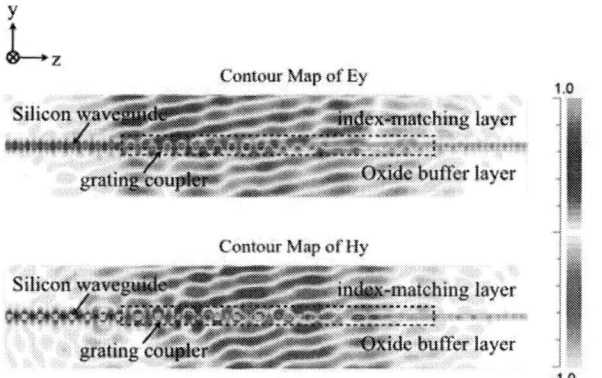

Fig.3 Simulation results by 2-D FDTD.

We have demonstrated that input grating couplers can be designed to have a desired coupling efficiency independent of the polarization of the incident light.

REFERENCES

[1] Robert Halir, Pavel Cheben, Siegfried Janz, Dan-Xia Xu, Íñigo Molina-Fernández, and Juan G. Wangüemert-Pérez, "Waveguide grating coupler with subwavelength microstructures" Opt. Lett., vol. 34 no. 9, pp. 1408-1410, 2009

[2] Günther Roelkens, Dries Van Thourhout and Roel Baets, "High efficiency Silicon-on-Insulator grating coupler based on a poly-Silicon overlay" Opt. Express, vol. 14 no. 24, pp. 11622-11630, 2006

[3] Dirk Taillaert, Peter Bientman, and Roel Baets, "Compact efficient broadband grating coupler for silicon-on-insulator waveguides," Opt. Lett., vol. 29, pp. 2749-2751, Dec. 2004

AN APPROACH FOR MODELING PHOTONIC CRYSTAL CIRCUITS

Yih-Bin Lin[1], Rei-Shin Chen[1], Ju-Feng Liu[2], and Han-Bin Lin[1]

Lunghwa University of Science and Technology[1], Taiwan, R.O.C.

China University of Science and Technology[2], Taiwan, R.O.C.

ABSTRACT

A novel approach for modeling photonic crystal circuits is proposed. In this approach, the effective length of photonic crystal waveguide between two devices is estimated. The frequency response of the photonic crystal circuit is modeled as a Fabry-Perot cavity. An example of double 90-degree bend is analyzed by the approach. The results have great agreement with those obtained by finite-difference time-domain method.

INTRODUCTION

Photonic crystal (PhC) structures have been the focus of research recently because of the potential ability to realize compact integrated optical circuits. Finite-difference time-domain (FDTD) method with optical pulse excitation [1] is used effectively to characterize device performance of PhC structures. However, the FDTD method is time-consuming when dealing with waveguide circuit which contains two or more devices. Some modelling methods [2][3] are proposed by effective equations or scattering matrix method. In this paper, we especially proposed a concept of frequency-depended waveguide length which is used to model a PhC circuit consisting of two waveguide bends.

In our method, some characteristics of the PC circuits must be calculated in advance and are saved as a database. For examples, the reflection coefficients of individual devices are obtained by FDTD method. The dispersion curve of waveguide is calculated by plane wave expansion (PWE) method. Besides, an effective waveguide segment contributed by the PhC device at each terminal is introduced, which is important for determining the cavity length between two devices. The length of the effective segment is estimated by simulating a circuit containing two identical devices connected by a section of waveguide. With considering the extra waveguide lengths contributed by the device, the resonant frequencies of a circuit can be predicted with good accuracies. An example of double bend is used to demonstrate the method

THEORY

The PhC circuit in Fig.1 consists of two 90° bends, a connection PhC waveguide of length L, and

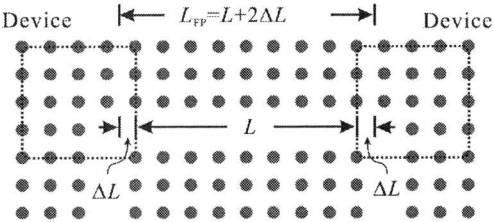

Fig. 1.   Layout of the double 90° waveguide bend.

input/output PhC waveguides. The propagating optical wave in the PhC waveguide encounters partial reflection at each of the bends. Therefore, the whole structure of the double 90° bends works like a Fabry-Perot (FP) cavity. From the theory of FP cavity, the cavity resonates and has minimum reflection while the round-trip optical path inside the cavity is an integer multiple of 2π. Then, the FP cavity length L_{FP} is given by

$$L_{FP} = \frac{m\pi}{k} \qquad (1)$$

where m is a positive integer and k is the propagation constant of the PhC waveguide. The value of k versus frequency, known as dispersion curve, is calculated

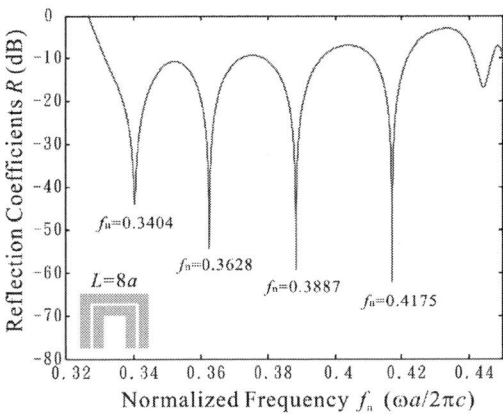

Fig. 2. Reflection spectrum of the double 90° waveguide bend simulated by FDTD for $L=8a$.

by PWE method. The reflection spectrum of the double 90° waveguide bends in Fig.1 is simulated by FDTD method in order to obtain information about resonant frequencies. The parameters used in Fig.1 are the same as in [1]. The simulation result is shown in Fig.2. The horizontal axis is the normalized frequency $f_n = \omega a/2\pi c$, where a is the lattice constant

978-1-4244-8926-8/10 $26.00 © 2010 IEEE

of the PhC structure. The reflection spectrum of the double 90° waveguide bends represents four resonant frequencies of the FP cavity. The values of normalized propagation constant $k_n = ka/2\pi$ at the resonant frequencies are obtained from the dispersion curve. From (1), by choosing suitable values of m, the corresponding lengths of the FP cavity are calculated and summarized in Table I.

Table I
Corresponding Data at Resonating Frequencies

f_n	k_n	m	L_{FP}	ΔL
0.3404	0.1676	4	11.9331a	1.9666 a
0.3628	0.2248	5	11.1210a	1.5605 a
0.3887	0.2777	6	10.8030a	1.4015a
0.4175	0.3296	7	10.6189a	1.3095 a

From Table I, we can see the FP cavity lengths L_{FP} calculated from (1) are not fixed values and thus are frequency depended. It can be explained that the effective FP cavity lengths corresponding to different waveguide modes are slightly differed from each others. The effective FP cavity lengths L_{FP}, therefore, are slightly longer than L. We define ΔL as the penetrating length at both ends of the FP cavity, that is

$$L_{FP} = L + 2\Delta L \qquad (2)$$

The values of penetrating length ΔL at resonant frequencies are shown in Table I. The penetrating lengths ΔL in Table I only have discrete values at the resonant frequencies. The other values of ΔL in the frequency range we are interested can be obtained by interpolation and extrapolation. If the cavity length of the double waveguide bends is changed, only the distance L is changed and the frequency depended penetrating length ΔL is still the same since ΔL is related to the individual waveguide bend. This concept is used to establish the model of the PhC circuit.

MODELING OF DOUBL 90° WAVEGUIDE BEND

From the theory of FP cavity, the reflection spectrum of a double 90° waveguide bend R' is given as

$$R' = \frac{4R\sin^2(\delta/2)}{(1-R)^2 + 4R\sin^2(\delta/2)} \qquad (3)$$

where R is the reflection spectrum of a single 90° waveguide bend and $\delta = 2kL_{FP}$ is the round-trip optical path within the cavity.

The model we proposed is based on Equation (3). After we have established a database containing R, k,

and ΔL, the reflection spectrum of a double 90° waveguide bend is obtained by some simple arithmetic steps instead of time-consuming FDTD method. Equation (3) is used to estimate the reflection spectrum of a double 90° waveguide bends for $L=18a$. The solid line in Fig. 3 is the reflection spectrum calculated from Equation (3) and the dash line is the reflection spectrum simulated by FDTD method for comparison.

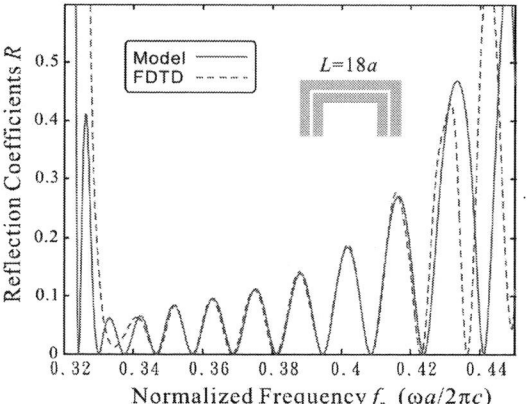

Fig. 3. Reflection spectrum of the double 90° waveguide bend for $L=18a$.

From the results in Fig.3, great agreement is obtained between the proposed model and the FDTD method within the middle part of the simulated frequency range. There exists more difference for $f_n > 0.42$ or $f_n < 0.34$ because that extrapolation is taken at these frequency range.

CONCLUSIONS

A model for photonic crystal circuits is proposed. The frequency-depended effective length of a connection waveguide is estimated. An example of double bends is analyzed by the model and gives great agreement with respect to the FDTD method.

REFERENCES

[1] A. Mekis, J. C. Chen, I. Kurland, S. Fan, P. R. Villeneuve, and J. D. Joannopoulos, " High Transmission through Sharp Bends in Photonic Crystal Waveguides," Physical Review Letters, vol. 77, no. 18, pp. 3787-3790, 1996.

[2] S. F. Mingaleev and Y. S. Kivshar, "Effective equations for photonic-crystal waveguides and circuits," Optics Letters, vol.27, no.4, pp. 231-233, 2002.

[3] S. F. Mingaleev and K. Busch, "Scattering matrix approach to large-scale photonic crystal circuits," Optics Letters, vol.28, no.8, pp. 619-621, 2003.

DESIGN OF HIGH TRANSMISSION BROADBAND 90-DEGREE BENDS FOR TWO DIMENSIONAL CUBIC PHOTONIC CRYSTALS

Yih-Bin Lin[1], Cheng-Ru Li[1], Rei-Shin Chen[1], Jung-Young Su[2]

Lunghwa University of Science and Technology[1], Taiwan, R.O.C.

Ta-Hwa Institute of Technology[2], Taiwan, R.O.C.

ABSTRACT

A simple design of 90-degree bend is proposed for cubic photonic crystals. By doubling the density of rods, new types of bends are created. The greatest proposed bend has transmission efficiencies above 99% for a wide frequency range.

INTRODUCTION

Photonic crystal structures have attracted a lot of interest recently because of their potential ability to control the propagation of light within a small area compared to traditional dielectric waveguides. Within various devices and components, 90-degree waveguide bends are fundamental components in cubic lattice structures. Some 90-degree bends [1-3] have been proposed for two dimensional cubic photonic crystals. In this work, we proposed a design of broadband 90-degree bend based on a lattice structure with double density of rods. In our design, the radii and index of the rods are all the same, which makes the design simple. A simple double-density bend has transmission efficiencies higher than 94%. Furthermore, a modified double-density bend has transmission efficiencies greater than 99% for a wide frequency range.

DESIGN AND SIMULATIONS

The bent waveguide shown in Fig.1 (a) is a traditional design having acceptable transmission for some frequencies within the band gap of the photonic crystals. The input and output waveguides are denoted along the (10) and (01) directions, respectively. The input and output waveguides are connected by a short waveguide section along the (11) direction. The structure in Fig.1 (a) is one of the bends proposed in [1]. However, the (11) waveguide itself is not symmetric. We extended the (11) waveguide in Fig.1 (a) as a long straight line-defect waveguide and then the FDTD simulation was performed. We found that the optical wave propagating within the (11) waveguide has a zigzag field distribution. Therefore, the guiding modes of the (11) waveguide should be more complicated than those of symmetric waveguides such as (10) or (01) waveguides.

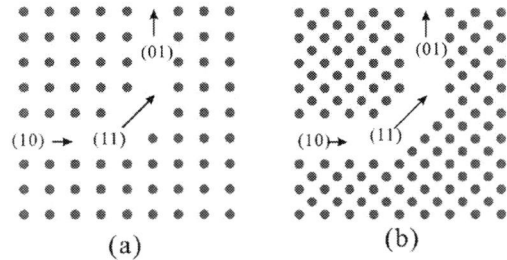

Fig. 1(a) The traditional 90° waveguide bend. (b) Proposed double-density 90° waveguide bend.

The lattice constant a is 1μm and the rod's radius r is 0.18μm. That means the r/a ratio is set to be 0.18. The refractive index of dielectric rod is 4.3 and the index of background is unity. The polarization of light wave is TM wave. Under these parameters, the transmission and reflection spectrum of the traditional bend is simulated and shown in Fig.2. The transmission efficiency is only 91% when the normalized frequency f_n ($=\omega a/2\pi c$) is 0.42.

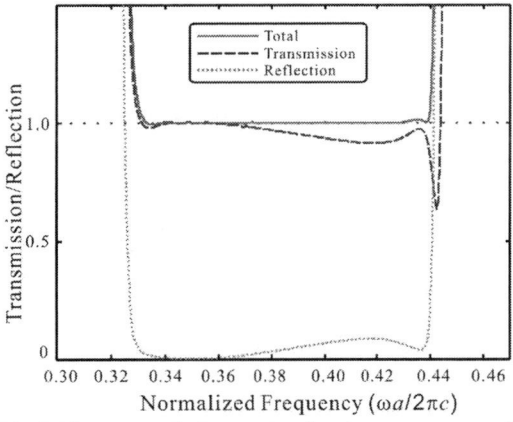

Fig.2 The transmission and reflection spectrums of the traditional 90° waveguide bend.

In this paper, we proposed a 90° waveguide bend with a density doubled lattice structure modified from the traditional bend in Fig.1 (a). The way to double the density of rods is by adding rods at the center of every square cell. The normal cubic lattice photonic crystals has a TM band gap which extends from f_n=0.30 to f_n=0.44. However, the double density lattice photonic crystals has a different TM band gap which has a range from f_n=0.25 to f_n=0.35.

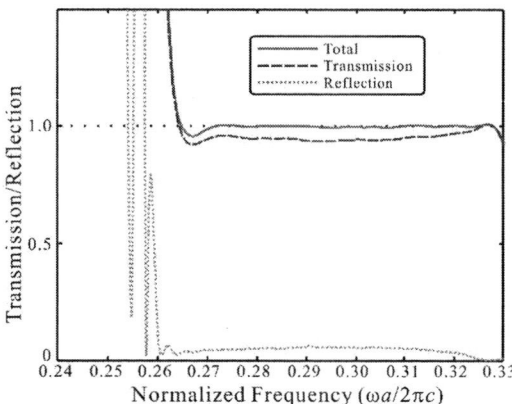

Fig.3 The transmission and reflection spectrums of the double density 90° bend in Fig.1 (b).

After modifying the lattice type from normal 2D cubic lattice into the double-density lattice, the traditional waveguide bend in Fig.1(a) becomes a double-density bend as shown in Fig.1(b). Unlike the traditional bend, the double-density bend has symmetric waveguides for (01), (10), and (11) waveguide sections. Therefore, the guiding modes of the three sections will have similar lateral field profile. The transmission and reflection spectrum of the bend in Fig.1(b) is simulated and the result is shown in Fig.3. The Transmission curve in Fig.3 is flat with respect to that in Fig.2. The transmission efficiency is about 94% for f_n =0.27 to f_n =0.31.

The dispersion relations for the (10) and (11) line-defect waveguides in Fig.1(b) is analysed by the plane wave expansion method. We found that the (10) waveguide supports two guiding modes but the (11) waveguide supports single mode. The (11) waveguide in Fig.1 (b) is a W2 line-defect waveguide with two lines of rods removed. If the (11) waveguide is a W3 line-defect waveguide with three lines of rods removed, it would support two guiding modes. We found the dispersion curve of the fundamental mode of a W3 line-defect (11) waveguide is close to that of a (10) waveguide, which means impedance match is achieved. The bend in Fig.4 is modified from Fig.1(b) by removing one line

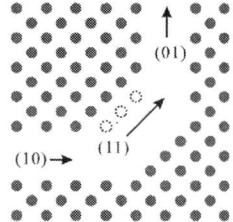

Fig.4 The 90° bend modified from Fig.1(b) by further removing one line of rods. The dashed circles indicate the removed rods.

of rods (total three rods denoted by dash circles). The (11) waveguide in Fig.4 is a W3 line-defect waveguide with three lines of rods removed.

The Transmission and reflection spectrums of the bend in Fig4 is simulated and shown in Fig.5. The results are extreme good. We can see that the transmission efficiencies are greater than 99% for frequencies from f_n=0.265 to f_n=0.30. The high transmission frequency range is useful for WDM technology in fiber communication systems.

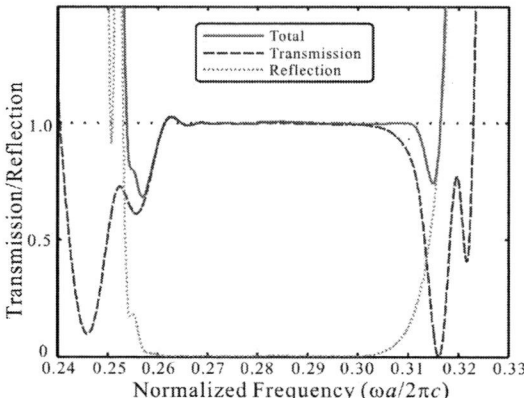

Fig.5 The transmission and reflection spectrums of the double-density 90° bend in Fig.4.

CONCLUSIONS

We proposed a design of high transmission broadband 90° bend in two dimensional cubic photonic crystals. The design of the proposed bend is achieved by double the density of rods first and then deleting one line of rods in the (11) waveguide. The simulation results show that the proposed bend has transmission efficiencies greater than 99% for the frequency range from f_n = 0.265 to f_n =0.30.

REFERENCES

[1] A. Mekis, J. C. Chen, I. Kurland, S. Fan, P. R. Villeneuve, and J. D. Joannopoulos, "High Transmission through Sharp Bends in Photonic Crystal Waveguides," Physical Review Letters, vol. 77, no. 18, pp. 3787-3790, 1996.

[2] J. Smajic, C. Hafner, and D. Erni, " Design and Optimization of an Achromatic Photonic Crystal Bend," Optics Express, vol.11, no. 12, pp. 1378-1383, 2003.

[3] Y. Naka and H. Ikuno, "Analysis of Two-Dimensional Photonic Crystal Sharply Bent Waveguides," 2004 URSI EMTS, International Symposium On Electromagnetic Theory, pp. 748-750, 2004

DESIGN AND FABRICATION OF DIELECTRIC NANOSTRUCTURED LUNEBURG LENS IN OPTICAL FREQUENCIES

Satoshi Takahashi, Chih-hao Chang, Se Young Yang, George Barbastathis
Massachusetts Institute of Technology, USA

ABSTRACT

A dielectric subwavelength Luneburg lens structure was designed using Hamiltonian ray tracing, and fabricated using electron beam lithography. Analysis from Hamiltonian ray tracing was in agreement with finite difference in time domain (FDTD) method with wavefront error between the two methods below $\lambda/8$, while an improvement in speed of approximately 100 times was observed. The fabricated Luneburg lens structure was tested with a fiber laser with a wavelength of $\lambda=1.55\mu m$, and proved its capabilities of focusing.

INTRODUCTION

Dielectric structures with feature sizes in the order of or smaller than the wavelength possess great potential for novel optical components, whether the structures are periodic [1,2] or non-periodic [3]. Especially in the latter case, structures can be engineered to have arbitrary and large variation in effective index, providing great design flexibility. These features open up the potential for characteristics that are very difficult or impossible with conventional gradient index (GRIN) manufacturing processes. One example of a GRIN structure that has been known but has been difficult to implement is the Luneburg lens [4]. The bulk (GRIN) Luneburg lens has a radially symmetric index variation of $n(r) = n_0\sqrt{2 - (r/R)^2}$, where n_0 is the index of refraction at the edge, and R is the lens radius. The Luneburg lens is known to produce a diffraction-limited focus at the edge of the structure opposite of the incident plane wave, having an f-number of zero.

DESIGN OF THE NANOSTRUCTURED LUNEBURG LENS WITH HAMILTONIAN RAY TRACING

For the design of nanostructured lenses, we start with a periodic lattice, and add gradual variation to the structure, such as the size or index of the structure in the periodic cell, lattice constant, etc. As long as the variation of the structure is slow enough to be considered periodic within the adiabatic length scale, Hamiltonian Optics can be used to analyze and design the geometrical ray paths. Particularly, the adiabatically changing structure's dispersion can be obtained assuming local periodicity as $\omega=\omega(\mathbf{p},\mathbf{q})$. The quantity ω is defined as the Hamiltonian [5,6]. Hence, the solution to the equations

$$\frac{d\mathbf{p}}{ds} = -\frac{\partial\omega(\mathbf{p},\mathbf{q})}{\partial\mathbf{q}}, \quad \frac{d\mathbf{q}}{ds} = \frac{\partial\omega(\mathbf{p},\mathbf{q})}{\partial\mathbf{p}} \quad (1)$$

yield the ray trajectories. We first create a library of dispersion relations for a range of variation to the periodic structure and then numerically solve the above Hamiltonian equations using the library to obtain ω at any given $(\mathbf{p},\mathbf{q})$ combination. In our numerical experiments, the Hamiltonian ray trace was two orders of magnitude faster than the FDTD when two-dimensional structures were calculated on the same computer, and further improvement should be expected in fully three-dimensional geometries. Note that the reduction in required computational power also allows for optimization of the structure using the Hamiltonian method, whereas with conventional methods such as finite difference in time domain (FDTD) analysis, that would not be practical.

In the case of the example Luneburg lens structure discussed below, we start with a four-fold-symmetric structure of cylindrical rods in 2D and add gradual variation in the radii of the rods. The refractive index of the rods is $n_{rod}=3.45$, and they are surrounded by air ($n=1$). The structure is as shown in Figure 1. The radii of the rods range from $0.27a$ to $0.42a$, where a is the lattice constant. In Figure 2, we show the Hamiltonian ray tracing results of this structure overlaid with FDTD analysis results. It can be seen from the figure that the two results match well, and the wavefront error between the two

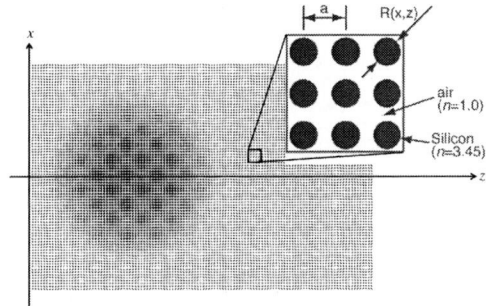

Figure 1. Schematic of dielectric nanostructured Luneburg lens

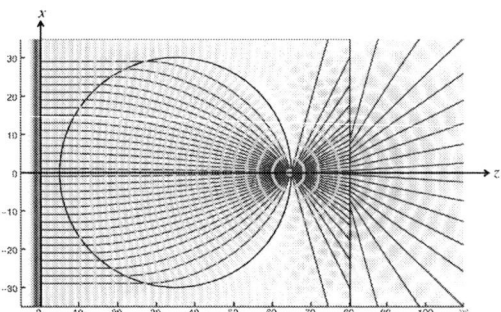

Figure 2. Hamiltonian ray tracing results (blue lines) overlaid with FDTD analysis (red and blue shading) for the Luneburg structure in figure 1.

methods is below $\lambda/8$. Further, it can be seen that the structure focuses a plane wave at the opposite edge, in agreement with the theory for the originally proposed bulk Luneburg lens.

FABRICATION

After confirming its capabilities through numerical analysis with the Hamitonian method, we fabricated the structure for experimental verification. In doing so, we used electron-beam (e-beam) lithography on a silicon-on-insulator (SOI) wafer with Hydrogen silsesquioxane (HSQ) as the resist material and mask for reactive ion etching (RIE) of the silicon structure layer. The e-beam exposure dose was precisely adjusted and calibrated to pattern the features with diameters in the order of 50nm with gradual variations, and RIE was done using HBr gas, which has high selectivity between the glass-like HSQ material and silicon. SEM image of the fabricated structure is shown in Figure 3. Although some proximity effects are still observed from the e-beam patterning, the structure is capable of focusing as shown in the next section.

EXPERIMENT AND RESULTS

An infra-red (IR) camera was used to track the

light path inside the structure by imaging light scattered at the rod edges, while a fiber laser was used to illuminate the structure through a lensed fiber coupled to a waveguide that was patterned on the substrate along with the Luneburg structure. The operating wavelength was $\lambda=1.55\mu m$.

A result captured by the IR camera is shown in Figure 4. A strong peak in intensity is observed, which corresponds to the focus created by the Luneburg lens structure. The slight shift in the position of the focal point is due to the fact that the input wave is not strictly a plane wave due to the fabrication tolerances of the input waveguides and also the slight shift in effective index of the lens structure due to fabrication errors. Further analysis of these errors and their influence towards the focus are under investigation.

REFERENCES

[1] E. Yablonovitch, "Inhibited Spontaneous Emission in Solid-State Physics and Electronics," Phys Rev Lett vol. 58, pp. 2059–2062, 1987
[2] J. D. Joannopoulos et al., Photonic Crystals Molding the Flow of Light Second Edition: Princeton, 2008
[3] R. Liu, C. Ji, J. J. Mock, J. Y. Chin, T. J. Cui, and D. R. Smith, "Broadband ground-plane cloak.," Science, vol. 323, pp. 366-369, 2009
[4] R. F. Rinehart, "A Solution of the Problem of Rapid Scanning for Radar Antennae," J. Appl Phys vol. 19, pp. 861-862, 1948
[5] Y. Jiao, S. Fan, and D. A. B. Miller, "Designing for beam propagation in periodic and nonperiodic photonic nanostructures: Extended Hamiltonian method," IEEE J. Quantum Electron Phys Rev E, vol. 70, pp. 036612-, 1999
[6] P. St. J. Russell and T. A. Birks, "Hamiltonian optics of nonuniform photonic crystals," J. Lightwave Technol., vol. 17, pp. 1982-1988, 1999

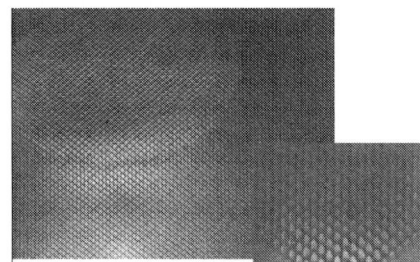

Figure 3. SEM image of the fabricated nanostructured Luneburg lens. The period is 194nm, and the overall structure size is about 20μm. The insert shows a magnified view.

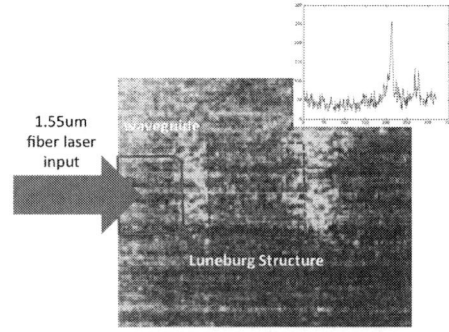

Figure 4. Results from the optical experiment of the nanostructured Luneburg lens. The insert shows the longitudinal cross-section at focus.

NONLINEAR KERR EFFECT APERIODIC LÜNEBURG LENS

Hanhong Gao, Satoshi Takahashi, Lei Tian and George Barbastathis

Massachusetts Institute of Technology, U.S.A.

ABSTRACT

A subwavelength-modulated dielectric nonlinear Lüneburg lens is proposed and numerically verified. In the linear limit, the structure focuses a plane wave to an ideal geometrical point image. In the presence of Kerr nonlinearity, focal shift to a Gaussian beam input is compensated at a tunable intensity value. Nonlinear Lüneburg lens is also capable of finite conjugate imaging. A modified aperiodic Lüneburg is also proposed to greatly reduce the aberration caused by the Kerr effect.

INTRODUCTION

The Lüneburg lens[1] is a spherically symmetric gradient-index (GRIN) lens which is able to focus an incoming plane wave into a perfect point at the opposite edge. However, exact implementation of the GRIN profile is not achievable in bulk media. Recently, methods to achieve this have been proposed using transformation optics[2] and metamaterials[3]. However, the former method is complex to implement. Besides, it breaks the circular symmetry of index profile and does not produce high quality focus. The latter method works only in the microwave domain. Here, Lüneburg lens is designed using aperiodic subwavelength dielectric structures which is feasible for nanofabrication and works in optical frequencies while preserving the circular symmetry.

Nonlinearity is widely used in photonic nanostructure devices[4]. The nonlinear Kerr effect, where refractive index changes in proportion to intensity, has been extensively studied in the context of self-focusing and self-trapping of beams[5, 6]. In this paper, the Kerr effect is applied to cancel the diffraction in imaging systems, compensating for the focal point shift in the Lüneburg lens caused by the diffraction of a Gaussian beam. Nonlinearity gives more design flexibility by adding one more parameter to control. The nonlinear Lüneburg lens is a generalized Lüneburg lens[7] which is capable of finite conjugate imaging (imaging objects from finite distances instead of infinity).

APERIODIC LÜNEBURG LENS

The bulk Lüneburg lens has an index profile of $n(r) = n_0\sqrt{2 - (r/R)^2}$, where n_0 is the index outside the lens, R is the radius, and r is the distance to the lens center. In this paper, subwavelength aperiodic dielectric structure bearing bulk media property is used to model an effective Lüneburg lens. The structure (Fig.1a) con-sists of dielectric rods ($n = 3.46$) spaced by a lattice constant $a_0 = \frac{1}{8}\lambda$, where λ is the free-space wavelength. The rod radius is gradually changing as $a(r) = a_1\sqrt{2 - (r/R)^2} + a_2$ where $a_1 = 0.367a_0$, $a_2 = -0.101a_0$ and $R = 30a_0$. Larger rod results in higher effective index. By controlling the radii of rods, refractive index can be controlled locally. Finite-difference time-domain (FDTD) simulation shows a good focal point at the edge when the input is a plane wave. (Fig.1b)

The radii of neighboring rods are almost the same so it satisfies the "locally periodicity" assumption[8]. Thus this aperiodic structure can also be tested with Hamiltonian ray-tracing method[8, 9]. Hamiltonian optics can provide a ray path picture and is computationally efficient compared with the FDTD method. With the dispersion relation $\omega = \omega(\mathbf{x}, \mathbf{k})$ known for each position, ray-tracing can be obtained by solving the following two Hamiltonian equations:

$$\frac{d\mathbf{x}}{d\sigma} = \frac{\partial\omega(\mathbf{x}, \mathbf{k})}{\partial\mathbf{k}}, \frac{d\mathbf{k}}{d\sigma} = -\frac{\partial\omega(\mathbf{x}, \mathbf{k})}{\partial\mathbf{x}}, \quad (1)$$

where ω is frequency, $\mathbf{x}$ and $\mathbf{k}$ are position and wave vector along the path and σ parameterizes the ray trajectories. Here ω is a function of both $\mathbf{x}$ and $\mathbf{k}$, since this structure is inhomogeneous and anisotropic. Hamiltonian ray-tracing also gives a good focus at the edge and matches the FDTD results. (Fig.1b)

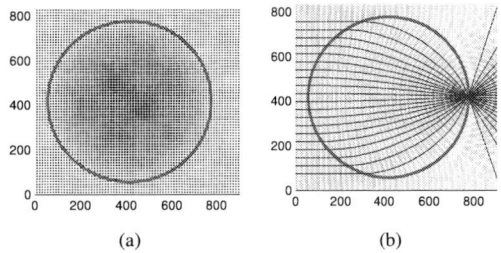

(a)   (b)

Figure 1: Aperiodic Lüneburg lens structure (a), with FDTD and Hamiltonian ray-tracing results (b).

NONLINEAR APERIODIC LÜNEBURG LENS

In practice, Gaussian beams are more realistic models than plane waves, by taking diffraction into account. For a Gaussian beam input, the focus, i.e. beam waist, is no longer perfect and shifts due to diffraction. (Fig.2a,b) The nonlinear Kerr effect could compensate for this diffraction effects and result in a good focus at the edge again. Kerr effect create an intensity-dependent refractive index expressed as $n = n_0 + n_2I$ where n_0 is

978-1-4244-8926-8/10 $26.00 © 2010 IEEE

the linear refractive index, I is the optical intensity and n_2 is the second-order refractive index. As in self-focusing[5, 6], Kerr effect is in competition with diffraction. Under appropriate conditions, it can be expected to cancel the focal shift and produce a minimum waist at the edge of the structures (i.e. at $r = R$). Here the index of each rod is increased according to the intensity and index change is larger for rods closer to the focus edge. Thus rays experience more index gradient and bend more, indeed canceling the focal shift as evident in Fig.2c,d.

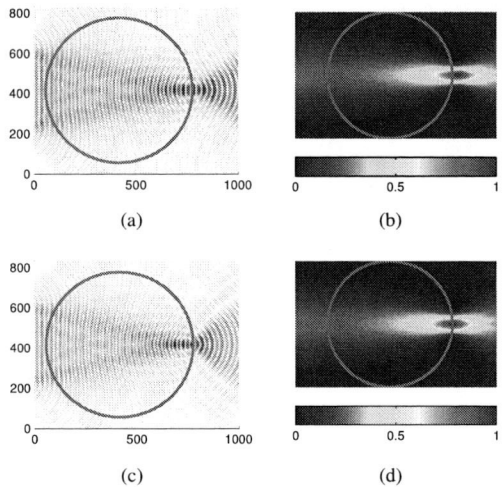

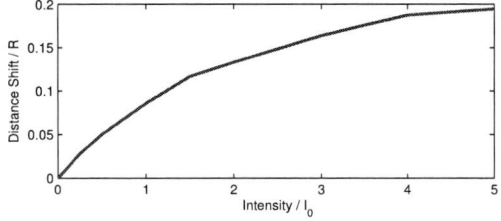

Figure 2: Aperiodic Lüneburg lens with Gaussian beam input, with beam waist $36a_0$. Linear case: FDTD (a) and intensity map generated from Hamiltonian ray-tracing (b) both show the waist shift from edge. Nonlinear case: FDTD (c) and Hamiltonian method (d) both show the shift is compensated by Kerr effect.

Focal shift is related to the intensity of the Gaussian beam. Larger intensity results in larger changes of refractive index, thus larger focal shift, as illustrated in Fig.3, where the normalized focal shift is shown with respect to normalized intensity.

Figure 3: Relationship between input intensity and shift of focal point.

One problem with the optical Kerr effect is that it introduces spherical aberration to the focal point as shown in Fig.4a. To reduce the aberration, a modified nonlinear aperiodic Lüneburg lens structure is designed where the rod radius gradient is modified from the Lüneburg prescription as shown in Fig.4c. The effective index

gradient of the outer part of Lüneburg is increased so the rays away from the optical axis are driven to bend more towards the optical axis, resulting reduced aberration. At the same time, the structure remains symmetric to ensure same focusing behavior from any incident direction. Hamiltonian ray-tracing results of the modified Lüneburg structure near the focal point are shown in Fig.4b. The aberration has been greatly reduced.

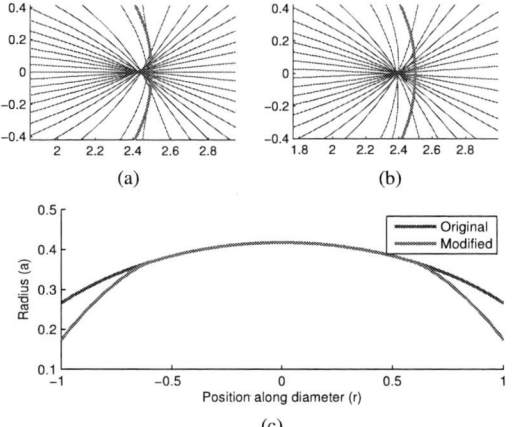

Figure 4: Hamiltonian ray-tracing near the focal point for original Lüneburg lens (a) and modified Lüneburg lens (b) with radius distribution along diameter (c). The waist of Gaussian beam is chosen as $125a_0$.

REFERENCES

[1] R. K. Lüneburg, Mathematical Theory of Optics, Providence: Brown U.P., 1944

[2] N. Kundtz and D. R. Smith, "Extreme-angle broadband metamaterial lens" Nat Mater, vol.9, 129-132, 2010

[3] Q. Cheng, H. F. Ma and T. J. Cui, "Broadband planar Luneburg lens based on complementary metamaterials" Appl.Phys.Lett., vol.95 no.18, 2009

[4] M. Takahashi and H. Goto, Progress in Nonlinear Optics Research, New York: Nova Science Publishers, 2008

[5] R. Y. Chiao, E. Garmire and C. H. Townes, "Self-Trapping of Optical Beams" Phys.Rev.Lett., vol.13 no.15, 479, 1964

[6] P. L. Kelley, "Self-Focusing of Optical Beams" Phys.Rev.Lett., vol.15 no.26, 1005, 1965

[7] S. P. Morgan, "General Solution of the Luneberg Lens Problem" J.Appl.Phys, vol.29 no.9, 1358-1368, 1958

[8] Y. Jiao, S. Fan and D. A. B. Miller, "Designing for beam propagation in periodic and nonperiodic photonic nanostructures: Extended Hamiltonian method" Phys Rev E., vol.70 no.3, 036612, 2004

[9] P. S. J. Russel and T. A. Birks, "Hamiltonian optics of nonuniform photonic crystals" Lightwave Technology, Journal of, vol.17 no.11, 1982, 1999

CONFIGURATION ANALYSIS OF SENSING ELEMENT FOR MICRO-CANTILEVER SENSOR USING DUAL NANO-RING RESONATOR

Bo Li[1], Fu-Li Hsiao[1,2] and Chengkuo Lee[1]*, *Member, IEEE*

[1] Dept. of Electrical & Computer Eng., National University of Singapore, Singapore 117576
[2] Graduate Institute of Photonics, National Changhua University of Education, Taiwan, ROC

ABSTRACT

Two photonic crystal rings of hexagonal lattice are lying transversely and longitudinally in silicon microcantilever at the junction of the microcantilever and the substrate. This unique dual nano-ring (DNR) resonator demonstrates channel drop filter characteristics. When DNR is used as sensing element of a microcantilever sensor, the wavelength shift of resonant peak measured at the backward drop terminal is a function of applied force at the microcantilever tip. The derived Q-factor is about 3000, and the minimum detectable force can be as small as 37nN for longitudinal case and 16.7nN for transverse case.

INTRODUCTION

Since the photonic band gap (PBG) first reported in 1987, many photonic crystal (PhC) resonators are reported. In recent years, the growing demands of applications in nano/microelectromechanical systems (NEMS/MEMS) have led to increased attention on development of new types of sensors. Among these reported sensors, microcantilever is a popular design because of its superior sensitivity and simpler fabrication process. Integration of PhC based nano-cavity resonator with microcantilever has reported as an intriguing approach [1]. In this paper, two types of microcantilever sensors integrated with PhC based hexagonal dual nano-ring (DNR) resonator is proposed and characterized as an optical nanomechanical sensor.

DEVICE CONFIGURATION AND SENSING MECHANISM

We proposed two types of microcantilever sensors with the DNR resonator arranged longitudinally (Type 1) and transversely (Type 2) at the junction edge of silicon cantilever and substrate as shown in the Fig.1. The microcantilever with length and width of 50 μm and 15 μm is patterned and released from a 220 nm thick silicon device layer of a silicon-on-insulator (SOI) substrate. The DNR resonator is created by forming two hexagonal defect rings in a two-dimensional (2-D) PhC of hexagonal lattice at the silicon cantilever, with 4 air hole separation in between the two rings. The lattice constant (a) is 410 nm and air hole radius (r) is 180 nm. This hexagonal DNR with 4 air-holes separation has been reported as a channel drop filter [2].

This work was supported in part by URC Tier 1 Fund R-263-000-475-112 at the National University of Singapore, and by the Agency for Science, Technology, and Research (A-STAR), SERC Grant No. 0921010049. *E-mail: elelc@nus.edu.sg.

For the type 1 sensor, the DNR is placed along the longitudinal direction of the cantilever, and it is along the transverse direction for the type 2 sensor. The hexagonal lattice gives the ring resonator a better photon confinement such that high quality factor (Q-factor) is achieved. The whole layout configuration including a, r and the separation between waveguides and hexagonal rings has been optimized in order to achieve resonant peak of higher Q-factor at the backward drop (BD) terminal.

Finite-element analysis (FEA) is deployed to obtain the deformation data of holes among PhC air-hole array under various force loads at the middle point of the edge of free standing cantilever tip end as shown in Fig. 1. The Young's modulus and Poisson's ratio of silicon is 130 GPa and 0.3 respectively. We further applied numerical 2-D finite-difference time-domain (FDTD) method to simulate the propagation of the electromagnetic waves in the deformed PhC DNR resonator structure under various applied force loads. In the FDTD modeling, we only consider the relative position shift of holes in the PhC structure. The composite air/Si/air structure is used to perform the FDTD analysis based on the effective refractive index method. The effective refractive index is derived as 2.7967 in TM mode, while the refractive indices of Si and air are 3.46 and 1 respectively. Apparently the ring deformation and change in the separation between two rings are different for the two types even these two sensors under the same force load. Therefore, the resonant wavelength of peak measured at BD is expected to be a function of applied force, while such dependence of resonant wavelength versus force loads is different for these two sensors due to the different layout arrangement.

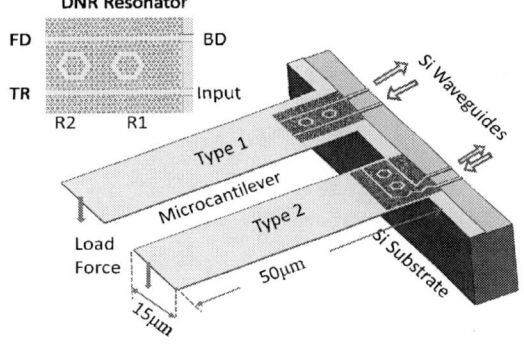

Fig.1. Schematic illustration of two types of PhC microcantilever sensor using a hexagonal dual nano-ring (DNR) resonator.

a)

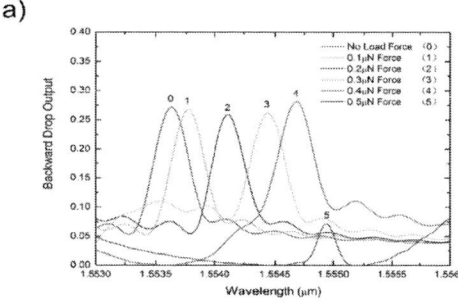

b)

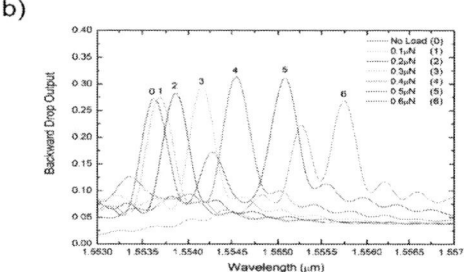

Fig. 2. Resonant wavelength peaks at the BD port under various force loads of type 1 sensor (a) and type 2 sensor (b).

RESULTS AND DISCUSSION

Fig. 2a and 2b show the resonant wavelength peak for two types of sensors derived at the BD terminal under different force loads in a step of 0.1 μN. The resonant peak which is located at 1.5536 μm with Q factor of about 3000 is referred as the no-load case. As force increases, both types of sensor show a red shift of the resonant wavelength and demonstrate well-defined resonant peak within the force range of 0.4 μN. However, in the case of the type 1 sensor, the intensity of resonant peak at BD terminal becomes relatively weak when the force reaches to 5 μN. It is suggested by a mechanism that the channel drop effect becomes weak such that limited amount of light coupled to the BD terminal via the DNR and most of the pumped energy goes to the transmission terminal. In the other words, the resonance of the DNR is slightly deviated, thus the coupling becomes worse. In the type 2 sensor case, both rings of DNR experience the same amount of deformation due to the cantilever bending, since the two rings are place in transverse direction of the cantilever. A minor peak of shorter wavelength except to the major peak is observed for type 2 sensor under force loads of 0.5 μN and 0.6 μN, because the deformed ring shape brings extra coupled energy which matches another resonance at BD terminal as well.

The change in the separation between two rings of DNR is considered as a contributing effect as well. There is sufficient large strain created in such ring separation in the type 1 sensor, i.e., the longitudinal placed rings. Thus we need to include ring shape deformation and strain of ring separation in account when we discuss the behavior of type 1 sensor, while we only consider ring shape deformation in the type 2 sensor due to the negligible change in the ring separation. Hence, the Q factors of the resonant peaks of type 2

sensor are holding better compared with the data of type 1 sensor. The only problem is the minor peak appears when the force load is larger than 0.4μN in the type 2 sensor. But the major peak of longer wavelength can be easily recognized, thus we this factor won't affect on the sensor characteristics.

Fig. 3 illustrates the relationship of the wavelength of resonant peak at BD terminal versus applied force of the both sensors. Linear relationship is observed for the type 1 sensor, while the slope is derived as 2.7nm/μN. It is suggested that the minimum detectable force will be 37nN, if we have the wavelength resolution of 0.1nm in the testing setup. However, a 2nd-order polynomial fitting curve has good agreement with the derived data point in the case of type 2 sensor. From the 2nd-order polynomial fitting curve, we get different sensitivity in different range of force loads for various applications. More specifically, the slope of the fitting curve for force less than 0.1μN is derived as 1 nm/μN, while such slope is 6 nm/μN in the force region of 0.5~0.6μN. As a result, the minimum detectable force is calculated as 100 nN and 16.7 nN in these two regions, respectively. Since the change of ring separation is neglected in the type 2 sensor, the ring shape deformation becomes the main root cause responsible for the 2nd-order polynomial relationship observed in data of the type 2 sensor. Therefore, the linearity of data of the type 1 sensor is mainly contributed to the change of the ring separation.

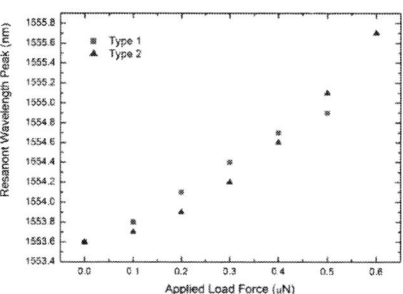

Fig. 3. Resonant wavelength as a function of applied force for both types of sensor.

CONCLUSIONS

We investigated two types of microcantilever sensors using DNR resonator. When the resonance is dominant by the change in the ring separation, linear relationship is observed in the curve of resonant wavelength versus force loads. In contrast, a 2nd-order polynomial relationship is observed in such curve when the ring shape deformation is the dominant coupling mechanism. The minimum detectable force is 37 nN and 16.7 nN for longitudinally and transversely arranged DNRs respectively.

REFERENCES

[1] C. Lee and J. Thillaigovindan, "Optical nanomechanical sensor using Si photonic crystals cantilever embedded with nanocavity resonator," *Appl. Opt.*, vol. 48, no. 10, pp. 1797–1803, Apr. 2009.

[2] F.-L. Hsiao and C. Lee, "Novel hexagonal nano-ring resonators," the 9th IEEE Conf. on Nanotechnology, IEEE-NANO 2009, p.303-306, Genoa, Italy, July 26-30, 2009.

978-1-4244-8926-8/10 $26.00 © 2010 IEEE

INVESTIGATION OF STRAIN SENSITIVITY OF PHOTONIC CRYSTAL NANOCAVITY FOR MECHANICAL SENSING

Bui Thanh Tung, Dzung Viet Dao and Susumu Sugiyama
Ritsumeikan University, Japan

ABSTRACT

This paper reports the theoretical investigation of the strain sensitive effect of a 2D photonic crystal (PhC) nanocavity resonator for mechanical sensing applications. By using finite element method (FEM) using ANSYS software and finite different time domain (FDTD) simulation using CrystalWave software, the strain sensitivity of high quality factor PhC nanocavity have been studied. Linear relationships between strain and shift of resonant wavelength of the cavity have been obtained. The sensitivities to longitudinal and transverse strains have been determined to be 1.67nm/mε and 1.4nm/mε, respectively.

INTRODUCTION

Photonic crystal (PhC) is known as a periodic distribution of one material in a different host material with different refractive index (RI). Silicon PhC structures based devices had been widely used in light flow control applications such as waveguide, photonic bandgap structures, resonator, etc. Because of its sensitive to small RI change, PhC structures had been considered for bio and chemical sensing by detecting the change of RI due to molecules immobilization in the PhC structures. However, mechanical sensing ability has not been investigated in-depth. Z. Xu *et al* proposed a displacement sensor which utilized the output light intensity as sensing signal [1]. D. Biallo *et al.* [2] and C. Lee *et al.* [3] proposed a nanomechanical sensors using silicon photonic crystal by detecting the resonant wavelength shift due to strain. In these studies, the refractive index of medium under stress/strain was not taken into account. Furthermore, these studies did not evaluate the influence of orientation of stress/strain to the sensitivity. In this paper, both RI changes in the material and geometric change of PhC structure due to mechanical strain will be considered in our analysis.

DESIGN OF TEST STRUCTURE

Fig. 1 shows the schematic of the test structure for analysis the effect. The test structure is a cantilever (15mm × 3mm × 0.75mm, length × width × thickness) which is fixed at one end, and PhC structure as a strain sensing element is placed at a suitable position near the fixed part of the cantilever. By applying force to the free-end of the cantilever, we can generate stress/strain at the position of the PhC structure. Input and output waveguides are placed above the fixed part of cantilever to avoid the displacement of the waveguide's inlet and outlet during application of force. The proposed test structure can be realized by using micro machining, such as electron beam lithography (EBL)

and cryogenic inductively coupled plasma (ICP) etching.

The sensing element in this study is PhC nanocavity resonator as shown schematically on fig.2. The nanocavity is formed by introducing point defects into a triangular lattice of holes in the high-index contrast Si/air waveguide. Cavity and waveguide are aligned on cantilever so that stress/strain will be parallel (longitudinal stress/strain) and perpendicular (transversal stress/strain) to the wave propagation direction.

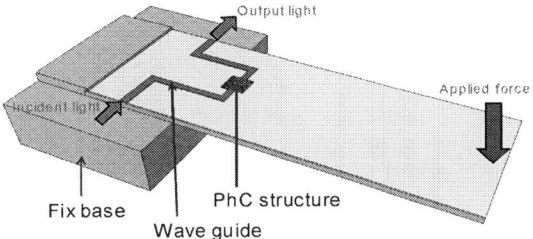

Figure 1. Test structure to investigate the strain sensitive effect.

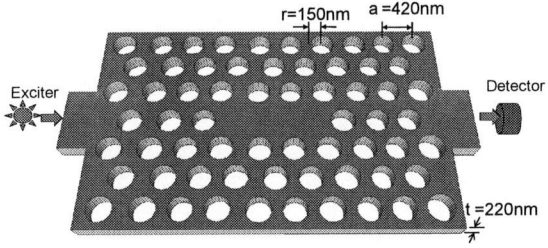

Figure 2. Schematic of 2D PhC cavity.

When the stress/strain applied to the PhC cavity, the resonant wavelength will be changed. This change is the result of refractive index change due to elasto-optic effect of silicon, and the change of PhC geometry. The elasto-optic effect of silicon and geometry change of PhC will be the input parameters for FDTD simulation.

STRUCTURAL SIMULATION

Under applied load, the stress/strain is generated at the position of PhC cavity, resulting the changing of geometry of PhC including positions and shapes of air holes. By finite element analysis using ANSYS, firstly, the position of nanocavity structure on cantilever is decided so that the strain in PhC is maximum. After that, the geometric changes are calculated based on diagram shown on fig.3-b,d. Figs.3-a,c show schematically the PhC structure under strain. The new positions and shapes of air holes are used as input parameters for FDTD simulation.

978-1-4244-8926-8/10 $26.00 © 2010 IEEE

FDTD OPTICAL SIMULATION

Besides the PhC geometry change as mentioned above, the changes in RI of material under stress/strain are also taken into account during 2D FDTD simulation using CrystalWave.

The RI changes due to elasto-optic effect in silicon can be expressed as [4]:

$$\Delta n_{\varepsilon_l} \approx -0.5 n_0^3 p_1 \varepsilon_l$$
$$\Delta n_{\varepsilon_t} \approx -0.5 n_0^3 p_2 \varepsilon_t \tag{1}$$

where ε_l, ε_t are longitudinal and transverse strains, n_0, n_{ε_l} and n_{ε_t} are the RIs of material at free strain state, under longitudinal strain, and transversal strain, respectively. p_1, p_2 are strain-optic constants. For Si, p_1 = -0.101, p_2 = 0.0094.

From these RI changes, the effective RI is calculated [5] for 2D FDTD simulation.

Fig.4 shows the transmission spectra of the nanocavity corresponding to different strain states. The quality factor of the cavity obtained to be about 1000. Fig. 5 shows the simulation results of cavity resonance wavelength shift due to longitudinal and transverse strains applied. In this study, both tensile and compressive strains have been investigated. Under longitudinal tensile strain, the resonant wavelength tends to shift to longer wavelengths, while it tends to shift to shorter wavelengths under transverse tensile strain. The wavelength shifts were 1.67 nm/mε and 1.4 nm/mε for longitudinal and transversal strains, respectively. The results obtained are about 2 times larger than that reported by C.Lee et al [3].

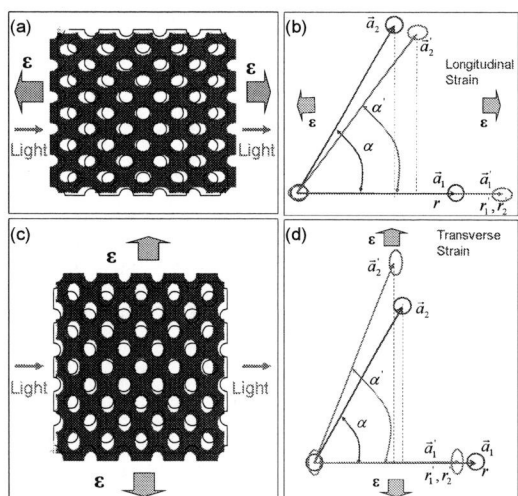

Figure 3. Deformation of PhC structure under applied stress. In (a) and (c), dashed and solid shapes show the intial and under-strain shape of PhC, respectively. The PhC lalttice geometry changes are shown schematically in (b) and (d).

With the minimum detectable wavelength of an up-to-date commercial optical spectrum analyzer is 10 pm,

we can obtain the minimum detectable longitudinal strain of 5.9×10^{-6}, and 7.1×10^{-6} for transverse strain. The difference of shifts between longitudinal and transversal effects is caused by the difference in elasto-optic constants and cavity lengths under these strains.

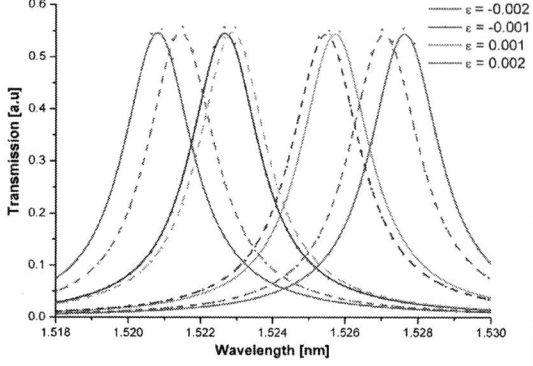

Figure 4. Transmission spectra shows the resonant wavelength under different strains. Solid lines and dashed lines show the results of longitudinal and transversal strains, respectively.

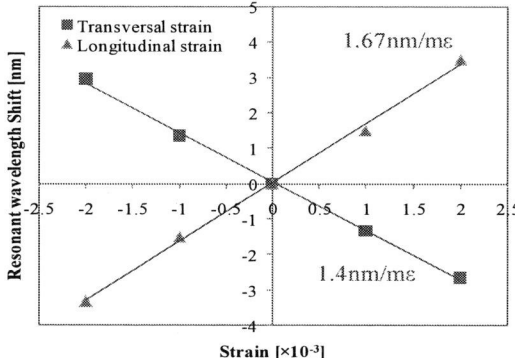

Figure 5. Resonant wavelength shift vs. strain

CONCLUSIONS

We have investigated the strain sensitivity of the resonant wavelength of Si PhC nanocavity resonators using FEM analysis and 2D FDTD simulation. Our results are expected to serve as practical guidance for detecting the mechanical strain by using PhC nanocavity. These results promise good ability to apply for PhC-based mechanical sensing devices.

REFERENCES

[1] Z. Xu, L. Cao, C. Gu, Q. He, and G. Jin, "Micro displacement sensor based on line-defect resonant cavity in photonic crystal", Optics Express, **14**, pp.298-305 (2006).

[2] D. Biallo *et.al* "High sensitivity photonic crystal pressure sensor", *J. Eur. Opt. Soc.* **2** pp.070171-070175(2007).

[3] C. Lee, J. Thillaigovindan "Optical nanomechanical sensor using a silicon photonic crystal cantilever embedded with a nanocavity resonator", *Appl. Opt.* **48**, pp.1797–1803 (2009)

[4] L. Pavesi, D. J. Lockwood "Silicon photonics", Springer-Verlag, 2004

[5] K. Okamoto, "Fundamentals of optical waveguide", Academic Press, 2000 pp.35.

Magnetic Response of Continuous Split Ring Structures at Visible Frequencies
Yi-Hao Chen, Alex F. Kaplan, and L. Jay Guo
The University of Michigan, Ann Arbor, MI, USA

ABSTRACT

Split ring resonators (SRRs) have been widely studied as fundamental elements of metamaterials showing magnetic response. In order to search for metamaterials with negative refractive index, we studied the electromagnetic properties of structures with modified SRRs. We proposed a structure composed of continuous SRRs showing both negative μ and negative ε. Experimental results are supported by numerical results to show negative refractive index at visible frequencies.

INTRODUCTION

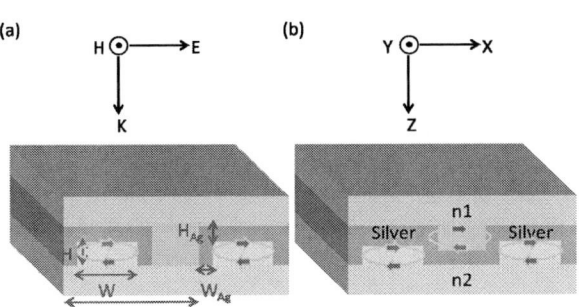

Fig. 1. (a) Discontinuous and (b) continuous SRRs. The light polarization shown is TM with H field in the transverse (y) direction.

SRRs have been extensively studied in the past years as negative μ metamaterials since it was first introduced [1]. The magnetic resonance wavelength scales proportionally with the size of SRRs in the microwave regime when metal acts as ideal metal. However, when the wavelength approaches the metal plasma oscillation wavelength, the metal can no longer be treated as ideal metal and the size scaling of SRRs with resonance wavelength break s down. Magnetic resonances in SRRs are limited to wavelengths above 800nm [2]. Nevertheless, not only μ but also ε need to be negative in order to achieve true left had materials [3]. To resolve this issue, we propose a continuous SRR structure that can exhibit negative permeability and permittivity simultaneously

As opposed to discontinuous SRRs shown in Fig. 1(a), the structure shown in Fig. 1(b) is formed by connected SRRs. The introduction of parallel capacitance reduces the total capacitance and increase the LC frequency [4]. The magnetic resonance can thus be pushed to visible wavelength. Moreover, the continuous metal layer provides negative ε similar to that of flat metal film. In this work, we optimize the

continuous SRRs based on numerical results and experimentally demonstrate that negative refractive index can be achieved at visible frequencies with high figure of merit.

NUMERICAL RESULTS AND ANALYSIS

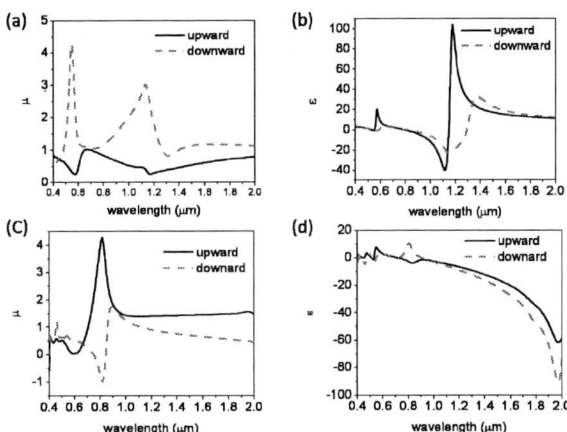

Fig. 2. (a) Retrieved real part of permeability and (b) permitivity of discontinuous SRRs depected in Fig. 1(a). (c) Retrieved real part of permeability and (d) permitivity of continuous SRRs depicted in Fig. 1(b). For both structures, n1 = 1, n2 = 1.55, Period=220nm, W=100nm, H=50nm, and $H_{Ag}=W_{Ag}$=10nm. Black solid and red dashed curves are retrieved from upward and downard propagation respectively.

We use finite difference time domain (FDTD) method to simulate the optical response of structures of interest. The FDTD simulation presented here is performed by using a commercial FULLWAVE package from the RSoft Design Group. Drude model is used to represent the bulk material permittivity of silver with plasma frequency ωpl=1.37*10^{16} rad/s and collision frequency ωcol=9·10^{13} rad/s.

With transmission and reflection coefficients obtained from FDTD simulation, we are able to retrieve the effective μ and ε. For SRRs sitting on dielectric substrate, n1= 1$\neq$ n2. Both structures are asymmetric along the propagation direction and we could expect the different effective μ and ε for light propagating in different directions as shown in Fig. 2. The effective μ is related to the direction and strength of induced magnetic field [1]. In continuous SRRs, induced field exists both inside and between the rings as opposed to discontinuous SRRs where induced field exists only inside the rings as illustrated in Fig. 1. The stronger induced field results in the stronger magnetic resonance for continuous SRRs compared to discontinuous SRRs and $\mu<0$ can be achieved

978-1-4244-8926-8/10 $26.00 © 2010 IEEE

around 825nm for in downward propagation. Unlike discontinuous, $\varepsilon<0$ can be achieved over a large bandwidth for continuous SRRs above electric resonance wavelength. However, simultaneously negative μ and ε can't be achieved.

In order to search for true left hand material, we tried to optimize the electromagnetic response by adjusting the structure parameters of continuous SRR. We found that simultaneous negative μ and ε can be achieved by adding a coating layer matching the index of substrate.

EXPERIMENTAL RESULTS

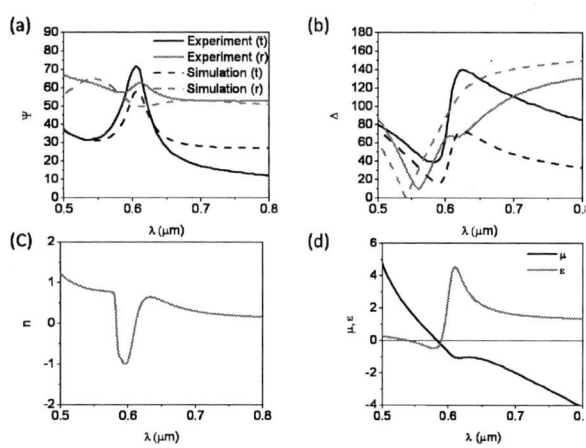

Figure 3. Experimental and simulated (a) Ψ and (b) Δ. Retrieved (a) n and (b) μ and ε from simulated optical responses. n1 = n2 = 1.45, Period=220nm, W=80nm, H=35nm, and H_{Ag}=20nm W_{Ag}=15nm. Black and red curves in (a) and (b) are from transmission and reflection

To demonstrate the real sample of negative index, the structure is fabricated with nano imprint lithography followed by procedures similar to Ref. [6]. Ellipsometry spectra were measured with a Woollam variable angle spectroscopic ellipsometer WVASE32. $\tan(\Psi)$ and Δ represents the amplitude ratio and the phase difference between TE and TM transmission or reflection coefficients. Fig. 3 (a) and (n) shows the experimental curves of Ψ and Δ have. By adjusting the structure parameters used in simulation, imulated Ψ and Δ are also shown and match the experimental results. The small

deviations between experiment and simulation may arise from the imperfections in fabrications such as inhomogeneous surface roughness. As both the amplitude and phase responses from experiment agree with simulation, the retrieved electromagnetic parameters from simulation would be valid to describe the real sample. The retrieved n shown in Fig. 3(c) shows that negative refractive index can be achieved around 590nm where retrieved μ and ε are simultaneously negative as shown in Fig. 3(d).

CONCLUSIONS

We showed that continuous SRRs can have strong magnetic resonance at visible wavelength. By optimizing the structure, negative index of refraction can be achieved at 590nm. To the best of our knowledge, this structure demonstrates the negative refraction at the shortest wavelength up to now.

REFERENCES

[1] J. B. Pendry, A. J. Holden, D. J. Robbins, and W. J. Stewart," Magnetism from conductors and enhanced nonlinear phenomena" IEEE Trans. Microwave Theory Tech., vol. 47 no. 11, pp. 2075-2084, 1999

[2] J. Zhou, Th. Koschny, M. Kafesaki, E. N. Economou, J. B. Pendry. C. M. Soukoulis, "Saturation of the magnetic response of split-ring resonators at optical frequencies " Phys. Rev. Lett., vol. 95, pp. 223902, 2005

[3] V. G. Veselago, "The electrodynamics of substances with simultanesouly negative values of ε and μ" Soviet Physics Uspekhi, vol. 10, pp. 509-514, 1986

[4] H. Schweizer, L. Fu, H. Gräbeldinger, H. Guo, N. Liu, S. Kaiser,and H. Giessen, "Negative permeability around 630nm in nanofabricated veritcal meander metamaterials", Phys. Stat. Sol. (a), vol. 204 no. 11, pp. 3886-3900, 2007

[5] D. R. Smith and S. Schultz, "Determination of effective permittivity and permeability of metamaterials from reflection and transmission coefficients" Phys. Rev. B, vol. 65, 195104, 2002

[6] A. F. Kaplan, Y. H. Chen, M. G. Kang, and L. J. Guo, "Subwavelength grating structures with magnetic resonances at visible frequencies fabricated by nanoimprint lithography for large area applications" J. Vac. Sci. Technol. B, vol. 26 no. 6, pp. 3175, 2009

An adaptive objective for optical motion correction in MRI

F. Schneider[1], J. Draheim[1], T. Burger[1], J. Maclaren[2], M. Herbst[2], M. Zaitsev[2], R. Bammer[3,4] and U. Wallrabe[1,4]

[1] University of Freiburg – IMTEK, Department of Microsystems Engineering, Germany
[2] University Hospital Freiburg, Department of Radiology, Germany
[3] Stanford University, Department of Radiology, USA
[4] University of Freiburg – FRIAS, Freiburg Institute for Advanced Studies, Germany

ABSTRACT

We present an adaptive membrane lens that is compatible with a magnetic resonance (MR) imaging system, as a part of an objective for in-bore optical motion tracking and correction. Compatibility tests with the high magnetic fields of a 3 T scanner and the shielding of the adaptive lens are discussed. For tracking of a marker an adaptive wide-angle objective has been designed and characterised. The measured mean distortions of the objective are between -2.05 and -2.6 %. The tracking system was verified in a MR scanner model.

INTRODUCTION

We develop an MR-compatible optical motion tracking system in order to monitor patient movements in a MR scanner. The tracking system has to be placed inside the bore of a 3 T magnet to ensure visibility of the tracking target. Due to different patient body dimensions as well as large uncontrolled spontaneous patient movements during scanning, the tracking target needs to be refocused. Hence, the autofocus optics and the camera chip have to withstand a constant magnetic field (up to 3 T), switching magnetic field gradients, and the RF fields that are used to generate signals from the imaged object [1]. Since the free space inside the bore is extremely limited, the entire tracking system has to be very compact, warranting a high degree of integration and the need to using micro components. Our adaptive lens [2, 3] is one part of an objective that allows for an "always-sharp" image of the tracking marker on the camera chip through fast refocusing.

WIDE-ANGLE OBJECTIVE

This work focuses on the motion correction of head movements. Typically, during an MR-scan, the head of a patient is placed inside a head coil within the magnet bore. A tracking marker can be fixed on the forehead of the patient and observed by the tracking system through the openings in the head coil (see Figure 1a).

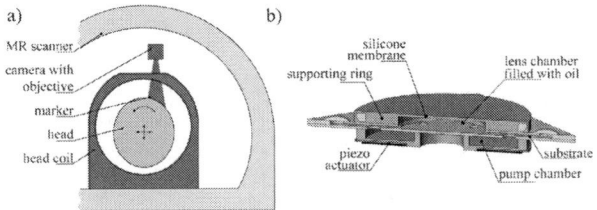

Fig. 1: Schematic of a) the tracking setup inside the magnet bore and b) the adaptive lens.

The adaptive lens consists of a lens and an actuator chamber (see Figure 1b) [2, 3]. The lens chamber, which is filled with oil, consists of a supporting ring and a membrane made of silicone. The actuator chamber consists

of a piezoelectric bending actuator (two active PZT layers), which is also embedded in silicone. Both pieces are connected to a substrate featuring several orifices for the fluid exchange. By applying a voltage to the actuator the focal length of the adaptive lens is tuneable.

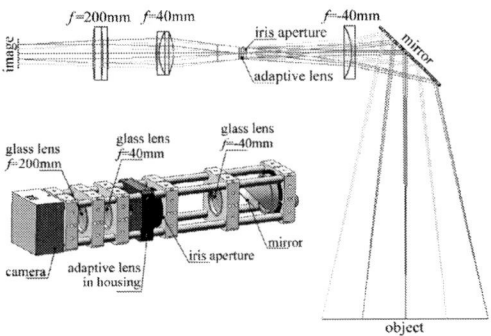

Fig. 2: Ray-tracing of the adaptive objective for a object distance of 90 mm. Inset: 3D schematic of the adaptive objective

The adaptive lens with a diameter of 10 mm must be integrated into an objective. The boundary conditions for the wide-angle objective are the object size of 60×45 mm^2, the image size at the camera chip, and the object distance of 90±20 mm. For the optical system design the ray-tracing software ZEMAX was used (see Figure 2). Due to the limited construction height between the bore and the head coil the light is reflected into the horizontal plane by a mirror. Behind the mirror a plane-concave lens collects the light. An iris aperture with a diameter of 4.5 mm is placed directly in front of the adaptive lens. For simplification the membrane shape of the adaptive lens is modelled by a sphere. Afterwards, two achromatic lenses are used for the sharp projection on the image plane. The working distance of the objective can be varied between 70 and 110 mm by the focal length variation of the adaptive silicone-membrane lens between -3.5 and +10.2 m. The mean distortions are between -1.07 and -1.65 %. After the optical design, we constructed the objective using mounting plates and aluminium bars (see inset Figure 2).

978-1-4244-8926-8/10 $26.00 © 2010 IEEE

MR-COMPATIBILITY AND OPTICAL PERFORMANCE

Firstly, the MR-compatibility of the materials used in the adaptive lens was verified. We fixed all lens sub-components (actuator chamber, lens chamber, bottle of lens liquid), one after the other, directly on the top of the head coil. The different materials did not influence the imaging of a phantom (liquid filled in a plastic bottle), which was placed in the head coil and scanned with a gradient-echo sequence (see Figure 3). The dark structure at the edge of the MR-images is an air bubble in the liquid-filled phantom.

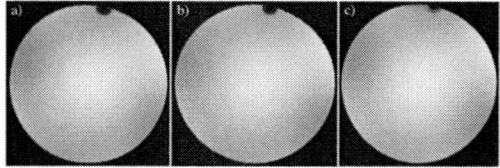

Fig. 3: MR-scan of a phantom with a) the actuator chamber, b) the lens chamber and c) a bottle of lens liquid on the top of the head coil

Secondly, we put the adaptive lens in an aluminium housing, to shield it against the scanner RF field and the switching gradients. During the MR scan we applied a voltage of ±40 V to the adaptive lens located on the head coil. The lens focused successfully inside the scanner. Furthermore, no interference from the lens was visible in the phantom images.

After the compatibility tests we focused on the optical characterisation of the adaptive objective. The resolution of the objective is measured with an USAF 1951 test target. In the initial state, without voltage applied to the actuator (90 mm object distance) the resolution is 5.95 lp/mm. Interpolating this across the whole object size a resolution of 685 × 547 pixels is obtained, which is better than the specified resolution for motion tracking.

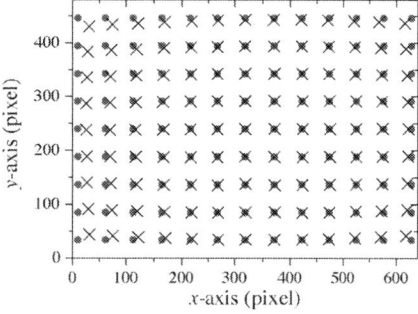

Fig. 4: Field distortions at an object distance of 90 mm. Red point: reference position, black cross: measured position

Field distortions are important for the accuracy of the optical tracking. The distortions were measured with a point array test target (3.75 mm lattice spacing) at nine object distances ranging for 70 to 110 mm. We used Labview Vision for the tracking of the points and the analysis of the distortions. Figure 4 shows the distortions of the adaptive objective at an object distance of 90 mm. The aberration of the objective has a barrel shape, which is typical for wide-angle objectives. A mean distortion of -2.22 % and a maximum of -7.02 % were measured. The reduction of the object distance decreased the mean distortion, which can be traced back to the higher magnification of the objective (see Figure 5).

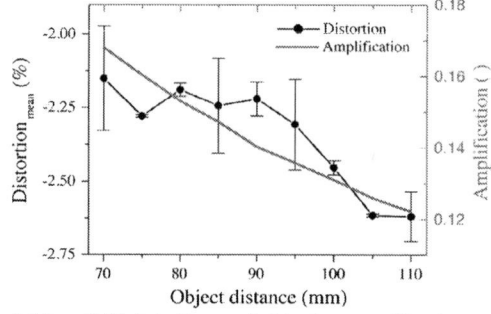

Fig. 5: Mean field distortions and objective magnification as a function of the object distance.

DUMMY MR SCANNER

For the optimisation of the adaptive objective and the testing of the tracking software we have constructed a model of a MR scanner (see Figure 6a). Additionally, a servo-motor actuated artificial head with 2 DOF was developed. Both motors are connected to a micro controller, which controls the two movements separately or simultaneously. The setup allows a head-rotation of ±25 ° and a lifting-height of 45 mm.

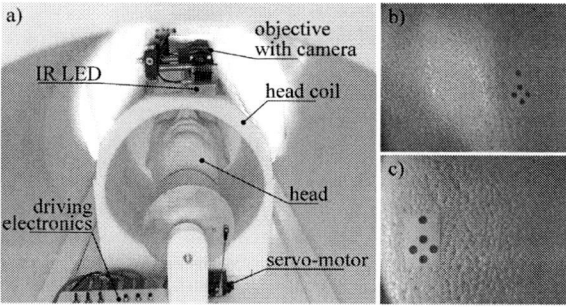

Fig. 6: a) Model of a MR scanner including a head coil and artificial head. b) Tracking marker focused in lower head position and c) in upper head position.

We use Labview Vision for the object tracking. A 5-point marker is detected and refocused by a contrast autofocus continuously. Additionally, the illumination of the marker with a IR-LED is controlled. Figure 6b and 6c show the focused marker in the upper position of the head as well as in the lower position.

CONCLUSIONS

MR-compatibility tests of an adaptive lens as a part of an in-bore motion tracing system has been successfully carried out. We have characterized the optical distortions and resolution of the objective. In a next step, we will verify the accuracy of the optical tracking system. Finally, the tracking and autofocus software of the model should be transferred to the MR tracking computer for real-time correction of the MR data.

REFERENCES

[1] M. Zaitsev et al., *Neuroimage* **31**, pp. 1038-1050, 2006.
[2] F. Schneider et al., *Proc. Optical MEMS 2009*, pp. 39-40, 2009.
[3] F. Schneider et al., *Optics Express* **17**, pp. 11813-11821, 2009.

978-1-4244-8926-8/10 $26.00 © 2010 IEEE

MEMS-BASED X-RAY OPTICS FOR FUTURE ASTRONOMICAL MISSIONS

Yuichiro Ezoe[1], Ikuyuki Mitsuishi[2], Kensuke Ishizu[1], Teppei Moriyama[1], Kazuhisa Mitsuda[2],
Noriko Y. Yamasaki[2], Takaya Ohashi[1], Mitsuhiro Horade[3], Susumu Sugiyama[3],
Raul E. Riveros[4], Taylor Boggs[4], Hitomi Yamaguchi[4], Yoshiaki Kanamori[5],
Nicholas T. Gabriel[6], Joseph J. Talghader[6], Kohei Morishita[7],
Kazuo Nakajima[7], Ryutaro Maeda[8]

[1] Tokyo Metropolitan University, Japan, [2] ISAS/JAXA, Japan, [3] Ritsumeikan University, Japan,
[4] University of Florida, US, [5] Tohoku University, Japan, [6] University of Minnesota, US,
[7] Kyoto University, Japan, [8] AIST, Japan

ABSTRACT

X-ray optics based on MEMS technologies can provide future astronomical missions with ultra light-weight and high-performance optical systems. Curvilinear micropores vertical to a thin wafer are made by using DRIE (Deep Reactive Ion Etching) or X-ray LIGA. The side walls are smoothed by using magnetic field assisted finishing and annealing technologies in order that the walls can reflect X-rays. Two or four such wafers are bent to spherical shapes with different curvature of radii and stacked, to focus parallel X-rays from astronomical objects by multiple reflections. In this paper, the concept and recent advances of the MEMS X-ray optics are reviewed.

INTRODUCTION

Future space astronomical missions need light-weight and high-performance (angular resolution and effective area) X-ray optics. Since X-rays are difficult to be collected refractively, grazing-incidence optical systems are widely used in X-ray astronomy. However, a high angular resolution generally means stiff thick mirror substrates and a large area needs a huge number (thousands) of mirrors. To dramatically reduce resources for the space X-ray optics, we proposed a novel X-ray optical system based on MEMS technologies.

MEMS X-RAY OPTICS

Figure 1 shows the concept of what we call MEMS X-ray optics [1]. Curvilinear micropores with widths of a few tens μm are made by using DRIE or X-ray LIGA techniques. Side walls of the micropores, made of silicon (DRIE) or nickel (X-ray LIGA), are used for X-ray mirrors. Since the surface roughness of the side walls must be as smooth as X-ray wavelength, i.e., nm or less, additional smoothing processes are necessary. Annealing and/or magnetic field assisted finishing are utilized for this purpose. Then, the wafer is bent to a spherical shape and stacked in combination with another wafer which is bent with a different curvature of radii. The final optics obtained in this way can produce a high quality image of astronomical objects.

Because the wafer is thin on the order of 100 μm and also a number of micropores, i.e., X-ray mirrors, can be made at once, this type of optics can be lightest and lowest-cost. The aspect ratio of the micropore and the surface roughness at micro scales are important for the effective area, while the flatness of the side wall and the bending accuracy are for the angular resolution. Figure 2 shows the expected performances of the three next generation optics including MEMS compared to those of the telescopes in the past missions.

To date, X-ray reflection upon the smoothed side walls has been verified using miniature optics or mirror chips [1,2]. An optical imaging test has been conducted for a single-stage silicon optical system [3]. Recent results on an X-ray imaging test is described in [4]. Figure 3 shows the silicon optics sample and its optical image. A parallel optical light is used instead of X-rays because of experimental easiness. Photons are reflected on the side walls of 20 μm line & space micropores. The wafer thickness is 300 μm and the focal length is 0.5 m. A fine focus (FWHM 2 arcmin) was confirmed, although, due to the optical diffraction within the micropores, a halo-like structure and blurring of the focus were seen.

OPTICS DESIGNS FOR FUTURE MISSIONS

Table 1 summarizes requirements on X-ray telescopes of three space missions that we are aiming at; the Micro-X rocket experiment, the small satellite mission DIOS, and the Jupiter orbiter EJSM JMO. To fulfill these requirements, we have started optimizing wafer parameters such as the pore size, pitch, and wafer thickness, considering the fabrication process easiness. Figure 4 shows an example of the simulated on-axis effective area for a two-stage 4-inch silicon optical system aiming at EJSM JMO as a function of X-ray energy. The field of view is r2.9°. Twenty μm line & space micro pores and side wall roughness of 1 nm are assumed, while the wafer thickness are variable (0.3, 0.5, 1.0 mm). The area is largest when the thickness is 1.0 mm, although such high-aspect micropores must be difficult to be fabricated. It is also clear that telescope arrays are necessary to satisfy the requirement (60 cm^2) if the usable wafer diameter is 4 inch. Considering these results, we will optimize the optics design for these three missions.

REFERENCES

[1] Y. Ezoe, et al., "Ultra light-weight and high-resolution X-ray mirrors using DRIE and X-ray LIGA techniques for Space X-ray Telescopes" Microsystem Tech., 2009, ISSN 0946-7076.
[2] H. Yamaguchi, et al., "Magnetic field assisted finishing for micropore X-ray focusing mirrors fabricated by deep reactive ion etching" CIRP Annals - Manufacturing Technology, in press
[3] I. Mitsuishi, et al., "Optical Image Analysis of the Novel Ultra-Lightweight and High-Resolution MEMS Optics" Proc. of Optical MEMS, Florida, US, Aug 17-20 2009, pp.123-124.
[4] I. Mitsuishi, et al., in these proceedings.

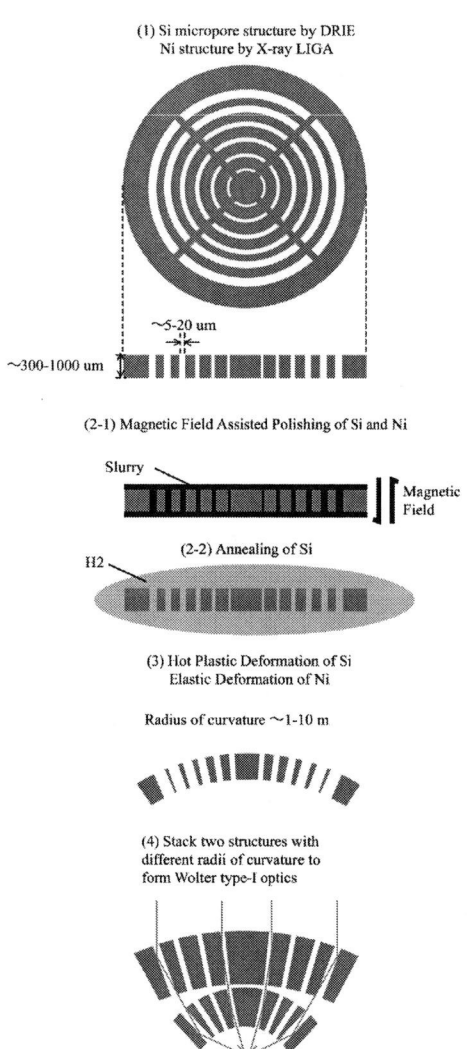

(1) Si micropore structure by DRIE
Ni structure by X-ray LIGA

~5-20 um

~300-1000 um

(2-1) Magnetic Field Assisted Polishing of Si and Ni

Slurry

Magnetic
Field

(2-2) Annealing of Si

H2

(3) Hot Plastic Deformation of Si
Elastic Deformation of Ni

Radius of curvature ~1-10 m

(4) Stack two structures with
different radii of curvature to
form Wolter type-I optics

Figure 1: Concept of the MEMS X-ray optics.

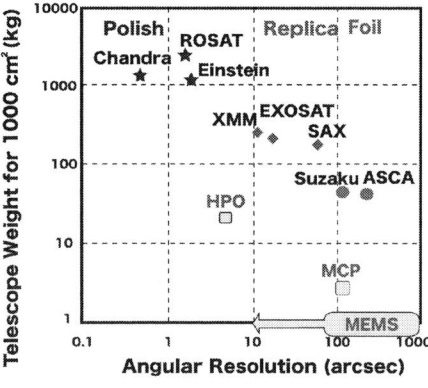

Figure 2: Expected performances of the MEMS X-ray optics compared to the other two next generation optics (HPO and MCP) and the telescopes used in the past missions. Stars, diamonds, and circles correspond to different mirror fabrication techniques.

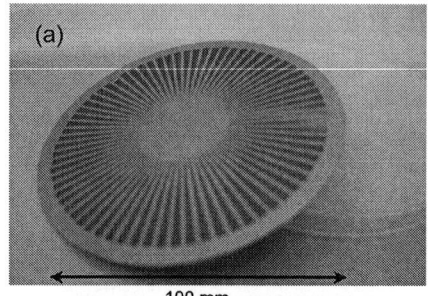

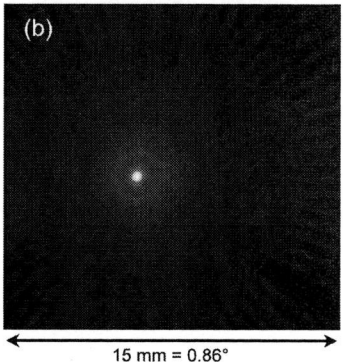

Figure 3: (a) A single-stage 4-inch Si X-ray optical system. (b) A CCD image of the optical focused light.

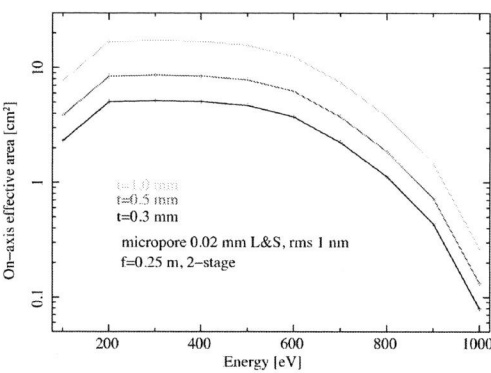

Figure 4: The estimated on-axis effective area of a two-stage 4-inch silicon optical system designed for EJSM JMO as a function of X-ray energy.

Table 1: Requirements on X-ray telescopes.

mission	Micro-X	DIOS	EJSM JMO
planned launch year	2011,12,13,14	2016	2022
energy range (keV)	0.2–3.0	0.3–1.5	<0.6
effective area (cm²)	200	100*	60
resolution (arcmin)	1	4	6
focal length (m)	2.2	0.7	0.25
field of view (deg)	0.19×0.19	1×1	r3
configuration	2-stage	4-stage	2-stage

* Defined by grasp (deg² cm²) for DIOS.

LARGE ELECTROSTATICALLY AND ELECTROMAGNETICALLY ACTUATED MIRROR SYSTEM FOR SPACE APPLICATIONS

Dara Bayat[1], Caglar Ataman[1], Benedikt Guldimann[2], Sebastian Lani[1], Wilfried Noell[1],
Nico F. de Rooij[1]

[1]Ecole Polytechnique Fédérale de Lausanne (EPFL), Switzerland
[2]European Space Agency (ESA-ESTEC), Netherlands

ABSTRACT

We present the design and microfabrication technology of a tip-tilt 2 degree of freedom mirror system made out of SOI wafers with two movable mirrors that have diameters of 1cm. The system is intended for precise beam positioning applications and tracking. Having two mirrors allows one mirror to have large static mechanical scan angles (±3.5°) and the other to have fast fine pointing capabilities within ±0.2° static mechanical scan angle. The first mirror is magnetically actuated and has a resonance frequency of 150Hz. The second mirror is actuated electrostatically and has a resonance frequency of 1 KHz. This mirror enables rapid fine-pointing. Potential space applications are robotic 3D vision, imaging LIDARS, docking sensors and inter-satellite laser communications.

INTRODUCTION

MEMS-based optical-beam steering scanners are in use in various systems these days. The dimensions of these mirrors vary from 10µm to 1mm [1, 2]. They are usually used in scanning mode (resonance and static) or in the "on – off" state [1]. Operating with laser beams at distances of a kilometer [2] and further requires laser beam diameters of several millimeters.

The current project aims at developing a compact microfabricated scanning mirror system of relatively large size (1 cm diameter). Comparable speeds and higher angular ranges are envisaged for this device in comparison with the category of the abovementioned devices. However, the non-resonant scan angle of a single mirror decreases if its response time is increased. This is due to the inverse relation of these two system attributes. In order to overcome this problem the system was separated into to mirrors. Analogously, this principle is used in many adaptive-optics systems for telescopes and imaging systems.

A schematic cross section of the system is shown in Figure 1. The device is based on a 3-mirror configuration, with two movable mirrors. This system consists of an electrostatically actuated mirror, shown on the left part of Figure 1 that acts as a fine tuning position finder that has a small range in the order of ±0.2° mechanical and a short response time in the order of 1ms this mirror acts as a fine tuning position finder. The other mirror that is magnetically actuated, shown on the right part, acts as a coarse position finder with a large deflection angle in the order of ±3.5° mechanical and a long response time in the order of 10ms. The fixed mirror that is positioned above the two moving mirrors reflects the light reflected from the electrostatically actuated mirror on to the magnetically actuated mirror. This configuration separates the large angular range and fast addressing issues from each other.

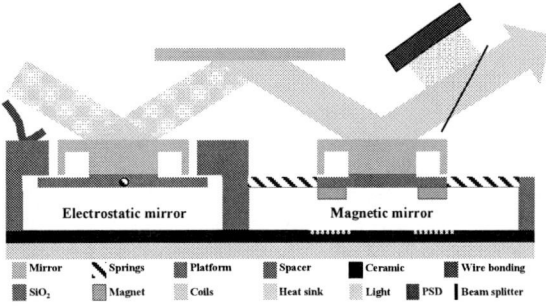

Figure 1: Schematic view of the electrostatically actuated (left) and electromagnetically actuated (right) mirrors of the system.

ELECTROMAGNETICALLY ACTUATED MIRROR

The magnetically actuated mirror uses four 2mm diameter NdFeB permanent magnets that are placed under the mirror as shown in Figure 1. The microfabricated copper coils that are positioned under the magnets create the actuation force. A spacer defines the gap between the magnets and the coils and the coils are positioned on a heat sink. The system uses a "moving magnet – stationary coil" configuration that allows a heat sink to be directly attached under the coils. This increases cooling efficiency of the magnetic mirror system since heat conduction is the only efficient cooling process in space and heat convection is nonexistent. Figure 2 shows the microfabricated coil that was microfabricated by Hightec MC AG, Switzerland. A cross section view of the coil shown in Figure 2 consists of two copper layers on top of each other with a wire width of 50µm. the gaps between two adjacent wires and the two layers are separated from each other with polyimide. The coils are fabricated on an Aluminum oxide substrate. Variations of the generated force due to coil-magnet distance are measured and computed using an empirical formula for the force; $F = \alpha I/(\beta+z)$, α and β are constants, I is the electrical current and z is the distance. Comparison with FEM simulations of the force shows a good agreement between measurements and FEM simulations as can be seen in Figure 2. An injected current density of $J=1.4\times10^9$A/m can produce a force of 7mN at 500µm coil-magnet distance which is the needed distance for a 3.5° rotation of a 1cm mirror.

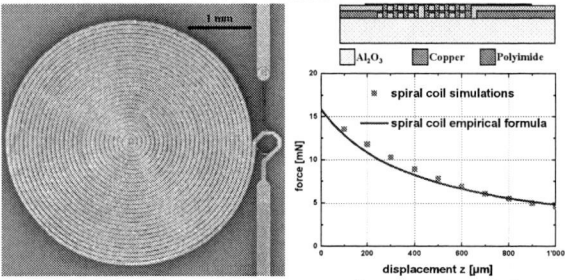

Figure 2: Microfabricated coil on ceramic substrate and force curve obtained with an NdFeB magnet (2mm diameter).

Figure 3 shows the schematic cross section of the magnetically actuated mirror. The microfabrication process is based on eutectic bonding of two SOI wafers. One of the wafers is formed into a platform where the magnets are placed and the second wafer is formed into the mirror. The advantage of bonding is the ease to shape the wafers as necessary without any mutual constraints. Also, by variation of the bonding pattern stress accommodation could be improved to keep the surface of the mirror as flat as possible.

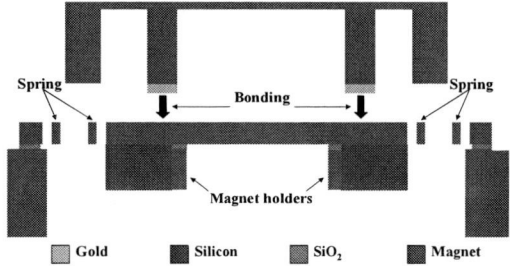

Figure 3: Schematic view of the magnetic actuator + mirror made out of two SOI wafers and bonded together.

ELECTROSTATICALLY ACTUATED MIRROR

The device presented here is 5mm in diameter and was fabricated to test the proposed process flow of the final electrostatically actuated mirror.

The most important part of the process is forming the vertical combs by etching them on the device layer of an SOI wafer and continuing the etching into the handle layer. This implies the removal of every other comb on the backside, i.e. handle layer, to enable the movement of the vertical combs. This process creates two set of combs that are completely self-aligned with a constant gap of 6µm in between each other. The major fabrication steps are shown in Figure 4:
1- DRIE of the device layer. 2- Buried Oxide (BOX) removal using vapor phase hydrofluoric acid (HF VPE). 3- DRIE into the handle layer through the device layer. 4- DRIE with a delayed mask process on the handle layer to trench out the unwanted combs. 5- HF VPE release of the unwanted combs.

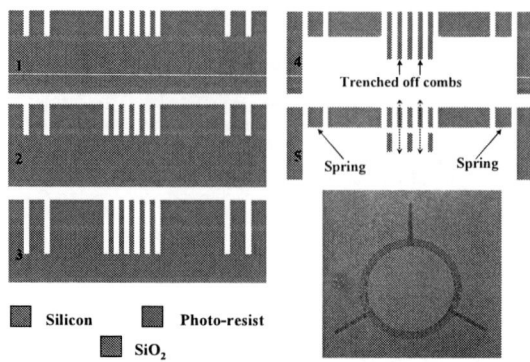

Figure 4: Process flow and top view of the electrostatic mirror depicting one-side DRIE for self aligned vertical comb drives.

Figure 5 shows the tilted view of the wafer's backside. The combs that were formed on the backside of the mobile part of the structure were released by HF VPE and removed, creating the height difference of the vertical combs. The frequency response $f_r = 760$Hz of the structure corresponding to the first excitation mode (piston movement) is very closely estimated by simulations $f_{r(simulation)} = 830$Hz.

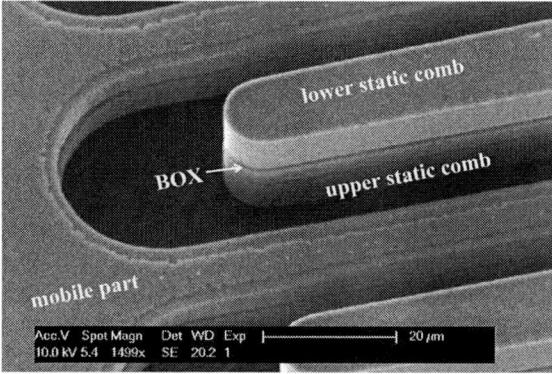

Figure 5: Backside view of the electrostatic vertical comb drive. Unwanted combs above mobile part drop out using HF-VPE.

CONCLUSION

Two mirrors were presented that work synchronically and divide the mutually contradicting scan angle and response time attributes of the mechanical system to achieve both a high angular range and a fast response. Future work concentrates on the feedback and control of the system.

ACKNOWLEDGMENT

This project is financially supported by the European Space Agency, ESA-ESTEC.

REFERENCES

[1] M. R. Douglass, in Reliability Physics Symposium Proceedings, 1998. 36th Annual. 1998 IEEE International, 1998, p. 9.

[2] T. Mukai, S. Abe, N. Hirata, R. Nakamura, O. S. Barnouin-Jha, A. F. Cheng, T. Mizuno, K. Hiraoka, T. Honda, H. Demura, R. W. Gaskell, T. Hashimoto, T. Kubota, M. Matsuoka, D. J. Scheeres, and M. Yoshikawa, Advances in Space Research 40 (2007) 187.

978-1-4244-8926-8/10 $26.00 © 2010 IEEE

Micromirror arrays designed and tested for space instrumentation

Frederic Zamkotsian[1], Michael Canonica[2], Kyrre Tangen[3], Patrick Lanzoni[1], Emmanuel Grassi[1],
Rudy Barette[1], Christophe Fabron[1], Severin Waldis[2], Wilfried Noell[2], Nico de Rooij[2],
Laurent Marchand[4], Ludovic Duvet[4]

[1] Laboratoire d'Astrophysique de Marseille, CNRS, 38 rue Frederic Joliot Curie,
13388 Marseille Cedex 13, France
[2] Ecole Polytechnique Fédérale de Lausanne, Jaquet Droz 1, CH-2002 Neuchâtel, Switzerland
[3] Visitech, Kjellstadveien 5, Lier, P.O.Box 616, N-3003 Drammen, Norway
[4] European Space Agency, Keplerlaan 1, 2200 AG, Noordwijk, The Netherlands

E-mail address: frederic.zamkotsian@oamp.fr

Abstract: Next-generation infrared astronomical instrumentation for ground-based and space telescopes requires MOEMS-based programmable slit masks for multi-object spectroscopy. We made a full space evaluation of Texas Instruments DMD chips, including tests at cold temperature and in vacuum, life tests, radiations, and vibrations and shocks. These results do not reveal any show-stopper concerning its ability to meet environmental space requirements. In parallel, a $100 \times 200 \mu m^2$ micro-mirror array was successfully designed for cryogenic temperature, fabricated and tested at 92K. Large micromirror arrays of 20'000 micromirrors have also been fabricated. These tests demonstrate the full ability of this type of components for space instrumentation.

Keywords: *MOEMS, micromirror array, multi-object spectrograph, space instrumentation, astronomical instrumentation.*

INTRODUCTION

Next-generation infrared astronomical instrumentation for ground-based and space telescopes could be based on MOEMS programmable slit masks for multi-object spectroscopy (MOS). This astronomical technique is used extensively to investigate the formation and evolution of galaxies. In order to avoid spectral overlap and background emission the slit-mask is placed at the focal plane of the telescope and thus permits to select hundreds of objects in parallel. For the near infrared spectrograph (NIRSpec) of the James Webb Space Telescope (JWST), a MEMS-based programmable microshutter array has been developed and fabricated by NASA [1]. We present in this paper a commercial array tested in a space evaluation program and a silicon-based micro-mirror array designed for cryogenic temperatures and tested at 92K.

DMD SPACE EVALUATION

In a MOS, the high precision spectra measurements could be obtained using Digital Micromirror Devices (DMD); these devices would act as object selection reconfigurable masks. ESA has engaged with Visitech and LAM in a technical assessment of using a DMD from Texas Instruments for space applications (for example in ESA EUCLID mission). The DMD features 2048 x 1080 mirrors on a 13.68μm mirror pitch. For MOS applications in space, the device should work in vacuum, at low temperature, and each MOS exposure would last for typically 1500s with micromirrors held in a static state (either ON or OFF) during that duration.

A specific thermal / vacuum test chamber has been developed for test conditions down to -40°C at 10^{-5} mbar vacuum (Fig. 1). Imaging capability for resolving each micromirror has also been developed for determining degradation in any single mirror. Dedicated electronics and software holds any pattern on the device for a duration of up to 1500s (Fig. 2). Data pipeline for data reduction has also been developed for revealing degradation in performance of any mirror. We have adopted three mirror degradation definitions: - the **blocked mirror** when the mirror is stuck, - the **lossy mirror** when the throughput is decreased by more than 20%, and – the **weak mirror** when the throughput is decreased between 10% and 20%.

Our first tests reveal that the DMD remains fully operational at -40°C and in vacuum. Then, a 1038 hours life test in space survey conditions (-40°C and vacuum), has been successfully completed. The device was operating continuously with typical MOS patterns, and optical measurements were done regularly. The number of affected mirrors remains identical through the whole test: 3 blocked mirrors, 7 lossy mirrors and 11 weak mirrors. Blocked mirrors are built-in failures while for the lossy and weak mirrors, a slight variation in number and location could occur but this remains very limited with a maximum variation of ± 2 micro-mirrors. However, all these numbers are very low compared to the 2 million mirrors of the array.

978-1-4244-8926-8/10 $26.00 © 2010 IEEE

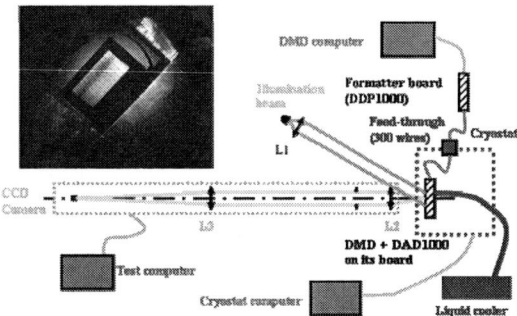

Fig. 1: Schematic of the DMD space evaluation set-up;
(insert: DMD image in the test chamber).

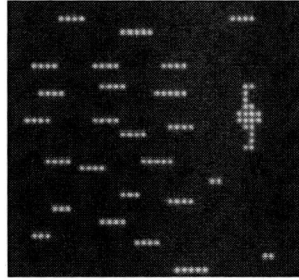

Fig. 2: Typical MOS pattern (individual mirrors are ON)

Total Ionizing Dose (TID) radiation tests have been completed, establishing between 10 and 15 Krads as the level of TID that the DMD can tolerate; at mission level, this limitation could likely be overcome by shielding the device. Finally, thermal cycling (500 cycles between room temperature and cold temperature, on a non-operating device) and vibrations and shocks tests have also been done; no degradation is observed from the optical measurements.

These results do not reveal any show-stopper concerning the ability of the DMD to meet environmental space requirements. Insertion of such devices into final flight hardware would however still require additional efforts such as development of space compatible electronics, and original opto-mechanical design of the instrument.

SILICON-BASED MICROMIRROR ARRAY

We are engaged in a European development of a micro-mirror array for generating reflective slit masks in future MOS. A prototype of micromirror arrays of 5 by 5 single-crystal silicon micromirrors was successfully designed and fabricated, based on a two chip technology where electrode chip and the mirror chip are processed separately and assembled consecutively. The 100 x 200 μm^2 micromirrors can be tilted by electrostatic actuation yielding 20° mechanical tilt-angle. Our MMA is designed such that all structural elements have a matched coefficient of thermal expansion (CTE) in order to avoid deformation or even flaking within the device when cooling it down to the operating temperature. The

10μm thick mirrors are covered with a 60nm thick gold layer for IR operation, whereas gold has a different CTE than silicon.

For characterising the surface quality and the performance of our MMA's at low temperature, we have developed a cryo chamber optically coupled to a high-resolution Twyman-Green interferometer. The interferometer provides a sub-nanometer accuracy, and the cryo-chamber allows pressure down to 10^{-6} mbar and temperatures down to 60 K. The MMA device is packaged in PGA chip carrier. The PGA is inserted in a ZIF-holder integrated on a PCB board (Fig. 3).

The micromirrors could be successfully actuated before, during and after cryogenic cooling at 92K. We could measure the surface quality of the gold coated micromirrors at room temperature, below 100K and being actuated: there is a slight increase of the deformation from 35 nm to 50nm PtV, due to CTE mismatch between silicon and gold layer. This small deformation is still well below the requirement for MOS application at IR. This value could be decreased if needed by using double-side coated mirrors, easily feasible in our process flow.

 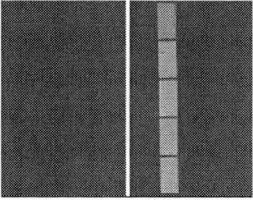

Fig. 3: Cryogenic set-up; functional testing of a
micromirror array at 92K (0V and 90V applied)

For applications in modern and future telescopes larger arrays are required. We are developing a new process where the mirror chip is bonded on top of the electrode chip and microfabricated pillars on the electrode chip provide the necessary spacing between the two parts [2]. The largest chip measures 25x22 mm^2 and is composed of 200x100 electrostatic actuated micromirrors.

In conclusion, these developments and tests demonstrate the full ability of this type of components for space instrumentation, especially in multi-object spectroscopy applications.

ACKNOWLEDGMENT

The authors would like to thank the SAMLAB at EPFL and the Service Essais at LAM. This work is partly funded by ESA, CNRS, Provence-Alpes-Cote d'Azur regional council and Conseil General des Bouches-du-Rhône county council.

REFERENCES

[1] M. J. Li; A. D. Brown; A. S. Kutyrev; H. S. Moseley; V. Mikula, " JWST microshutter array system and beyond", Proc. SPIE **7594**, San Francisco, USA, 2010

[2] M. Canonica, S. Waldis, F. Zamkotsian, P. Lanzoni, W. Noell, Nico de Rooij, " Realization and Characterization of MEMS-Based Programmable Slit Mask for Multi-Object Spectroscopy ", Proc. SPIE **7594**, San Francisco, USA, 2010

Commercialization of Self-Assembled Quantum-Dot Lasers: From Optical Communication to Consumer Electronics

M Sugawara
QD Laser, Inc.

E-mail: sugawara@qdlaser.com

Quantum-dot lasers, first proposed by Arakawa and Sakaki of University of Tokyo in 1982 [1], are semiconductor lasers with nano-sized semiconductor particles called quantum dots as light sources. Self-assembled quantum dots emitting light of 1.3 μm was found in 1995 [2], and room-temperature continuous lasing was achieved at the wavelength in 1999[3]. Based on subsequent research achievements under industry-academia collaboration in Japan, QD Laser, Inc. [4] was launched to commercialize quantum-dot lasers, which have a variety of superior performance to conventional lasers like temperature insensitivity, high temperature operation, low-power consumption, low-cost mass production and SHG-based visible light lasing including green to realize mobile laser projectors. This talk will provide QD laser activity, i.e., the organization, applications & markets from optical communication to consumer electronics, self-assembled quantum-dot lasers, the product portfolio and the business model.

References

[1] Y. Arakawa and H. Sakaki, Appl. Phys. Lett. 40, 939 (1982).
[2] K. Mukai, N. Ohtsuka, M. Sugawara, and S.Yamazaki, Jpn. J. Appl. Phys. 33, L1710-1712 (1995).
[3] K. Mukai, Y. Nakata, H. Shoji, M. Sugawara, K. Otsubo, N. Yokoyama, and H. Ishikawa, Electron. Lett. 34, 1588 (1998).
[4] http://www.qdlaser.com/

AUTHOR INDEX

Adato, Ronen 63
Ahn, Jaewook 55
Akai, Daisuke 125
Aksu, Serap 63
Alaca, Erdem 9
Aljasem, Khaled 87
Altug, Hatice 63
Amaya, Satoshi 27
Amemiya, Yoshiteru 101
An, Jae Yong 17
Aonuma, Takuro 93
Aoyagi, Seiji 101
Arakawa, Takahiro 83
Artar, Alp 63
Ataka, M. 111
Ataman, Caglar 191
Ayerden, N. Pelin 105
Baba, Yoshinobu 41
Bammer, R. 187
Barbastathis, George 37, 143, 177, 179
Barette, Rudy 193
Bauer, Ralf 115
Bayat, Dara 191
Bich, Andreas 61
Bish, Sheldon 3
Boeck, Robi 169
Boggs, Taylor 151, 189
Bonacina, Luigi 59
Brosnihan, Timothy 73
Brown, G. 57
Brown, Gordon 115
Burger, T. 187
Burger, Tobias 15
Burns, D. 57
Canonica, Michael 193
Carmon, Tal 51, 157, 159
Chalasani, S. 135
Chang, Chih-hao 177
Chang, Chun-Che 99
Chao, Keng-hsing 7
Chau, Fook Siong 21, 23, 69, 81, 89
Chen, Haiqing 89
Chen, Kelvin Wei Sheng 23
Chen, Nanguang 23
Chen, Rei-Shin 173, 175
Chen, Tzung-Ming 127
Chen, Yi-Hao 185
Chen, Ying-Chuan 141
Chen, Yue 43, 45
Cheo, Kelvin K.L. 21
Cheo, Kelvin Koon Lin 81
Chew, Xiongyeu 69
Chi, J.Y. 167
Chia, Bonnie Tingting 123
Chiou, Eric Pei-Yu 43

Chiou, Jin Chern 121, 139
Chiou, Pei Yu 45, 47
Chiu, Chun-Chia 171
Chiu, Wei-Chao 99
Choi, Hyung-ryul Johnny 37
Choi, Jun-Hyuk 55
Choi, Yongje 55
Chou, Chao-Min 145
Chrostowski, Lukas 71, 169
Chu, Hoang Manh 113
Chuang, Huang Yun-Ju 141
Dao, Dzung Viet 27, 183
De Rooij, Nico F 59, 61, 77, 191, 193
Deng, Jie 69
Deterre, Martin 37
Di Carlo, Dino 43
Draheim, J. 187
Draheim, Jan 15
Du, Yu 21, 81
Duvet, Ludovic 193
Extermann, Jerome 59
Ezoe, Yuichiro 151, 189
Fabron, Christophe 193
Feng, Hanhua 23, 81
Fijol, John 73
Fujita, H. 75
Fujita, Hiroyuki 107, 109
Fukuyama, Masataka 101
Gabriel, Nicholas T. 189
Gandhi, Jignesh 73
Gao, Hanhong 179
Gessei, Tomoko 83
Gomes, J. 57
Gopal, Ashwini 3, 49
Grasshoff, Thomas 25, 105
Grassi, Emmanuel 193
Grutter, Karen E. 33
Gu, Zhengbin 165
Gueissaz, Francois 77
Guldimann, Benedikt 191
Guo, L. Jay 185
Hagood, Nesbitt 73
Halfman, Mark 73
Hane, K. 79, 97
Hane, Kazuhiro 113, 149, 163
Hayashi, Takayuki 151
Herbst, M. 187
Herzig, Hans Peter 161
Higo, Akio 107, 109
Holmstrom1, Sven TS 105
Horade, Mitsuhiro 151, 189
Hori, Masaru 137
Hoshino, Kazunori 49
Hou, Kuan Chou 139
Howe, Roger T. 35

AUTHOR INDEX

Hsiao, Fu-Li 181
Hsu, S.T. 167
Hu, F. 79
Hu, Fangren 163
Huang, Ding-Wei 171
Huang, Min 63
Huang, Sheng-Wen 145
Huang, Yu-Yen 49
Hung, Chen-Chun 121
Hung, Kuo Yung 141
Hur, Ji Hwan 17
Hur, Sung-Young 43
Ikeda, T. 97
Ishida, Makoto 125
Ishimaru, Ichiro 5
Ishizu, Kensuke 151, 189
Itabashi, Gen 83
Jaeger, Nicolas A. F. 71
Jaeger, Nicolas A.F. 169
Jeong, Ki-Hun 91
Jeong, Soon-Jong 129
Jeong1, Ki-Hun 55
Jiang, Hongrui 65, 131
Jin, Kyung-Hwan 55
Joo, Jae Young 147
Jung, Jae Hak 155
Jutzi, Fabiio 59
Jutzi, Fabio 77
Kamberger, Robert 15
Kanamori, Y. 97
Kanamori, Yoshiaki 151, 189
Kang, Jingran 89
Kaplan, Alex F 185
Kavakli, Halil 9
Kawai, Gou 101
Kelly, A. 57
Kim, Doogwook 165
Kim, In-Sung 129
Kim, Min-Soo 129
Kim, Myun-Sik 161
Kiselev, Denis 59
Kobayashi, Takeshi 119
Koh, Kah How 119
Kubo, Hironori 39
Kudo, Hiroyuki 83
Kumagai, Shinya 39, 93, 137
Kuo, H.C. 167
Kwong, Dim-Lee 53
Lan, R.L. 167
Lani, Sebastian 191
Lani, Sebastien 59
Lanzoni, Patrick 193
Lee, ChaBum 149
Lee, Chengkuo 119, 181
Lee, Daesung 35

Lee, Jong-Hyun 17
Lee, Ming-Chang M. 99, 145
Lee, Sung Kil 17
Lee, Sun-Kyu 147, 149
Lee, Taelim 107
Lee, W. Scott 35
Lewis, Steve 73
Li, Bo 181
Li, Cheng-Ru 175
Li, Chensha 65
Li, Li 115, 117
Liao, Bo-Ting 123
Liao, Chun-da 7
Liao, Chun-Wei 145
Lin, Chun-Ying 121
Lin, G. 167
Lin, G.R. 167
Lin, Han-Bin 173
Lin, Lih Y. 67
Lin, Pi-Yao 145
Lin, Yih-Bin 173, 175
Lin, Yu-Sheng 153
Liu, Jinsong 115
Liu, Ju-Feng 173
Liu, Ye 131
Lo, C.Y. 75
Lo, Chi-Wei 65
Lo, Patrick 53
Lockhart, R. 103
Loke, Yee Chong 69
Lu, H.H. 167
Lubeigt, W. 57
Lugo, Katherine 67
Ma, Cheng-Wen 123
Maclaren, J. 187
Maeda, Ryutaro 151, 189
Maeda, Yoshitomo 151
Marchand, Laurent 193
Maruyama, Satoshi 109
Masson, Jonathan 61
Mastrangelo, C. H. 135
Masuno, Katsuya 137
Miao, Xiaoyu 67
Min, Bonggi 165
Mita, M. 109, 111
Mita, Makoto 151
Mitra, Subhasish 35
Mitsubayashi, Kohji 83
Mitsuda, K. 151
Mitsuda, Kazuhisa 189
Mitsuishi, Ikuyuki 151, 189
Mizuno, T. 111
Mönch, Wolfgang 13
Morishita, Kohei 151, 189
Morita, Hironobu 133

AUTHOR INDEX

Moriyama, Teppei 151, 189
Mu, Xiaojing 23
Müller, Philipp 13
Nakada, M. 109
Nakajima, Kazuo 151, 189
Nakano, Yoshiaki 1, 107, 109
Nargul, Sezin 9
Nishioka, Kouji 101
Noell, Wilfried 59, 61, 77, 191, 193
O'Sullivan, Ciara K. 11
Oda, Kentaro 5
Ohashi, Takaya 189
Oohira, Fumikazu 5, 29
Ozaki, Katsuya 125
Park, C. 155
Park, Chinho 165
Park, Hyeon-Cheol 91
Park, Sang-Gil 55
Park, Soojung Claire 43
Park, Sun Sub 147
Park, Sung-Yong 45
Patra, Susant K. 33
Payne, Richard 73
Peng, P.C. 167
Rieke, Fred 67
Riveros, Raul E. 151, 189
Sameshima, H. 79
Sandner, Thilo 25, 105
Sasaki, Minoru 39, 93, 137
Sasaki, Osamu 29
Sato, Takuro 151
Savitski, V. 57
Sawada, Kazuaki 125
Scharf, Toralf 161
Schenk, Harald 25, 105
Schneider, F. 187
Schneider, Florian 15, 127
Seifert, Andreas 85, 87
Seren, Huseyin R. 105
Sharma, Jaibir 105
Shen, Lan 165
Shi, Wei 71
Shieh, Jia-Min 99
Shieh, Jia-Ming 145
Shih, Sun-Chih 123
Shih-Hao, Chen 141
Singh, Janak 23
Song, Cheol 91
Song, Jae-Sung 129
Spengler, Nils 13
Stanley, R.P. 103
Stewart, George 117
Steyn, J. Lodewyk 73
Su, Jung-Young 175
Sugawara, M 195

Sugiyama, Susumu 27, 151, 183, 189
Suzuki, Masato 101
Suzuki, Takaaki 5, 29
Suzuki, Yuki 83
Takahashi, Daishi 83
Takahashi, K. 109
Takahashi, Satoshi 177, 179
Takahashi, T. 109
Takahashi, Tomokazu 101
Takao, Hidekuni 5, 29
Talghader, Joseph J. 189
Tan, Chee Wei 23
Tanae, T. 79
Tang, Xiaosong 69
Tangen, Kyrre 193
Tashiro, Kohji 137
Terao, Kyohei 5
Terao, Kyouhei 29
Thursby, Graham 117
Tian, Lei 179
Timurdogan, Erman 9
Tomes, Matthew 51, 157, 159
Tormen, M. 103
Torres, Miguel Angel Guillen 71
Tortissier, G 75
Toshiyoshi, and Hiroshi 109
Toshiyoshi, H. 75, 111
Toshiyoshi, Hiroshi 107
Truong, N.T.N. 155
Truong, Quynh Nhu Nguyen 155
Tsai, Jui-che 7
Tung, Bui Thanh 183
Tunnell, James W 3
Urey, Hakan 9, 105
Uttamchandani, D. 57
Uttamchandani, Deepak 115, 117
Vafaei, Raha 71
Voelkel, Reinhard 61
Waldis, Severin 193
Wallrabe, U. 187
Wallrabe, Ulrike 15, 127
Wang, Yih-Ching 19
Wang, Yongjin 163
Wang, Youmin 3
Wang, Yuyan 49
Watanabe, Minoru 133
Weber, Niklas 85
Weber, Stefan M. 59
Weible, Kenneth J. 61
Wilkinson, James S 95
Wolf, Jean-Pierre 59
Wong, H. S. Philip 35
Woo, Do-Kyun 147
Wu, Jiun-Ming 99
Wu, Joyce 73

AUTHOR INDEX

Wu, Ming C. 33
Wu, Ting-Hsiang 43, 45
Xiao, Fan..47
Xie, Y...135
Xu, Chao-Nan 31
Xu, Yingshun 23
Yamaguch, Hitomi 151
Yamaguchi, Hitomi 189
Yamasaki, N Y 151
Yamasaki, Noriko Y...189
Yang, Chih-Cheng 19
Yang, Se Young..37, 177
Yang, Yao-Joe 123
Yang, Yao-Tsu 145
Yanik, Ahmet A..63
Ye, Jong-Chul 55
Yeh, Anthony M. 33
Yeh, J. Andrew 19, 153
Yi, Minwoo 55
Yokoyama, Shin 101
Yu, Aibin...23
Yu, Hongbin 21, 89
Zaitsev, M. 187
Zamkotsian, Frederic 193
Zappe, Hans..13, 85, 87
Zehnpfennig, John 51
Zeng, Y...135
Zhang, Baile...143
Zhang, Qingxin 81
Zhang, Wei 87
Zhang, Xiaojing...3, 49
Zhang, Xingyu 159
Zhou, Guangya...21, 23, 69, 81, 89